# Basic Radar Tracking

For a listing of recent titles in the
*Artech House Radar Series*,
turn to the back of this book.

# Basic Radar Tracking

Mervin C. Budge, Jr.
Shawn R. German

ARTECH
HOUSE

BOSTON | LONDON
artechhouse.com

**Library of Congress Cataloging-in-Publication Data**
A catalog record for this book is available from the U.S. Library of Congress.

**British Library Cataloguing in Publication Data**
A catalog record for this book is available from the British Library.

ISBN-13: 978-1-63081-335-2

**Cover design by John Gomes**


**685 Canton Street**
**Norwood, MA 02062**



10 9 8 7 6 5 4 3 2 1

## DISCLAIMER OF WARRANTY



For supplemental materials to this book, please visit http://us.artechhouse.com/Assets/downloads/budge_335.zip

# Contents

Preface ........ ix
Acknowledgments ........ xi
Chapter 1 Tracking Basics ........ 1
1.1 Introduction ........ 1
1.2 Tracker Types ........ 1
1.3 Book Outline ........ 3
Chapter 2 Control Theory Review ........ 7
2.1 Introduction ........ 7
2.2 Continuous Time Systems ........ 9
2.2.1 System Type and Steady State Error ........ 9
2.2.2 Root Locus and Transient Behavior ........ 13
2.3 Discrete Time Servos ........ 19
2.3.1 System Type and Steady State Error ........ 20
2.3.2 Root Locus and Transient Behavior ........ 22
2.4 Modeling Closed Loop Servos ........ 24
2.4.1 Analog Servo Modeling ........ 24
2.4.2 Digital Servo Modeling ........ 35
2.5 Exercises ........ 36
Chapter 3 Track Filters ........ 41
3.1 Introduction ........ 41
3.2 Kalman, α-β, and α-β-γ Track Filters ........ 41
3.2.1 Background ........ 41
3.3 The Prediction Equation ........ 44
3.3.1 Closed Loop Tracker Structure ........ 46
3.3.2 Filter Stability and Variance Reduction ........ 48
3.4 Benedict-Bordner Method for α-β Filter Design ........ 53
3.5 Polge-Bhagavan Method for α-β-γ Filter Design ........ 56
3.6 Calculation of α for the Benedict-Bordner and Polge-Bhagavan Filters 57
3.7 Responses of the Optimal α-β and α-β-γ Filters ........ 58
3.8 Control Theory Approach ........ 59
3.8.1 Type 1 Servo ........ 59
3.8.2 α-β Tracker ........ 61

3.8.3 α-β-γ Tracker .... 65
3.9 Linear Kalman Filter .... 70
3.10 Example .... 72
3.11 Exercises .... 76
Appendix 3A: Stability Triangle and Variance Ratio .... 79
Appendix 3B: Derivation of (3.60)—Benedict and Bordner α-β Relation .... 84

Chapter 4 Closed Loop Range Tracking .... 91

4.1 Introduction .... 91
4.2 Sampling Gate Range Discriminator .... 92
4.2.1 LFM Pulse .... 98
4.2.2 Other Waveforms .... 101
4.3 Summing Gate Range Discriminator .... 101
4.3.1 Unmodulated Pulse .... 103
4.3.2 LFM Pulse .... 107
4.3.3 Barker Coded Pulse .... 107
4.3.4 Digital Matched Filter Implications .... 107
4.4 Direct Range Measurement .... 109
4.5 Range Tracker Modeling .... 111
4.5.1 Signal Model .... 112
4.5.2 Noise Model .... 115
4.5.3 Scaling the Signal and Noise .... 120
4.5.4 Signal and Noise Generation Algorithm .... 121
4.6 Signal Processor Considerations .... 121
4.7 Examples .... 121
4.7.1 Example 1: Sampling Gate Discriminator and α-β Filter .... 122
4.7.2 Example 2: Summing Gate Discriminator and α-β Filter .... 128
4.7.3 Example 3: Direct Range Measurement and α-β-γ Filter .... 129
4.8 Functional Level Error Model .... 133
4.9 Exercises .... 134
Appendix 4A: Derivation of $v_e$ when $q < ½$ and $q\tau_p < |\Delta\tau| < (1-q)\tau_p$ .... 137

Chapter 5 Closed Loop Angle Tracking .... 139

5.1 Introduction .... 139
5.1.1 Chapter Outline .... 142
5.2 Types of Monopulse Sensing .... 142
5.3 Phase Comparison Monopulse .... 146
5.4 Amplitude Comparison Monopulse .... 154
5.5 Monopulse Combiners .... 160
5.5.1 Magic Tee .... 160
5.5.2 Rat Race .... 161
5.5.3 3-dB Coupler .... 163
5.6 Monopulse Receivers .... 167
5.6.1 Three-Channel Monopulse Receiver .... 168
5.6.2 Two-Channel Monopulse Receiver .... 170
5.6.3 Simultaneous Multiple Beams/Digital Beam Forming .... 175

5.7 Conical Scan......176
5.8 Monopulse Processors......181
5.8.1 Exact Processor......182
5.8.2 Modified Exact Processor......187
5.8.3 Log-Based Processor......190
5.9 Example 1: Angle Tracker with Different Monopulse Processors......193
5.10 Example 2: Combined Angle and Range Tracker......198
5.11 Functional Level Error Model......202
5.12 Exercises......203

Chapter 6 Closed Loop Doppler Tracking......207
6.1 Introduction......207
6.2 CW Doppler Discriminator......208
6.3 Pulsed Doppler Discriminator......214
6.4 Direct Doppler Measurement......217
6.5 Doppler Tracking in Low PRF Pulsed Radars......218
6.6 Example 1: CW Doppler Tracker......226
6.7 Example 2: Low PRF tracker......232
6.8 Functional Level Error Model......239
6.9 Exercises......240
Appendix 6A: Derivation of the Correlation Coefficient of the BPF Outputs......241

Chapter 7 Simulation Examples......245
7.1 Introduction......245
7.2 RCS Fluctuation......245
7.2.1 Example 1......247
7.3 Dual Target Tracking......251
7.3.1 Background......251
7.3.2 Example 2: Ideal Angle Tracker—Demonstration of Dichotomous Tracking......252
7.3.3 Example 3: A Variation on Example 2......255
7.3.4 Example 4: Example 3 with a Realistic Antenna......257
7.4 Crossing Target Examples......258
7.4.1 Example 5: Equal Doppler Frequencies and Target Sizes......260
7.4.2 Example 6: Different Doppler Frequencies and Target Sizes......262
7.4.3 Example 7: A Different Geometry—Collinear Target Trajectories......264
7.5 Crossing Target Examples—Fluctuating Target RCS......266
7.5.1 Example 8: Same Set Up as Example 5 with Fluctuating RCSs......266
7.5.2 Example 9: Same Set Up as Example 7 with Fluctuating RCSs......269
7.6 Multipath Examples......269
7.6.1 Specular Multipath Modeling......271
7.6.2 Target Model for Multipath......272

7.6.3 Example 10: Tracking in the Presence of Specular Multipath ... 274

7.7 Exercises ... 278

Chapter 8 Acquisition and Track Initiation ... 285

8.1 Introduction ... 285

8.2 Background ... 286

8.2.1 Acquisition Volume Design ... 286

8.2.2 Acquisition and Track Initiation Process ... 287

Acronyms and Abbreviations ... 295

Variables ... 297

About the Authors ... 307

Index ... 309

# Preface

This book is an outgrowth of a course in radar tracking taught at the University of Alabama, in Huntsville. The motivation for the course, and this book, was to provide an introduction to how to build, analyze, and simulate closed loop trackers that might be applicable to modern, multi-function, phased array radars. To that end, this book considers not only the topic of track filters [alpha-beta ($\alpha$-$\beta$), alpha-beta-gamma ($\alpha$-$\beta$-$\gamma$), Kalman], but how the signals that drive these filters are generated. In particular, we consider the topics of range discriminators, Doppler discriminators, and monopulse processors. The intent is to cover the topics in sufficient detail to support the construction and simulation of range, Doppler and angle trackers, and to explore how these trackers work together to support the overall track function. A significant portion of the book is devoted to the simulation of range, angle and Doppler trackers, and to integrated combinations of these trackers. We use the integrated trackers to investigate, through simulation and theoretical predictions, how trackers will operate in a multiple target environment.

Chapter 1 begins with a definition of the elements of closed loop trackers, and contains discussions of the basic difference between closed and open loop trackers. Since closed loop trackers are basically servomechanisms, some of the properties of servomechanisms related to tracking are discussed in Chapter 2. We begin the development of simulations in this chapter by discussing the structure of closed loop trackers and how to model them.

Chapter 3 contains discussions of $\alpha$-$\beta$ and $\alpha$-$\beta$-$\gamma$ filters, and how to design them to provide a desired closed loop tracker bandwidth. It also contains discussions of variance reduction and the Benedict-Bordner and Polge-Bhagavan approaches to designing $\alpha$-$\beta$ and $\alpha$-$\beta$-$\gamma$ filters. The chapter also contains a discussion of Kalman-type filters that could be used in closed loop trackers.

Chapters 4, 5, and 6 contain discussions of how a radar forms the signals needed to drive the track filters of closed loop trackers. Chapter 4 is devoted to deriving signals for range trackers, Chapter 5 is devoted to deriving signals for angle trackers, and Chapter 6 is devoted to deriving signals for Doppler trackers. All of these chapters contain simulation examples, and we begin integrating the various tracker types in Chapters 5 and 6.

The topic of Chapter 7 is closed loop tracking of two targets. In particular, we use the various trackers discussed in Chapters 4, 5, and 6 to investigate, through simulation and theoretical development, how the interaction between the returns from two targets affects tracking. We specifically study the impact of relative target Doppler frequency and radar cross section variation on the behavior of range and angle trackers in a two-target environment. The chapter also contains a brief discussion of tracking in the presence of specular multipath, and contains simulation examples of such tracking when radar cross section fluctuation is considered.

In the simulation examples of Chapters 4 through 7, the trackers are initialized with either perfect target information, or target information with small errors. In Chapter 8, we briefly address the issue of target acquisition and track initiation by providing an acquisition and track initiation process a track radar might need to include.

We hope you find this book useful, and we welcome your feedback.

# Acknowledgments

We would like to express our thanks to several people who contributed to this book. First, we thank David Barton, our Artech reviewer, for his meticulous review of the chapters and his many valuable suggestions on how to improve the material in the book. We also thank Stacy Thompson, Shawn's sister, for reviewing the various chapters and fixing our grammatical errors. We would also like to thank others who reviewed portions of the book and/or offered technical ideas and materials, including Dr. B. K. Bhagavan, Dr. Craig Newborn, Alan Volz, Johnathan Andrews, Alexandria Carr, Cooper Barry, and Belinda Byron.

On the publishing side, we owe a special thanks to the Dynetics Creative Media Solutions department, especially Joyce Walters and Virginia Elmer for preparing the manuscript.

Finally, we thank Carmie, Merv's wife, and Karen German, Shawn's mother, for their unwavering encouragement and support during the preparation of this book.

Merv Budge
Shawn German
book@thebudges.com

# Chapter 1

## Tracking Basics

### 1.1 INTRODUCTION

The IEEE dictionary defines tracking as "The process of following a moving object or a variable input quantity" [1]. In the context of this book, the "object" would be the target, and the "input quantities" would be target metrics, or parameters, such as range, range-rate, angle, x-y-z position, velocity, acceleration, etc. In the simplest form of tracking, we could use sensor measurements such as range, Doppler frequency, and antenna pointing angles as a tracking mechanism. However, because of noise, measurements are not accurate enough for most tracking applications. A tracker filters, or smoothes, measurements to improve the accuracy of target metric estimates. In some instances, the tracker also performs coordinate transformations needed to use measurements in one coordinate system (e.g., range, Doppler frequency, angle) to track in a different coordinate system (e.g., x-y-z positions and rates).

### 1.2 TRACKER TYPES

We can define two types of trackers: closed loop trackers and open loop trackers. Closed loop trackers, which are sometimes termed continuous trackers [2, 3], are generally used in tracking radars or the track function of multifunction radars. Open loop trackers are used in search radars or the search function of multifunction radars. More specifically, open loop trackers are used in the track-while-scan (TWS) [4, 5] function of these radars. In the TWS process, the target metrics obtained during search are used to establish and maintain target tracks. When detection logic reports a detection, target metrics (e.g., angle, range, Doppler frequency) associated with the detection are estimated and recorded. These metrics are then used to update an existing target track, or initialize a new track if the measured metrics cannot be associated with a current track.

Generic block diagrams of open and closed loop trackers are contained in Figure 1.1. Both tracker types consist of a sensor and a track filter. The sensor is what we normally think of as the radar. In the open loop tracker, the sensor measures target parameters such as range, angle, and Doppler frequency. This is why we use the term *measurement sensor* in Figure 1.1. In a closed loop tracker, the sensor can measure target parameters, or it can measure the difference, or error, between target parameters and the tracker's estimate of them. The fact that the sensor can measure error in a closed loop tracker is why we use the term *error sensor* in Figure 1.1. Whether the sensor in a closed loop tracker measures the parameters or errors depends on the specifics of the tracker implementation. If the track update rate is such that the tracker's estimate of a target's parameter is close to the actual target parameter, the tracker could use an

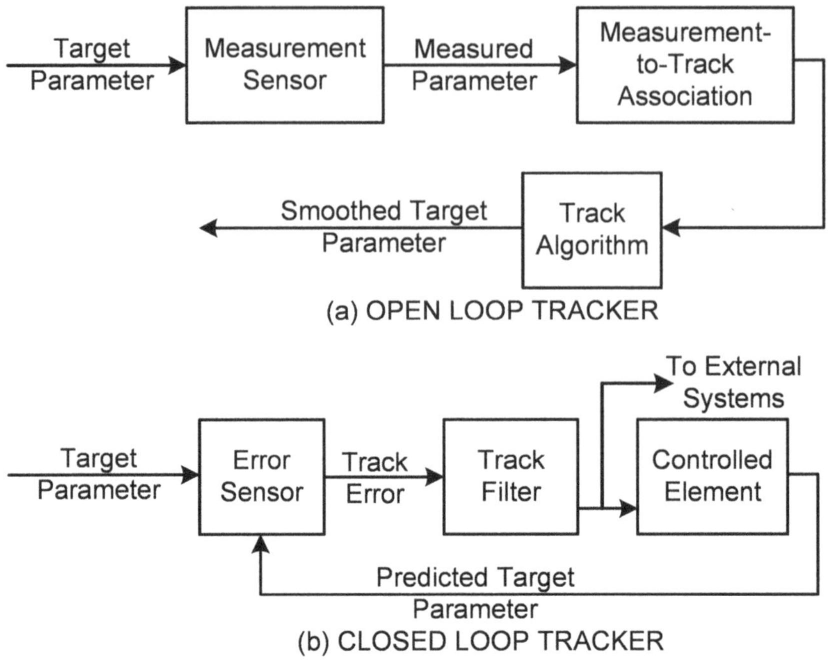


**Figure 1.1** Open and closed loop trackers.

error sensor. However, if the update rate is such that there could be a considerable difference between the actual target parameter and the tracker's estimate, the tracker would need to use a measurement sensor. An example of a closed loop tracker that uses a measurement is discussed in Chapters 4 and 6.

As with the sensor, both types of trackers include some type of track algorithm or track filter. The track algorithm and track filter have similar implementations (e.g., Kalman, α-β or α-β-γ). However, the specifics of the implementations differ in open and closed loop trackers.

In an open loop tracker, the input to the track algorithm is the measured parameter (possibly after a coordinate transformation), and the output is the filtered, or *smoothed*, version of the *parameter*. In a closed loop tracker, the input to the track filter can also be the parameter measurement, however, in this case the output of the track filter is the *predicted* value of the *parameter*, which is used by the sensor to measure the parameter on the next track update. We discuss this further in Chapter 3, where we consider the different types of implementations.

The track filter input in closed loop trackers can also be the error between target parameter and the predicted parameter value from the track filter, if that is what the sensor provides. However, the output is still the predicted value of the parameter.

One difference between closed and open loop trackers is the inclusion of a controlled element in the closed loop tracker. The controlled element prepares the sensor for the next measurement. An example is the beam steering computer and phase shifters of a phased array antenna or the control system that positions a mechanically scanned antenna. Another example of a controlled element is the timing circuit that positions the range gates of the sensor.

The fact that the closed loop tracker uses the output of the track filter as feedback to the controlled element is what makes it a closed loop tracker. Because of the feedback, we can use classical servomechanism techniques, such as system type and root locus, to help analyze closed loop trackers [6–8].

In a sense, the track algorithm in an open loop tracker is also a closed loop tracker. However, it closes its loop through itself rather than through the controlled element and the sensor.

In open loop trackers, there is no feedback from the track algorithm to the sensor. Because of this, an open loop tracker needs some means of associating target parameter measurements with target tracks. This is the purpose of the measurement-to-track association block of Figure 1.1. The measurement-to-track association algorithm uses measurement variances, and covariances from the track algorithm, to determine the most likely association between a particular measurement and various, existing target tracks. The algorithms in the measurement-to-track association block can range from an elementary nearest neighbor technique, to more sophisticated techniques such as probabilistic data association or multiple hypothesis tracking [9–14].

## 1.3 BOOK OUTLINE

Since closed loop trackers are servomechanisms, some of the properties of servomechanisms related to tracking are discussed in Chapter 2. In particular, the chapter contains discussions of system type and root locus, and how they can be used to characterize the properties and expected behavior of closed loop trackers [6–8]. This chapter also contains discussions of the structure of closed loop trackers and how to model them. The models are used in track loop simulation examples discussed throughout the book.

Chapter 3 contains a discussion of α-β (g-h) and α-β-γ (g-h-k) filters and how to design them to provide a desired closed loop bandwidth and damping ratio [6, 8, 15–18]. It also contains discussions of variance reduction and specifically the Benedict-Bordner approach to designing α-β track filters [19] and the Polge-Bhagavan approach to designing α-β-γ track filters [20]. Both of these design techniques attempt to satisfy the dual requirement of providing good tracking behavior and maximizing the variance reduction offered by the filter. Finally, we discuss the structure of Kalman type filters that could be used in closed loop trackers [21–23].

Chapters 4, 5, and 6 contain discussions of error and measurement sensors for range (Chapter 4), angle (Chapter 5), and Doppler frequency (Chapter 6). We also discuss how the sensors are integrated into closed loop trackers.

The discussions of range error sensors include sampling gate discriminators as well as integrating gate discriminators. For the measurement sensor, we discuss the effect of sample spacing and how to structure the measurement sensor. The chapter also contains simulation examples of closed loop trackers.

The angle sensor discussions of Chapter 5 focus mainly on monopulse techniques [24–27]. The angle sensor discussions include different types of monopulse processors, such as the log processor, the exact discriminator, and a modified exact discriminator. As with Chapter 4, this chapter contains simulation examples of closed loop angle trackers. It also contains a simulation example of an integrated, closed loop, range and angle tracker.

We begin the discussion of Doppler sensors by discussing continuous wave (CW) Doppler discriminators. We then discuss discriminators that might be used in pulsed Doppler radars and radars that use long, modulated pulses. As with range, we also discuss direct Doppler measurement. The chapter contains example block diagrams of integrated range, Doppler, and angle trackers. The chapter concludes with simulation examples of a Doppler tracker, and an integrated range and Doppler tracker.

The topic of Chapter 7 is closed loop tracking of two targets. In particular, we investigate, through simulation and a theoretical development, how the interaction between the returns from two targets affects tracking. We specifically discuss the impact of relative target Doppler frequency and radar cross section (RCS) fluctuation [28–29]. The chapter also contains a brief discussion of tracking in the presence of specular multipath and contains simulation examples of such tracking when RCS fluctuation is considered.

In the simulation examples of Chapters 4 through 7, the trackers were initialized with either perfect target information, or target information with small errors. In Chapter 8, we address the issue of target acquisition and track initiation.

## References

[1] F. Jay, Ed., *IEEE Standard 100, IEEE Standard Dictionary of Electrical and Electronics Terms*, Fourth ed., New York: The Institute of Electrical and Electronics Engineers, Inc., 1988.

[2] M. I. Skolnik, *Introduction to Radar Systems*, 2nd ed., McGraw-Hill, 1980.

[3] S. A. Hovanessian, *Radar System Design and Analysis*, Norwood, MA: Artech House, 1984.

[4] S. A. Hovanessian, *Radar Detection and Tracking Systems*, Dedham, MA: Artech House, 1973.

[5] D. R. Billetter, *Multifunction Array Radar*, Norwood, MA: Artech House, 1989.

[6] C. H. Houpis and S. N. Sheldon, *Linear Control System Analysis and Design with MATLAB*, 6th ed., Boca Raton, FL: CRC Press, 2013.

[7] W. R. Evans, "Graphical Analysis of Control Systems," *Transactions of the American Institute of Electrical Engineers,* vol. 67, no. 1, pp. 547-551, January 1948.

[8] R. C. Dorf and R. H. Bishop, *Modern Control Systems*, 13th ed., Pearson Education, Inc., 2016.

[9] S. S. Blackman and R. Popoli, *Design and Analysis of Modern Tracking Systems*, Norwood, MA: Artech House, 1999.

[10] S. S. Blackman, *Multiple-Target Tracking with Radar Application*, Norwood, MA: Artech House, 1986.

[11] S. S. Blackman, "Multiple hypothesis tracking for multiple target tracking," *IEEE Aerospace and Electronic Systems Magazine,* vol. 19, no. 1, pp. 5-18, 2004.

[12] J. C. McMillan and S. S. Lim, *Data Association Algorithms for Multiple Target Tracking*, Report No. 1040, Ottowa: Defence Research Establishment of Ottowa, 1990.

[13] Y. Bar-Shalom, Ed., *Multitarget-Multisensor Tracking: Applications and Advances*, vol. II, Norwood, MA: Artech House, 1992.

[14] S. H. Rezatofighi, A. Milan, Z. Zhang, Q. Shi, A. Dick and I. Reid, "Joint Probabilistic Data Association Revisited," in *2015 IEEE International Conference on Computer Vision (ICCV)*, Santiago, 2015.

[15] J. Sklansky, "Optimizing the dynamic parameters of a track-while-scan system," *RCA Review,* vol. 18, 1957.

[16] B. K. Bhagavan and R. J. Polge, "Performance of the g-h Filter for Tracking Maneuvering Targets," *IEEE Transactions on Aerospace and Electronic Systems,* Vols. AES-10, no. 6, pp. 864-866, Nov. 1974.

[17] H. Simpson, "Performance measures and optimization condition for a third-order sampled-data tracker," *IEEE Transactions on Automatic Control,* vol. 8, no. 2, pp. 182-183, April 1963.

[18] D. T. a. T. Singh, "Optimal design of α-β-(γ) filters," in *American Control Conference, 2000. Proceedings of the 2000*, Chicago, IL.

[19] T. Benedict and G. Bordner, "Synthesis of an optimal set of radar track-while-scan smoothing equations," *IRE Transactions on Automatic Control,* vol. 7, no. 4, pp. 27-32, 1962.

[20] R. J. Polge and B. K. Bhagavan, "A Study of the G-H-K Tracking Filter," UAH Research Report No. 176, MICOM Report No. RE-CR-76-1, DTIC ADA021317, 1975.

[21] S. Neal, "Parametric relations for the α-β-γ filter Predictor," *IEEE Transactions on Automatic Control,* vol. 12, no. 3, pp. 315-317, June 1967.

[22] R. E. Kalman, "A new approach to linear filtering and prediction problems," *Transactions of the ASME, Journal of Basic Engineering, Series D,* vol. 82, no. 1, pp. 35-45, March 1960.

[23] R. E. Kalman and R. S. Bucy, "New results in linear filtering and prediction theory," *Transactions of the ASME, Journal of Basic Engineering,* vol. 83, no. 3, pp. 95-107, December 1961.

[24] D. K. Barton, *Radars*, Volume 1, Monopulse Radar, Dedham, MA: Artech House, 1975.

[25] A. I. Leonov and K. I. Fomichev, *Monopulse Radar*, Artech House, 1986.

[26] S. M. Sherman and D. K. Barton, *Monopulse Principles and Techniques*, 2nd ed., Norwood, MA: Artech House, 2011.

[27] E. Brookner, *Aspects of Modern Radar*, Norwood, MA: Artech House, 1988.

[28] J. Andrews, "Examination Of Dichotomous and Centroid Tracking of A Monopulse Angle Tracking Unit," The University of Alabama in Huntsville Master's Thesis, Huntsville, 2017.

[29] J. Andrews, M. Budge, L. Joiner. "An Examination of Dichotomous and Centroid Tracking of a Monopulse Angle Tracking Unit." Proceedings IEEE SoutheastCon 2018, Apr. 2018.

# Chapter 2

# Control Theory Review

## 2.1 INTRODUCTION

As indicated in Chapter 1, we will make use of system type and root locus to help characterize closed loop trackers and open loop track filters. We use system type to help characterize steady state behavior of the tracker and root locus to help characterize its expected transient behavior [1; 2; 3, p. 147].

The closed loop trackers in radars are unity feedback servomechanisms, or servos [1]. We call them unity feedback because the predicted target parameter has the same units as the actual target parameter. As indicated in the block diagrams presented in Figure 2.1, analog and digital trackers have the same components, but the signals, track filters, and controlled elements are in different domains. In the analog tracker, the signals are in the time domain, and the track filter and controlled element transfer functions are in the *s*-domain [1]. In the digital tracker, the signals are in the discrete time domain, and the track filter and controlled element transfer functions are in the *z*-domain [4]. In some cases, the tracker can be hybrid where, for example, the track filter might be digital, and the controlled element might be analog. As a note, the type of receiver and signal processor (analog, digital, or hybrid) does not enter into determining whether the tracker is analog, digital, or hybrid.

An example of an analog tracker would be the angle tracker in a single-target track radar with a reflector antenna [5, 6]. In this case, the controlled element would be the controllers, motors and mechanical structure that move the reflector, thus the controlled element would be analog. We could also treat the dynamics of the reflector controller as part of the track filter since they influence the closed loop response. The rest of the track filter would consist of some op-amp-based R-C (resistor-capacitor) networks and would also be analog. Thus, the overall tracker would be considered an analog tracker, and we would analyze it using differential equations and *s*-domain transfer functions.

An example of a digital tracker would be the angle tracker in a radar that uses a phased array antenna and tracks multiple targets, or performs multiple functions such as track, search, missile guidance, and so forth [7, 8]. Because of this multiplexing, the radar produces error signals, for any one target, at discrete instances of time. Also, in such radars the track filter would most likely be implemented in a digital computer, possibly as a $\alpha$-$\beta$, $\alpha$-$\beta$-$\gamma$, or Kalman filter [9–17]. The controlled element would be the beam steering computer and the phase shifters in the antenna, which move the beam at discrete instances of time. Since events in this tracker are occurring at discrete instances of time, we would consider it a digital tracker and analyze it using difference equations and *z*-transforms.

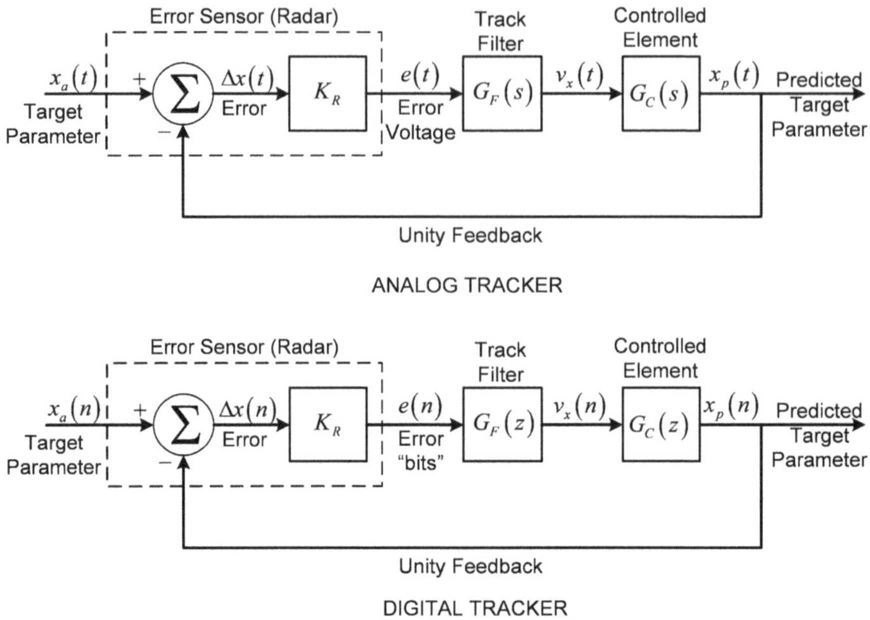


**Figure 2.1** Analog and digital closed loop tracker block diagrams.

An example of a hybrid tracker might be the angle tracker in a seeker that implements the track filter in a digital computer as a *z*-transfer function or a α-β, or other types of track filter. However, the antenna controller might be analog if the seeker has a reflector or flat plate antenna. Also, we consider that the seeker is tracking a single target and is not performing multiple functions. The fact that the track filter is implemented as a digital filter means that we should analyze it using difference equations and *z*-transforms. However, the fact that the controller for the antenna is analog means that we should analyze it using differential equations and *s*-transforms. Actually, what we want to do is recast the *z*-transfer functions to the *s*-domain or recast the *s*-transfer functions to the *z*-domain and analyze the overall system in a single domain. That decision is usually determined by designer/analyst preferences. However, it could be determined by the overall operation of the hybrid tracker. For example, if the tracker updates the antenna controller at fixed, discrete instances of time, the analyst might be inclined to treat it as a digital tracker since events are happening at discrete instants of time.

In the block diagrams of Figure 2.1, the signal-related components of the radar (i.e., the transmitter, antenna, receiver, signal processor, etc.) are represented as a difference and a gain, $K_R$. That is, it forms the error between the target parameter and its estimate and outputs a voltage or number that is proportional to the error. In practice, the gain, $K_R$, is a nonlinear curve that is termed a *discriminator curve*. When we study how the radar forms the error signal, we will see how this nonlinear curve is arises.

In many cases, the controlled element can be considered a unit converter. Its input is a voltage or number, and its output is an estimate of the parameter being tracked. This holds even if the controlled element is a hardware device such as a motor that moves the antenna, or a timing circuit that positions a range gate. In both cases, the input is a voltage or number, and the output is the angle for the antenna or time for the timing circuit.

As implied by Figure 2.1, for continuous-time (analog) trackers we use the Laplace transform and write transfer functions (i.e., $G_F(s)$ and $G_C(s)$) in the $s$-domain. For discrete-time (digital) trackers, we use the $z$-transform and write transfer functions (i.e., $G_F(z)$ and $G_C(z)$) in the $z$-domain. We discuss characteristics of the various transfer functions in terms of their *poles* and *zeros*. [3, p. 41] The poles of a transfer function are the values of $s$ or $z$ where the value of the transfer function goes to infinity, and the zeros are the values of $s$ and $z$ where the value of the transfer function goes to zero. If the transfer function is written as a ratio of polynomials in $s$ or $z$, which is the most common form, the poles are the roots of the denominator polynomial, and the zeros are the roots of the numerator polynomial.

We use *system type*, [3] to characterize the steady state behavior of the closed loop tracker. It tells us whether or not the tracker will experience a steady state bias error. The system type is determined by the location of the poles of the open loop transfer function for the tracker. In terms of Figure 2.1, the open loop transfer functions are the product $G_F(s)G_C(s)$ and $G_F(z)G_C(z)$.

*Root locus* [3, p. 41] is a plot of the poles and zeros of the overall, or *closed loop*, tracker transfer function as we vary the *open loop gain*, $K_R$, of the tracker. The root locus gives an indication of the transient behavior of the tracker. It provides an indication of how fast the tracker will respond to changes in target motion, and whether or not the tracker response will oscillate as it heads toward steady state.

## 2.2 CONTINUOUS TIME SYSTEMS

Since most analysts are more familiar with continuous time systems than discrete time systems, we will begin the discussion of system type and root locus by first considering continuous time systems. The particular model we use is illustrated in Figure 2.2 where we have combined all of the blocks of the continuous time diagram of Figure 2.1 into a single transfer function we term the *open loop transfer function* and denote as $G_{OL}(s)$.

### 2.2.1 System Type and Steady State Error

Most analog trackers are Type 1 or Type 2 servomechanisms [18, pp. 236–255]. A Type 1 servo tracks a constant (step) input with zero steady state error, and a Type 2 servo tracks linearly varying (ramp) and constant inputs with zero steady state error. These features are attractive for radar trackers in that we would like to have zero steady state tracking error. Analog trackers do not normally use higher than Type 2 servos because it is difficult to build stable analog trackers that are Type 3 or higher.

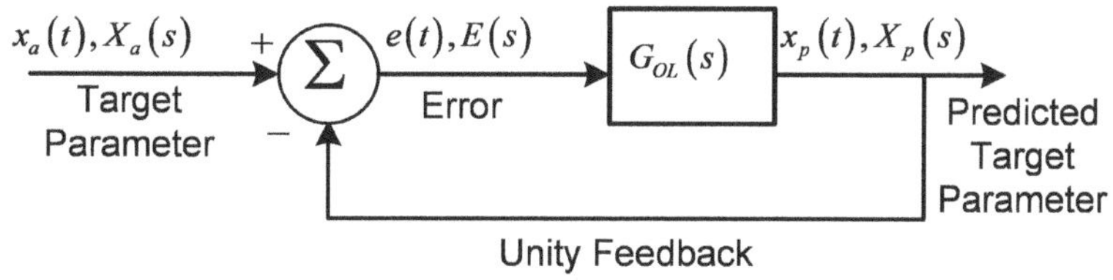


**Figure 2.2** Analog tracker block diagram used in system type and root locus analyses.

The form of $G_{OL}(s)$ that we use to study system type is

$$G_{OL}(s)=\frac{G(s)}{s^N} \tag{2.1}$$

where $G(s)$ has no poles or zeros at the origin ($s = 0$). In the notation of (2.1), the system type is $N$. If $N = 1$, $G_{OL}(s)$ has one pole at the origin, and the system is Type 1. If $N = 2$, $G_{OL}(s)$ has two poles at the origin, and the system is Type 2.

Since we are interested in the steady state error, we will be concerned with the term $e(t)$ in Figure 2.2. Specifically, we are interested in $e_{ss}$, the steady state value of $e(t)$. That is, we are interested in

$$e_{SS} = \lim_{t\to\infty} e(t). \tag{2.2}$$

We will work in the $s$-domain. From Figure 2.2 we have

$$E(s)=X_a(s)-X_p(s) \tag{2.3}$$

or

$$X_p(s)=X_a(s)-E(s). \tag{2.4}$$

We further note that

$$X_p(s)=G_{OL}(s)E(s). \tag{2.5}$$

Combining (2.4) and (2.5) gives

$$G_{OL}(s)E(s)=X_a(s)-E(s) \tag{2.6}$$

and solving for $E(s)$ in terms of $X_a(s)$ gives

$$E(s)=\frac{1}{1+G_{OL}(s)}X_a(s)=\frac{s^N}{s^N+G(s)}X_a(s) \tag{2.7}$$

where we made use of (2.1).

We are interested in $e_{ss}$ for different values of $N$ and inputs of the form

$$x_a(t)=A\frac{t^M}{M!}U(t) \tag{2.8}$$

where $M$ is an integer, and $U(t)$ is the unit step function which is defined as

$$U(t)=\begin{cases}0 & t\le 0\\ 1 & t>0\end{cases}. \tag{2.9}$$

$x_a(t)$ is a step, or constant, input for $M = 0$, a ramp, or constant velocity, input for $M = 1$ and a quadratic, or constant acceleration, input for $M = 2$.

The Laplace transform of (2.8) is

$$X_a(s)=\frac{A}{s^{M+1}}. \tag{2.10}$$

Substituting (2.10) into (2.7) yields

$$E(s)=\frac{s^N}{s^N+G(s)}\frac{A}{s^{M+1}}. \tag{2.11}$$

To find $e_{ss}$, we invoke the final value theorem of Laplace transforms and write [1]

$$e_{ss}=\lim_{t\to\infty}\left[e(t)\right]=\lim_{s\to 0}\left[sE(s)\right]=\lim_{s\to 0}\left[\frac{s^{N+1}}{s^N+G(s)}\frac{A}{s^{M+1}}\right]. \tag{2.12}$$

We will consider a few examples then generate a table of steady state errors for Type 0, 1, and 2 servos with step, ramp, and quadratic inputs. For the first example, we consider a Type 0 servo with a step input. For this case, we have $N = 0$ and $M = 0$. With this, (2.12) becomes

$$e_{ss}=\lim_{s\to 0}\left[\frac{s^{0+1}}{s^0+G(s)}\frac{A}{s^{0+1}}\right]=\lim_{s\to 0}\left[\frac{A}{1+G(s)}\right]=\frac{A}{1+G(0)} \tag{2.13}$$

which tells us that a Type 0 servo with a constant input (step input) will have a steady-state error. Further, the steady state error will be directly proportional to the magnitude of the step input ($A$) and inversely proportional to the DC gain of the open loop transfer function, with the poles at the origin removed (i.e., $G(0)$). The consequence of the last statement is that the steady state error can be controlled by increasing the DC gain of the servo. Unfortunately, this can have a deleterious effect on the transient performance of the servo.

If we consider a Type 0 servo with a ramp input, we would have $N = 0$ and $M = 1$. This would give a steady state error of

$$e_{ss}=\lim_{s\to 0}\left[\frac{s^{0+1}}{s^0+G(s)}\frac{A}{s^{1+1}}\right]=\lim_{s\to 0}\left[\frac{A}{(1+G(s))s}\right]=\infty. \tag{2.14}$$

In other words, the steady state error would be infinite for this case.

As still another example, we consider a Type 1 servo with a step input. For this case, we would have $N = 1$ and $M = 0$, and

$$e_{ss} = \lim_{s\to 0}\left[\frac{s^{1+1}}{s^1 + G(s)}\frac{A}{s^{0+1}}\right] = \lim_{s\to 0}\left[\frac{sA}{s+G(s)}\right] = 0 \tag{2.15}$$

or zero steady state error. That is, a Type 1 servo will, in steady state, track a constant input with zero steady state error.

We can generalize the above discussions to generate Table 2.1 which is a tabulation of steady state errors for Type 0, 1, 2, and 3 servos with step, ramp, and quadratic inputs.

In general, Type 0 analog trackers should be avoided. Even if the target is stationary (a constant input), the tracker will experience a bias error, which is generally undesirable. If the target is moving, which it usually is, the error could grow to the point where the tracker loses track.

Type 1 servos might be acceptable for angle trackers in ground based radars because angle rates in these are generally low, which means the bias errors may be tolerable. The exception to this would be very short range radars with crossing targets. In such situations, the angle rate can become large, which may cause a bias error that is large enough to cause the radar to lose track. Also, it is suspected that Type 1 angle trackers in seekers or airborne radars may have problems because the seeker or aircraft motion might cause large enough angle rates to create unacceptably large bias errors.

In range trackers, the range is constantly changing. This means a Type 1 range tracker might experience bias errors that are large enough to cause the range tracker to lose track. A Type 2, or maybe Type 3, tracker might be a better choice for range trackers. While Type 2 trackers can be made robust from a stability perspective, Type 3 servos tend to be less robust.

The above discussions provided a characterization of the steady state error. If the tracker is too sluggish, it could lose track before it reaches steady state. For example, in a short-range radar with crossing targets, the angle tracker may not be able to follow the short-term changes in position and velocity as the target flies in a tangential direction. Similarly, if a target executes a maneuver, the change in range may be faster than the range tracker can accommodate. The short-term behavior of a tracker is related to the bandwidth and transient characteristics of the tracker. Because of this, we also need a mechanism for characterizing the transient behavior of a closed loop tracker. The root locus provides this mechanism.

**Table 2.1**

Steady State Error versus Input and Analog Servo Type

| *Servo Type* | *Input Type*<br>*Step* | <br>*Ramp* | <br>*Quadratic* |
|---|---|---|---|
| 0 | $\frac{A}{1+G(0)}$ | $\infty$ | $\infty$ |
| 1 | 0 | $\frac{A}{G(0)}$ | $\infty$ |
| 2 | 0 | 0 | $\frac{A}{G(0)}$ |
| 3 | 0 | 0 | 0 |

### 2.2.2 Root Locus and Transient Behavior

The easiest way to get a qualitative idea of the transient behavior of an analog tracker is via the root locus, which gives an idea of how the closed loop poles of the tracker vary as the open loop gain varies from zero to infinity [1, 2]. In general, it provides an idea of whether the tracker will go unstable and a general idea of whether the closed loop transfer function will have all real poles or some complex poles. Complex poles are a requirement if we want to have an underdamped transient response, which is generally the case for radar trackers.

We will restrict ourselves to unity feedback systems since this is the type of servo that we have indicated radar trackers use.

To set the stage for a root locus algorithm, we want to find the closed loop transfer function of the servo block diagram of Figure 2.2. Specifically, we want to find

$$G_{CL}(s) = \frac{X_p(s)}{X_a(s)}. \tag{2.16}$$

If we combine (2.4) and (2.7), and perform the appropriate manipulations, we get

$$\begin{aligned} X_p(s) &= X_a(s) - \frac{1}{1+G_{OL}(s)} X_a(s) \\ &= \frac{G_{OL}(s)}{1+G_{OL}(s)} X_a(s) \end{aligned} \tag{2.17}$$

and, using (2.16),

$$G_{CL}(s) = \frac{G_{OL}(s)}{1+G_{OL}(s)}. \tag{2.18}$$

To study root locus, we write $G_{OL}(s)$ in the form

$$G_{OL}(s) = \frac{K_{OL}(s-z_1)(s-z_2)\cdots(s-z_m)}{(s-p_1)(s-p_2)\cdots(s-p_n)} \tag{2.19}$$

where $K_{OL}$ is the open loop gain, the $z_m$ are the zeros of $G_{OL}(s)$, and the $p_n$ are the poles of $G_{OL}(s)$.

If we substitute (2.19) into (2.18), we get

$$G_{CL}(s) = \frac{K_{OL}(s-z_1)(s-z_2)\cdots(s-z_m)}{(s-p_1)(s-p_2)\cdots(s-p_n) + K_{OL}(s-z_1)(s-z_2)\cdots(s-z_m)}. \tag{2.20}$$

It will be noted that, as $K_{OL}$ varies, the pole locations change. The plot of the pole locations as $K_{OL}$ is varied is termed a *root locus*. The name derives from the fact that the plot is the *locus* of the *roots* of the denominator of $G_{CL}(s)$.

As $K_{OL}$ approaches zero, the poles of the closed loop transfer function ($G_{CL}(s)$) approach the poles of the open loop transfer function ($G_{OL}(s)$), and as $K_{OL}$ approaches infinity, the poles of $G_{CL}(s)$ approach the zeros of $G_{OL}(s)$. This means the poles of $G_{CL}(s)$ will start at the poles of $G_{OL}(s)$ and terminate on the zeros of $G_{OL}(s)$, as $K_{OL}$ varies from zero to infinity. If $n > m$, that is, $G_{OL}(s)$ has more poles than zeros, (which is the usual case), we assume that there are $n - m$ zeros of $G_{OL}(s)$ at infinity. This means some of the poles of $G_{CL}(s)$ go to infinity.

If we restate the above in root locus terminology, the branches of the root locus start on the poles of $G_{OL}(s)$ and end on the zeros of $G_{OL}(s)$. If $n > m$, some of the branches of the locus go to infinity.

We note that the zeros of $G_{OL}(s)$ are also the zeros of $G_{CL}(s)$.

The discussion above implies that it is necessary to form $G_{CL}(s)$ and use a root solver to compute the pole locations as $K_{OL}$ is varied. While this is the rigorous way to plot the root locus, it is also cumbersome. As an alternate approach, control system analysts have developed algorithms for sketching a root locus. One variant of those algorithms is[1]:

1. Draw an $s$-plane where the horizontal axis is the real part of $s$, and the vertical axis is the imaginary part of $s$.
2. Plot the poles of $G_{OL}(s)$ as ×'s and the zeros of $G_{OL}(s)$ as o's.
3. Plot the real-axis portion of the root locus using the rule that a point will be on the locus if the total number of poles and zeros to the right of the point is an odd number.
4. The locus always migrates away from the poles of $G_{OL}(s)$ and towards the zeros of $G_{OL}(s)$.
5. If $G_{OL}(s)$ has more poles than zeros, branches of the locus will go to infinity.
6. If two real-axis branches of the locus meet, the root locus will branch away from the real axis. Sometimes the locus heads toward infinity, and other times it returns to the real axis.
7. If the locus branches away and goes to infinity, it will approach infinity along asymptotes that are located at angles of

$$\theta_k = (2k-1)\frac{180}{K}\text{deg} \quad k = 1,2\cdots K \tag{2.21}$$

   where $K = n - m$ is the difference of the number of poles and zeros of $G_{OL}(s)$.
8. The asymptotes intersect the real axis at the location

$$\sigma_{asy} = \frac{\sum_{i=1}^{n} p_i - \sum_{j=1}^{m} z_j}{K} . \tag{2.22}$$

---

[1] Derivations of these root locus algorithms can be found in many text books on control theory [1; 3, p. ch 6; 21]

To illustrate how to construct a root locus, we consider several root locus examples.

2.2.2.1 Example 1

For the first case, we consider

$$G_{OL}(s) = \frac{K_{OL}}{s+a} \tag{2.23}$$

where $a \geq 0$. $G_{OL}(s)$ has one pole at $s = -a$ and no zeros. Since $K = n - m = 1$, there will be one asymptote at an angle of (see Step 7)

$$\theta_1 = (2-1)\frac{180}{1}\text{deg} = 180^\circ . \tag{2.24}$$

Since there are an even number of poles and zeros to the right of all $s > -a$ (we assume zero is an even number), there will be no locus on the real axis for $s > -a$. However, since there are an odd number of poles and zeros (one) to the right of all $s < -a$, there will be a locus in this region (by definition, the point at $s = -a$ is also part of the locus). Finally, the locus starts on the pole at $s = -a$ and goes to infinity along the asymptote at an angle of $\theta_1 = 180^\circ$. The resultant root locus is shown in Figure 2.3.

The root locus of Figure 2.3 tells us the closed loop response for a step input will be of the form

$$x(t) = c + de^{-\alpha t} \tag{2.25}$$

where $\alpha$ depends on the location of the closed loop pole. Since $\alpha$ increases with increasing $K_{OL,}$ the settling time of $x(t)$ will decrease with increasing open loop gain. Since the closed loop pole, $-\alpha$, always remains in the left half of the $s$-plane, the tracker will always be stable.

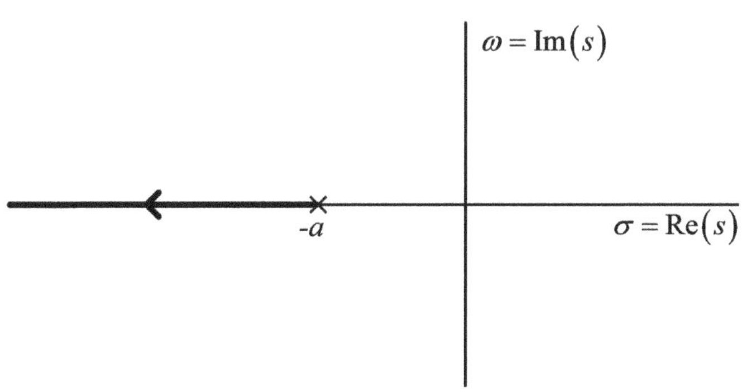


**Figure 2.3** Root locus for $G_{OL}(s)$ of (2.23).

### 2.2.2.2 Example 2

For a second example, we consider the open loop transfer function

$$G_{OL}(s) = \frac{K_{OL}}{(s+a)(s+b)} \tag{2.26}$$

where $b > a \geq 0$. A sketch of the root locus for this case is shown in Figure 2.4. Since there are an odd number of poles and zeros to the right of the region between the two open loop poles, the locus will exist in this region. For all other points on the real axis of the $s$-plane, there are an even number of poles and zeros to the right, thus there is no locus at these points. Since the locus must move away from the poles, segments of the locus start at $s = -a$ and $s = -b$ and meet somewhere between the two poles. The locus then branches away from the real axis and approaches infinity along asymptotes located at angles of

$$\theta_k = (2k-1)\frac{180}{2}\text{deg} \quad k = 1,2, \text{ or } 90^\circ \text{ and } 270^\circ. \tag{2.27}$$

The two asymptotes intersect the $\sigma$ axis at

$$\sigma_{asy} = \frac{\sum_{i=1}^{2} p_i - 0}{2} = \frac{-a-b-0}{2} = \frac{-(a+b)}{2} \tag{2.28}$$

or the center of the two pole locations. There are methods of computing the exact point where the locus branches away from the real axis, but the equations are somewhat tedious and not worth computing for qualitative analyses.

The root locus of Figure 2.4 tells us that, for low values of $K_{OL}$, $G_{CL}(s)$ will have two real poles. This means the time, or transient, response will be *overdamped*. One of the poles will be located between the pole at $s = -a$ and the breakaway point, and the other will be located between $s = -b$ and the breakaway point [1; 3, p. 216].

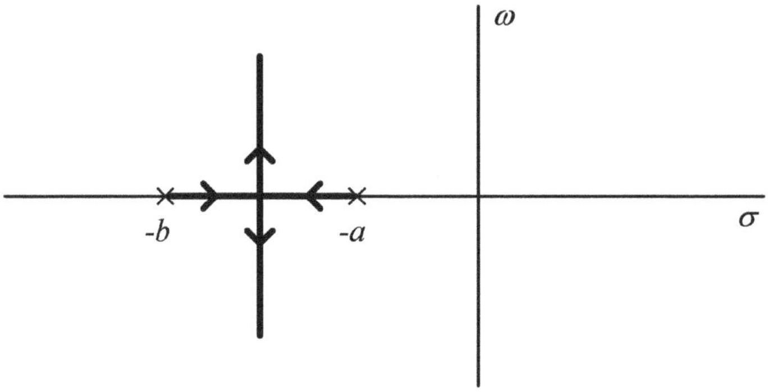


**Figure 2.4** Root locus for the $G_{OL}(s)$ of (2.26).

At some value of $K_{OL}$, the poles will become equal. When this happens, we say the transient response is *critically damped*. Finally, at larger values of $K_{OL}$ the poles will become complex, and the transient response will become *underdamped*. As $K_{OL}$ continues to increase, the frequency of the underdamped oscillation will increase because the imaginary parts of the poles will become larger. However, the damping, or decay, time will remain fixed because the real parts of the poles stay constant at some value between $-a$ and $-b$ (the breakaway point).

Examples of the three closed-loop pole locations and their associated time responses are contained in Figure 2.5. As the response becomes more overdamped (the top figure), the time it takes for the tracker to settle increases. In the limit, as $K_{OL} \rightarrow 0$, the settling time is governed mainly by the pole at $s = -a$. The critically damped case (the center figure) offers the fastest settling time, without overshoot. The transient time for the underdamped case appears to be much shorter than the other two cases because of the oscillations. The actual settling time, the time for the response to get within 10% of its final value, is actually the same as for the critically damped case [3, p. 138].

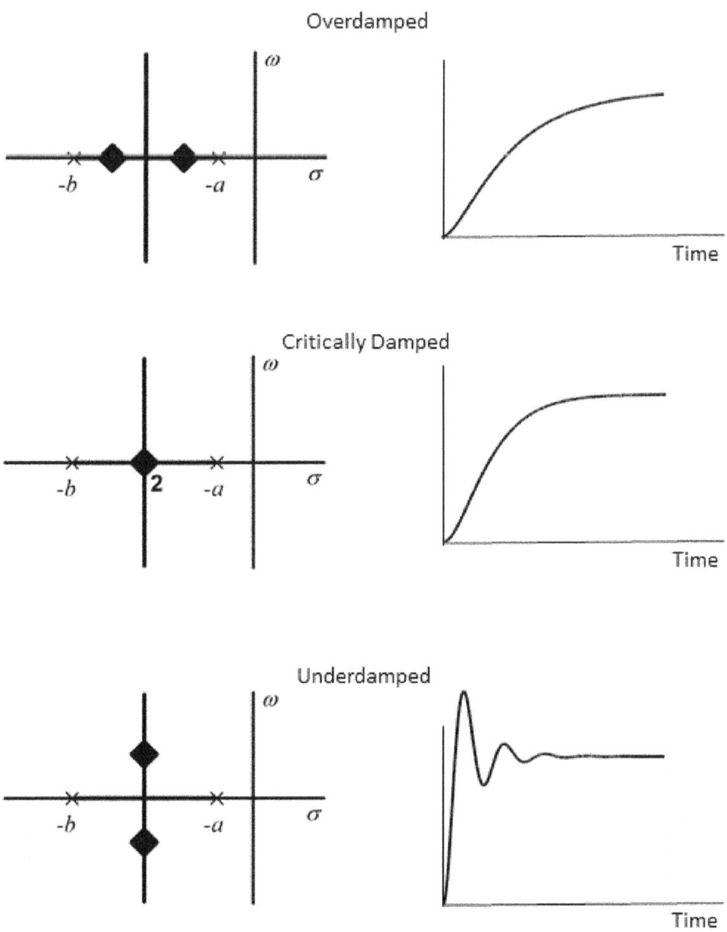


**Figure 2.5** Illustration of three types of damping (closed loop poles denoted by black diamonds).

### 2.2.2.3 Example 3

For a third example, we consider a Type 1 servo with an open loop transfer function of

$$G_{OL}(s) = \frac{K_{OL}(s+b)}{s(s+a)}. \tag{2.29}$$

We consider two cases for the relative sizes of $a$ and $b$. For the first, we let $a > b > 0$. The resulting root locus is shown in Figure 2.6. As can be seen, there will be locus points on the real axis between the pole at the origin (at $s = 0$) and the zero at $s = -b$, and for $s \leq -a$. Since $K = n - m = 2 - 1 = 1$, there will be one asymptote at an angle of 180 degrees. As $K_{OL}$ increases, one of the closed loop poles moves toward the open loop zero, and the other moves toward infinity. Since the two closed loop poles are real, we would expect the closed loop time response to be overdamped. However, the presence of the closed loop zero at $s = -b$ (recall that the closed loop zeros are the open loop zeros) will affect the time response and could make it appear to be underdamped.

For the second case, we let $b > a > 0$. A sketch of the resulting root locus is shown in Figure 2.7. In this case, the locus will exist on the real axis between the poles at the origin and $s = -a$, and to the left of the zero at $s = -b$. However, in this case, the locus branches leave the poles at the origin and $s = -a$, and come together somewhere between the two poles. The locus then branches away from the real axis and returns to the real axis somewhere to the left of the zero at $s = -b$. Finally, the locus travels to the zero, and to infinity along the asymptote at an angle of 180 degrees.

This case is interesting because it says the two closed loop poles are real and unequal for small values of $K_{OL}$, and the transient response is overdamped (except for the possible influence of the closed loop zero at $s = -b$). For intermediate values of $K_{OL}$, the poles become complex, which yields an underdamped response. As the response becomes underdamped, the frequency of oscillation will increase then decrease (as the poles travel around the circular part of the locus) while the settling time will continue to decrease. As the gain is further increased, the poles again become real, and the response becomes overdamped again.

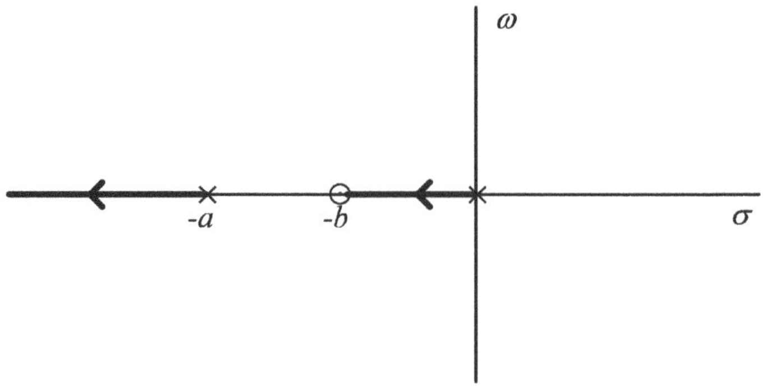


**Figure 2.6** Root locus for the $G_{OL}(s)$ of (2.29) with a > b > 0.

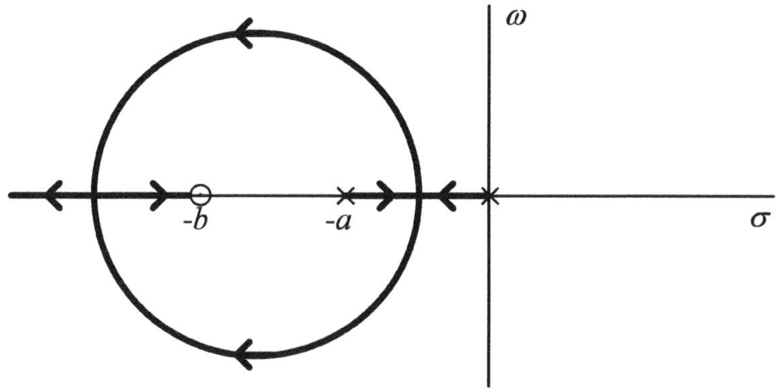


**Figure 2.7** Root locus for the $G_{OL}(s)$ of (2.29) with b > a > 0.

The fact that the locus follows a circle between leaving and re-entering the real axis is notional. To determine the actual shape of the locus, we would need to compute and plot the poles of $G_{CL}(s)$ as we vary $K_{OL}$. However, for qualitative analyses, this is not necessary.

2.2.2.4 Example 4

As a final example, we consider the open loop transfer function

$$G_{OL}(s) = \frac{K_{OL}}{s(s+a)(s+b)} \tag{2.30}$$

with $b > a > 0$. A sketch of the resultant root locus is shown in Figure 2.8. The real axis locus exists between the pole at the origin and the pole at $s = -a$ and to the left of the pole at $s = -b$. Since $K = n - m = 3$, the locus will approach infinity along three asymptotes located at 60, 180, and 300 degrees. Further, the locus approaches the asymptotes at 60 and 300 degrees after it branches away from the real axis somewhere between the origin and $s = -a$. We can conclude that the response will start as an overdamped response for small values of open loop gain then become underdamped for larger values of open loop gain. As the open loop gain continues to increase, two of the closed loop poles move into the right half of the s-plane, which means the tracker will become unstable.

## 2.3 DISCRETE TIME SERVOS

We now want to consider discrete time servos, which we also call digital servos [4]. As indicated by Figure 2.9, the block diagram for a digital servo looks the same as an analog servo, except the transfer functions are a function of $z$, and the time responses are a function of $n$, a discrete time, "digital", time variable.

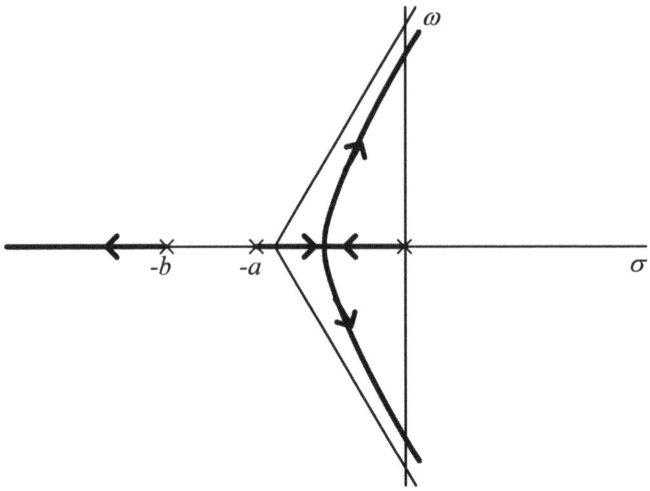


**Figure 2.8** Root locus for the $G_{OL}(s)$ of (2.30).

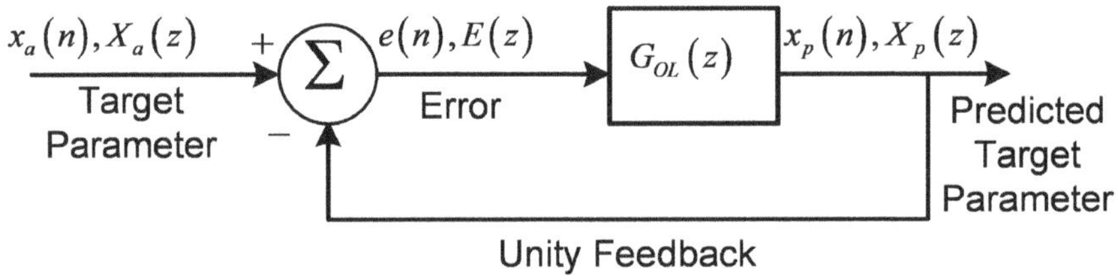


**Figure 2.9** Digital servo block diagram.

### 2.3.1 System Type and Steady State Error

The transfer function manipulations for digital servos are the same as analog servos. Thus, the closed loop transfer function of the servo of Figure 2.9 is

$$G_{CL}(z) = \frac{G_{OL}(z)}{1 + G_{OL}(z)}. \tag{2.31}$$

As with analog servos, we can have Type 0, 1, 2, etc. digital servos. A Type $N$ digital servo will have $N$ poles at $z = 1$. That is, the open loop transfer function can be written as

$$G_{OL}(z) = \frac{G(z)}{(z-1)^N} \tag{2.32}$$

where $G(z)$ has no poles or zeros at $z = 1$. The steady state response properties of analog servos also apply to digital servos. That is, a Type 0 digital servo will follow a step input with a constant steady state error. However, its error will become unbounded for ramp and higher order inputs. A Type 1 digital servo will follow a step input with zero steady state error, and a ramp input with a constant steady state error, and so forth.

The equation for finding steady state error from the $z$-transform is different for digital servos than for analog servos. Specifically, the steady state error is given by [4]

$$e_{ss} = \lim_{n\to\infty}\left[e(n)\right] = \lim_{z\to 1}\left[(z-1)E(z)\right]. \tag{2.33}$$

If we use a development similar to that of Section 2.2.1, we arrive at

$$e_{ss} = \lim_{z\to 1}\left[(z-1)\frac{1}{1+G_{OL}(z)}X_a(z)\right] = \lim_{z\to 1}\left[\frac{(z-1)^{N+1}}{(z-1)^N + G(z)}X_a(z)\right]. \tag{2.34}$$

In general, the form of $X_a(z)$ is not as simple as for the analog servo. However, specific forms for step, ramp, and quadratic inputs are relatively simple as indicated in Table 2.2. The table also contains an expression for higher order inputs [19, 20].

As for the analog case, we consider a few specific examples then present a summary table like Table 2.1. For the first example, we consider a Type 0 servo with a step input. For this case, we have $N = 0$ and use the $F(z)$ from Table 2.2 that corresponds to a step input. With this, (2.34) reduces to

$$e_{ss} = \lim_{z\to 1}\left[\frac{(z-1)}{1+G(z)}\frac{Az}{(z-1)}\right] = \frac{A}{1+G(1)} \tag{2.35}$$

which, as with the analog case, tells us that a Type 0 servo with a constant input (step input) will have a steady-state error. Further, the steady state error will be directly proportional to the magnitude of the step input, $A$, and inversely proportional to the DC gain of the open loop transfer function, with the poles at the origin removed [i.e., $G(1)$]

If we consider a Type 0 servo with a ramp input, we would have $N = 0$ and use the $F(z)$ of Table 2.2 that corresponds to a ramp input. This would give a steady state error of

**Table 2.2**

$z$-Transforms of Some Standard Inputs

| $f(n)$ | $F(z)$ |
|---|---|
| $Au(n)$ *(step)* | $\frac{Az}{z-1}$ |
| $Anu(n)$ *(ramp)* | $\frac{Az}{(z-1)^2}$ |
| $\frac{An^2}{2}u(n)$ *(quadratic)* | $\frac{Az(z+1)}{2(z-1)^3}$ |
| $\frac{An^n}{m!}u(n)$ *($n^{th}$ order)* | $A\lim_{\alpha\to 0}\frac{(-1)^n}{n!}\frac{\partial^n}{\partial\alpha^n}\left[\frac{z}{z-e^{-\alpha}}\right]$ |

$$e_{ss} = \lim_{z \to 1} \left[ \frac{(z-1)}{1+G(z)} \frac{Az}{(z-1)^2} \right] = \lim_{z \to 1} \left[ \frac{Az}{(z-1)(1+G(z))} \right] = \infty . \tag{2.36}$$

In other words, the steady state error would be infinite for this case.

As still another example, we consider a Type 1 servo with a step input. For this case, we would have $N = 1$and use the $F(z)$ from Table 2.2 that corresponds to a step input. The result is

$$e_{ss} = \lim_{z \to 1} \left[ \frac{(z-1)^2}{(z-1)+G(z)} \frac{Az}{(z-1)^2} \right] = \lim_{z \to 1} \left[ \frac{Az(z-1)}{(z-1)+G(z)} \right] = 0 \tag{2.37}$$

or zero steady state error. That is, a Type 1 servo will, in steady state, track a constant input with zero steady state error.

We can generalize the above discussions to generate Table 2.3 which is a tabulation of steady state errors for Type 0, 1, 2, and 3 servos with step, ramp, and quadratic inputs.

As with analog trackers, Type 0 digital trackers should be avoided. Even if the target is stationary (a constant input) the tracker will experience a bias error, which is generally undesirable. If the target is moving, which it usually is, the error could grow to the point where the tracker loses track.

As we will see in the next chapter, α-β filters, which are used with angle trackers, and some range trackers, are Type 2 servos. α-β-γ filters, which are used with some range trackers, are Type 3 servos. Unlike Type 3 analog trackers, trackers that contain α-β-γ filters are reasonably easy to make stable.

### 2.3.2 Root Locus and Transient Behavior

The procedure for constructing root loci for digital servos is the same as for analog servos [4]. As an example, we draw the root locus for a digital servo analogous to the analog servo of Example 3 (Section 2.2.2.3). The open loop transfer function is

$$G_{OL}(z) = \frac{K_{OL}(z-b)}{(z-1)(z-a)} . \tag{2.38}$$

**Table 2.3**

Steady State Error versus Input and Digital Servo Type

| | | *Input Type* | |
|---|---|---|---|
| *Servo Type* | *Step* | *Ramp* | *Quadratic* |
| 0 | $\frac{A}{1+G(1)}$ | $\infty$ | $\infty$ |
| 1 | 0 | $\frac{A}{G(1)}$ | $\infty$ |
| 2 | 0 | 0 | $\frac{A}{G(1)}$ |
| 3 | 0 | 0 | 0 |

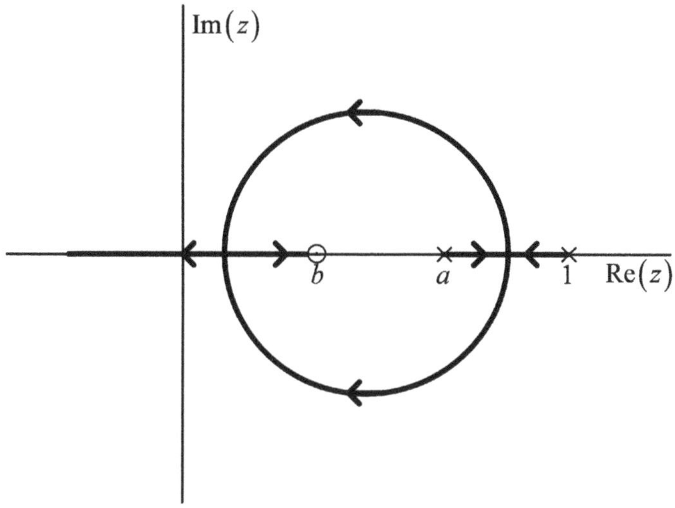


**Figure 2.10** Root locus for the $G_{OL}(s)$ of (2.38) for $0 < b < a < 1$.

Analogous to the analog equivalent, $G_{OL}(z)$ has a zero at $z = b$ and a pole at $z = a$, where the zero and pole locations satisfy $0 < b < a < 1$. It has a second pole at $z = 1$, which is equivalent to the pole at $s = 0$ for the analog equivalent. From the system type discussions of Section 2.3.1, the servo is Type 1 because it has one pole at $z = 1$.

A sketch of the root locus is shown in Figure 2.10. As can be seen, the sketch looks similar to Figure 2.7, but shifted so that the circle is in the right half plane. As with Figure 2.7, the circular path of Figure 2.10 is notional. To find the actual shape of the locus, we would need to compute the poles of $G_{CL}(z)$ as $K_{OL}$ varies from zero to infinity.

The locations of the poles, zeros, and the circles, in Figure 2.7 and Figure 2.10 prompt a discussion of how regions of the $s$- and $z$-planes relate to each other, and to servo behavior. For analog servos, the $s$-plane regions of interest are those bounded by the imaginary, or $\omega$, axis. If all poles of $G_{CL}(s)$ lie in the left half of the $s$-plane, the response will remain bounded for a bounded input (the system is asymptotically stable). If any poles lie in the right half of the $s$-plane, the response increases without bound (the system is unstable). If any poles lie on the imaginary axis, the response could increase without bound or remain bounded (the system is marginally stable).

For digital servos, the $z$-plane regions of interest are those bounded by a unit circle in that plane. If any poles of $G_{CL}(z)$ lie outside of the unit circle, the response increases without bound (the system is unstable). If all of the poles lie inside of the unit circle, the response will remain bounded for a bounded input (the system is asymptotically stable). However, if any of the poles lie inside the left half of the unit circle, the response will oscillate from sample to sample [21]. If any poles lie on the unit circle, the response could increase without bound or remain bounded (the system is marginally stable). As indicated in Figure 2.11, the point $z = 1$ is analogous to the origin of the s-plane. That is why the number of poles at $z = 1$ determines the system type.

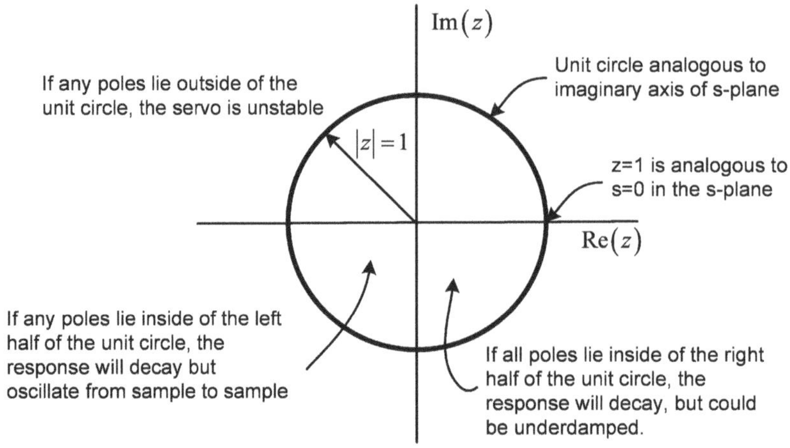


**Figure 2.11** Description of the significant parts of the $z$-plane.

When we create a frequency response for a continuous time system represented by $G(s)$, we compute $|G(s)|$ or $|G(s)|^2$ for $s = j\omega = j2\pi f$. When we create a frequency response for a discrete time system represented by $G(z)$, we compute $|G(z)|$ or $|G(z)|^2$ for $z = e^{j\theta}$, where $\theta$ varies from 0 to $2\pi$ or $-\pi$ to $\pi$, In other words, we compute $|G(z)|$ as $z$ varies around the unit circle of the z plane. While this is strictly legal, this approach makes it difficult to relate the frequency response to frequency, $f$. To get around this problem, we prefer to compute $|G(z)|$ or $|G(z)|^2$ for $z = e^{j2\pi fT}$ where $f$ varies from 0 to $1/T$ or $-1/(2T)$ to $1/(2T)$. Of course, this means we must know the sample period, $T$. In trackers, we know this because it is the track update period.

## 2.4 MODELING CLOSED LOOP SERVOS

Now that we have a technique for determining the general properties of analog and digital servos, we want to consider how to model them.

### 2.4.1 Analog Servo Modeling

We will discuss two methods of modeling analog servos. With one method, we derive a state variable representation of $G_{OL}(s)$ and use a differential equation solver like Runge-Kutta[2], predictor-corrector or the MATLAB® ode45 to solve the resulting matrix differential equation [22–34]. The second method, which we term the $z$-transform method, is more direct but requires a bilinear transform routine and the MATLAB filter function [34–37].

#### 2.4.1.1 State Variable Method

To use the state variable method, we need to review how to find a state variable representation of an $s$-domain transfer function. We consider the transfer function

---

[2] The Runge–Kutta methods are named after German mathematician and physicist Carl Runge and German mathematician Martin Kutta [23, 24].

$$G(s) = \frac{b_0 s^m + b_1 s^{m-1} + b_2 s^{m-2} + \cdots + b_{m-1}s + b_m}{s^n + a_1 s^{n-1} + a_2 s^{n-2} + \cdots + a_{n-1}s + a_n} = \frac{Y(s)}{E(s)} \tag{2.39}$$

which is $G_{OL}(s)$ with $K_{OL}$ incorporated into the numerator coefficients (the $b$'s). In this equation, we stipulate that $m < n$ which will apply to the cases of interest to us. If $m = n$, we would need to add an extra step of dividing the numerator by the denominator before applying the technique below. If $m > n$, we have a situation where the transfer function has derivatives, which we avoid because they are physically unrealizable.

There are several state variable representations that will result in the transfer function of (2.39). The particular form we will use is the *phase variable* form because it is the easiest to derive [32, 33, 38].

For the first step, we divide (2.39) into the two equations

$$\frac{W(s)}{E(s)} = \frac{1}{s^n + a_1 s^{n-1} + a_2 s^{n-2} + \cdots + a_{n-1}s + a_n} \tag{2.40}$$

and

$$\frac{Y(s)}{W(s)} = b_0 s^m + b_1 s^{m-1} + b_2 s^{m-2} + \cdots + b_{m-1}s + b_m . \tag{2.41}$$

We solve (2.40) for $E(s)$ in terms of $W(s)$ and (2.41) for $Y(s)$ in terms of $W(s)$ to get

$$\left(s^n + a_1 s^{n-1} + a_2 s^{n-2} + \cdots + a_{n-1}s + a_n\right)W(s) = E(s) \tag{2.42}$$

and

$$Y(s) = \left(b_0 s^m + b_1 s^{m-1} + b_2 s^{m-2} + \cdots + b_{m-1}s + b_m\right)W(s). \tag{2.43}$$

We next use the Laplace transform of derivatives to convert (2.42) and (2.43) to differential equations. The particular Laplace transform of interest is [19, 36]

$$\mathcal{L}\left[\frac{d^n f(t)}{dt^n}\right] = s^n F(s) - \sum_{l=0}^{n-1} s^{n-l-1} \left.\frac{d^l f(t)}{dt^l}\right|_{t=0} \tag{2.44}$$

where the summation captures the initial conditions on all of the derivatives of $f(t)$. Since the transfer function of a system is defined as the ratio of the Laplace transform of the output divided by the Laplace transform of the input, assuming the initial conditions are zero, we can ignore the initial condition terms of (2.44) when finding the differential equations. With this, we can transform (2.42) and (2.43) to differential equations as

$$w^{(n)}(t)+a_1w^{(n-1)}(t)+a_2w^{(n-2)}(t)+\cdots+a_{n-1}w'(t)+a_nw(t)=e(t) \tag{2.45}$$

and

$$b_0w^{(m)}(t)+b_1w^{(m-1)}(t)+b_2w^{(m-2)}(t)+\cdots+b_{m-1}w'(t)+b_mw(t)=y(t). \tag{2.46}$$

In these equations, $w^{(k)}(t)=d^kw(t)/dt^k$ .

To transform this to a state variable representation, we define the state variables as

$$\begin{bmatrix} x_1(t) \\ x_2(t) \\ \vdots \\ x_n(t) \end{bmatrix} = \begin{bmatrix} w(t) \\ dw(t)/dt \\ \vdots \\ d^{n-1}w(t)/dt^{n-1} \end{bmatrix}. \tag{2.47}$$

We next take the derivative of both sides of (2.47) to give

$$\begin{bmatrix} x_1'(t) \\ x_2'(t) \\ \vdots \\ x_n'(t) \end{bmatrix} = \begin{bmatrix} dw(t)/dt \\ d^2w(t)/dt^2 \\ \vdots \\ d^nw(t)/dt^n \end{bmatrix} = \begin{bmatrix} x_2(t) \\ x_3(t) \\ \vdots \\ d^nw(t)/dt^n \end{bmatrix} \tag{2.48}$$

where the apostrophe (′) denotes the derivative. To get the right vector of (2.48), we made use of (2.47).

We next need to consider the last element of the right vector. From (2.45) we have

$$w^{(n)}(t)=-a_1w^{(n-1)}(t)-a_2w^{(n-2)}(t)-\cdots-a_{n-1}w'(t)-a_nw(t)+e(t) \tag{2.49}$$

or, with relations of (2.47),

$$\begin{aligned} w^{(n)}(t)&=d^nw(t)/dt^n \\ &=-a_1x_n(t)-a_2x_{n-1}(t)-\cdots-a_{n-1}x_2(t)-a_nx_1(t)+e(t) \end{aligned} \tag{2.50}$$

Substituting this into (2.48) gives

$$\begin{bmatrix} x_1'(t) \\ x_2'(t) \\ \vdots \\ x_n'(t) \end{bmatrix} = \begin{bmatrix} x_2(t) \\ x_3(t) \\ \vdots \\ -a_1 x_n(t) - a_2 x_{n-1}(t) - \cdots - a_{n-1} x_2(t) - a_n x_1(t) + e(t) \end{bmatrix} \quad (2.51)$$

which we write this in matrix form as

$$X' = AX + Be(t) \quad (2.52)$$

where

$$X = \begin{bmatrix} x_1(t) \\ x_2(t) \\ x_3(t) \\ \vdots \\ x_n(t) \end{bmatrix}, X' = \begin{bmatrix} x_1'(t) \\ x_2'(t) \\ x_3'(t) \\ \vdots \\ x_n'(t) \end{bmatrix}, A = \begin{bmatrix} 0 & 1 & 0 & \cdots & 0 \\ 0 & 0 & 1 & \cdots & 0 \\ \vdots & \vdots & \vdots & \vdots & \vdots \\ 0 & 0 & 0 & \cdots & 1 \\ -a_n & -a_{n-1} & -a_{n-2} & \cdots & a_1 \end{bmatrix}, B = \begin{bmatrix} 0 \\ 0 \\ 0 \\ \vdots \\ 1 \end{bmatrix} \quad (2.53)$$

We dropped the time dependency of $X$ and $X'$ as a notational convenience.

From (2.46) and (2.47) we can write

$$y(t) = b_m x_1(t) + b_{m-1} x_2(t) + \cdots + b_1 x_{m-1}(t) + b_0 x_m(t) \quad (2.54)$$

or

$$y(t) = C^T X \quad (2.55)$$

where

$$C^T = [b_m \quad b_{m-1} \quad \cdots \quad b_1 \quad b_0 \quad 0 \quad \cdots]. \quad (2.56)$$

$C^T$ has $n$ elements. Equations (2.52), (2.53), (2.55), and (2.56) constitute a state variable representation of the transfer function defined by (2.39).

#### 2.4.1.2 *z*-Transform Method

For the *z*-transform method, we find the *z*-domain equivalent of $G_{OL}(s)$ and use something like the MATLAB filter function to model the resulting *z*-domain transfer function, $G_{OL}(z)$. There are several methods of deriving $G_{OL}(z)$ from $G_{OL}(s)$ such as the impulse invariant technique, the step invariant technique, and variations on the bilinear transform. We will use one of the bilinear transform techniques. With this technique, we derive $G_{OL}(z)$ from $G_{OL}(s)$ by replacing $s$ with

$$s = \frac{2}{T}\frac{z-1}{z+1} \tag{2.57}$$

where $T$ is a sample period equivalent to the update period used in differential equation solvers such as Runge-Kutta. A bilinear transform function is included in the MATLAB Signal Processing and Controls toolboxes. For those who do not have these toolboxes, a simplified bilinear transform routine is included on the MATLAB files that accompanies this book. The details of how to use the $z$-transform technique will be illustrated via example in Section 2.2.4.2.

### 2.4.1.3 Simulation Examples

To illustrate the simulation techniques of Sections 2.4.1.1 and 2.4.1.2, we consider a specific example. We choose $G_{OL}(s)$ as

$$G_{OL}(s) = \frac{K_{OL}}{s(s+a)}. \tag{2.58}$$

Since $G_{OL}(s)$ has one pole at the origin, the closed loop servo is Type 1. Thus, we expect it will track a step input with zero steady state error and a ramp input with a constant steady state error. If we were to plot the root locus, we would conclude that we can adjust $K_{OL}$ to produce an underdamped transient response.

#### *2.4.1.3.1 State Variable Approach*

To derive the state variable representation of $G_{OL}(s)$, we need to write (2.58) in the form of (2.39). Specifically

$$G_{OL}(s) = \frac{K_{OL}}{s(s+a)} = \frac{b_0 s + b_1}{s^2 + a_1 s + a_0}. \tag{2.59}$$

From (2.59) we conclude that $b_0 = 0$, $b_1 = K_{OL}$, $a_1 = a$, and $a_2 = 0$. If we use these assignments in (2.53) and (2.56), we get

$$A = \begin{bmatrix} 0 & 1 \\ -a_2 & -a_1 \end{bmatrix} = \begin{bmatrix} 0 & 1 \\ 0 & -a \end{bmatrix},\ B = \begin{bmatrix} 0 \\ 1 \end{bmatrix},\ C = \begin{bmatrix} K_{OL} & 0 \end{bmatrix}. \tag{2.60}$$

The resulting state variable equation is

$$\begin{bmatrix} x_1' \\ x_2' \end{bmatrix} = \begin{bmatrix} 0 & 1 \\ 0 & -a \end{bmatrix}\begin{bmatrix} x_1 \\ x_2 \end{bmatrix} + \begin{bmatrix} 0 \\ 1 \end{bmatrix} e,\ x_p = \begin{bmatrix} K_{OL} & 0 \end{bmatrix}\begin{bmatrix} x_1 \\ x_2 \end{bmatrix}. \tag{2.61}$$

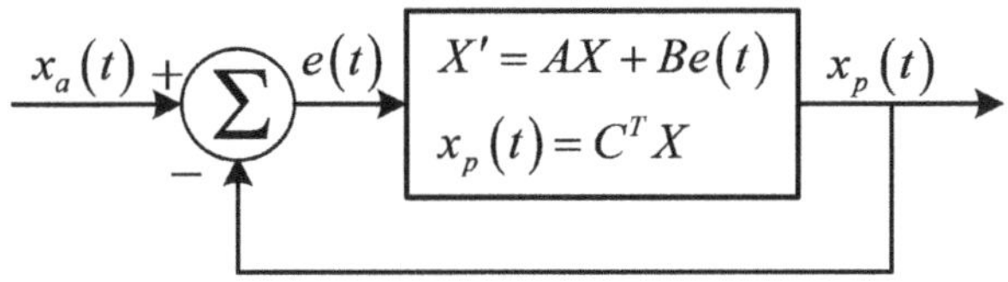


**Figure 2.12** Block diagram of the state variable example.

A block diagram of the resulting closed loop servo is contained in Figure 2.12, and a flow diagram of the computer code is shown in Figure 2.13.

In the computer code, we start by defining *A*, *B*, and *C* and the integration step size, *dt*. Care must be taken when setting *dt*. If it is too small, the simulation run time may be unacceptably large. If it is too large, the simulation results will be wrong because we will have violated the Shannon sampling theorem [39, 40][3]. A rule of thumb is to set the initial value of *dt* to 1/10 the smaller of

1. The reciprocal of the closed loop bandwidth and;
2. $2\pi$ divided by the magnitude of the largest open loop pole.

Once *dt* is defined, we can compute the time array and the input, $x_a(t)$. We also set the initial state, *Xold*, to zero. We also set the initial error, $e(0)$ and the initial output, $x_p(0)$, to zero.

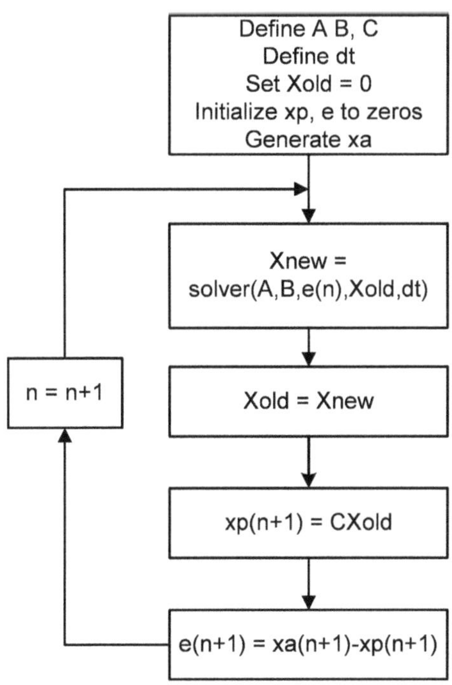


**Figure 2.13** Flow diagram of state variable computer code.

---

[3] Also known as the Nyquist sampling theorem (perhaps erroneously) after Harry Nyquis [40] In Russia, the equivalent theorem is known by the name Kotel'nikov after Vladimir Aleksandrovich Kotel'nikov (Владимир Александрович Котельников) [41] .

The first operation in the iterative loop is to compute the state one *dt* in the future. The block diagram shows this being done with a routine called solver. This could be a Runge-Kutta routine or some other differential equation solver. A specific example is the MATLAB ode45 routine. The inputs to the solver are the state variable equation, the integration step size, *dt*, the initial state, *Xold*, and value of the input at the current time, *e*. The output is the new state vector, *Xnew*.

In the next block, *Xold* is set to *Xnew* and will serve as the initial condition on the next pass through the loop.

The next two blocks compute the servo output, *xp*, and the error, *e*, for the next pass through the loop. As a note, the input used to compute *e* (from $e = xa - xp$) is computed at the next time step, *tnew*.

*2.4.1.3.2 z-Transform Approach*

For the *z*-transform approach, we start with (2.58) and apply the bilinear transform given by (2.57). This yields the result

$$G_{OL}(z) = G_{OL}(s)\Big|_{s=\frac{2}{T}\frac{z-1}{z+1}} = \frac{K_{OL}}{s(s+a)}\Bigg|_{s=\frac{2}{T}\frac{z-1}{z+1}} = \frac{K_{OL}}{\frac{2}{T}\frac{z-1}{z+1}\left(\frac{2}{T}\frac{z-1}{z+1}+a\right)}$$
$$= \frac{\left(T^2/4\right)K_{OL}(z+1)^2}{(z-1)\left(z-1+aT/2(z+1)\right)} = \frac{\left(T^2/4\right)K_{OL}\left(z^2+2z+1\right)}{(aT/2+1)z^2-2z+1-aT/2} \quad (2.62)$$

A block diagram of the closed loop servo for this approach is contained in Figure 2.14, and a flow diagram of a computer program that implements the block diagram is shown in Figure 2.15.

Again, the first part of the computer code is concerned with establishing the various parameters. The computation loop is similar to the state variable method except that the MATLAB filter function (or some other function capable of modeling *z*-domain transfer functions) is used to compute the output of $G_{OL}(z)$. As a note, the filter function requires an initial state vector (as did the state variable method) and returns the updated state vector. The lengths of these vectors equal the order of the denominator of $G_{OL}(s)$ (2 in this example). As with the state variable method, we start with an initial state vector of zero.

*2.4.1.3.3 Determining the Open Loop Parameters*

To complete the example, we want to use specific values for $K_{OL}$ and *a*, and generate step and ramp responses. As mentioned earlier, a plot of the root locus reveals that the closed loop response could be made to have a pair of complex poles and thus exhibit an underdamped response (see Section 2.2.2.2). From our previous work (see Section 2.2.2), we can write the closed loop transfer function as

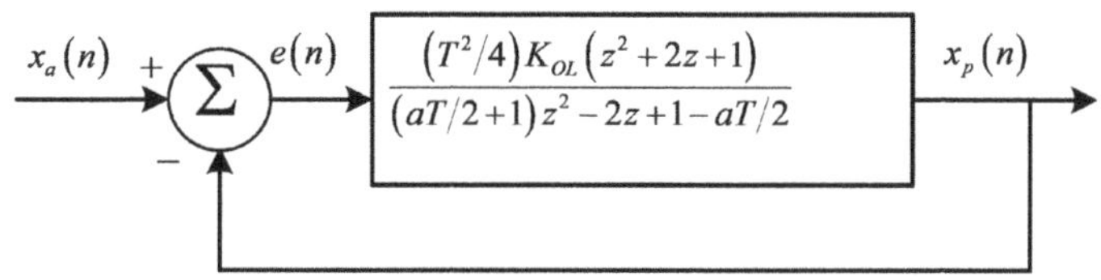


**Figure 2.14** Block diagram of *z*-transform example.

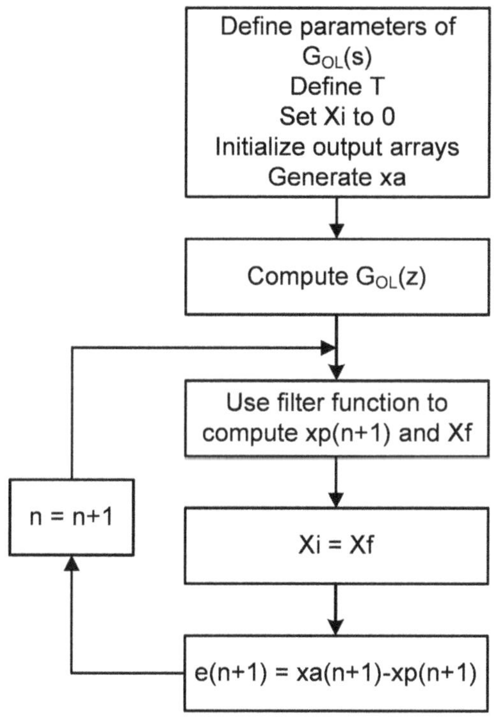


**Figure 2.15** Flow diagram of *z*-transform computer code.

$$G_{CL}(s) = \frac{G_{OL}(s)}{1+G_{OL}(s)} = \frac{\frac{K_{OL}}{s(s+a)}}{1+\frac{K_{OL}}{s(s+a)}} = \frac{K_{OL}}{s(s+a)+K_{OL}} = \frac{K_{OL}}{s^2+as+K_{OL}} \tag{2.63}$$

which we can write a standard quadratic form as

$$G_{CL}(s) = \frac{K_{OL}}{s^2+as+K_{OL}} = \frac{K_{OL}}{s^2+2\zeta\omega_n s+\omega_n^2}. \tag{2.64}$$

In the right side of (2.64), $\omega_n$ is termed the *undamped natural frequency,* and $\zeta$ is termed the *damping coefficient* or *damping ratio* [1; 3, p. 41; 38]. These parameters tell us something about the settling time and the oscillation frequency of the transient.

If we factor the denominator of the last term of (2.64), we obtain the roots

$$s_1, s_2 = -\zeta\omega_n \pm \omega_n\sqrt{\zeta^2 - 1}\,. \qquad (2.65)$$

If $\zeta > 1$, both roots are real, which means the transient response will be overdamped. If $\zeta = 1$, the two roots are equal, and the response is said to be critically damped. If $\zeta < 1$, the roots are complex conjugates, and the response is underdamped, which is the condition we want. In (2.65),

$$\omega_d = \omega_n\sqrt{1 - \zeta^2} \qquad (2.66)$$

is defined as the *damped frequency*. It, and not $\omega_n$, is the frequency of the ringing, or oscillation, in the transient response. The *time constant*, or *decay time constant*, is $1/\zeta\omega_n$. It determines how long the response will take to decay, or settle to a steady value. Generally, we select $\zeta$ to be a value between 0.5 and $1/\sqrt{2} \approx 0.707$ then choose $\omega_n$ to give a desired decay time. $\zeta = 0.707$ results in a step response that has one overshoot peak before reaching steady state. When $\zeta = 0.5$, the rise time of the response will be faster, but there will be more ringing as the response settles. As $\zeta$ gets smaller, the rise time decreases, and the ringing increases. As $\zeta$ becomes larger, the rise time increases, and the ringing disappears (the response becomes critically damped then overdamped).

In many cases, we find it intuitively appealing to characterize the closed loop behavior in terms of frequency rather than time. Specifically, we specify the damping ratio, $\zeta$, and a *closed loop bandwidth*, $\omega_c$ rad/s or $f_c = \omega_c/2\pi$ Hz. Since $G_{OL}(s)$ has a low pass response, we define the bandwidth as the value of $\omega_c$ at which $|G_{OL}(s)|^2$ is ½ its value at $s = 0$. In equation form, we want the $\omega_c$ such that

$$\left|G_{OL}(s)\right|^2_{s=j\omega_c} = \frac{1}{2}\left|G_{OL}(0)\right|^2 \qquad (2.67)$$

or, for this example,

$$\left|\frac{K_{OL}}{-\omega_c^2 + 2j\zeta\omega_n\omega_c + \omega_n^2}\right|^2 = \frac{1}{2}\left|\frac{K_{OL}}{\omega_n^2}\right|^2 \qquad (2.68)$$

which leads to

$$\left(\omega_n^2 - \omega_c^2\right)^2 + \left(2\zeta\omega_n\omega_c\right)^2 = 2\omega_n^4\,. \qquad (2.69)$$

Solving (2.69) for $\omega_n$ results in

$$\omega_n = \omega_c\sqrt{\sqrt{\left(1 - 2\zeta^2\right)^2 + 1} + 2\zeta^2 - 1}\,. \qquad (2.70)$$

Interestingly, if we choose $\zeta = 1/\sqrt{2}$, $\omega_n = \omega_c$ which says the undamped natural frequency is the closed loop bandwidth.

Once we determine $\zeta$ and $\omega_n$, we can use (2.64) to relate them to $a$ and $K_{OL}$ by equating like powers of $s$. This gives

$$a = 2\zeta\omega_n \tag{2.71}$$

and

$$K_{OL} = \omega_n^2 . \tag{2.72}$$

With this we get an open loop transfer function of

$$G_{OL}(s) = \frac{\omega_n^2}{s(s + 2\zeta\omega_n)} . \tag{2.73}$$

*2.4.1.3.3.1 A Specific Case*

As a specific example, we choose a closed loop bandwidth of $f_c = 1$ Hz and a damping ratio of $\zeta = 1/\sqrt{2}$. This gives $\omega_c = 2\pi f_c = 2\pi$ rad/sec and from (2.70)

$$\omega_n = \omega_c \sqrt{\sqrt{\left(1 - 2\zeta^2\right)^2 + 1} + 2\zeta^2 - 1} = \omega_c = 2\pi \tag{2.74}$$

With this we get

$$G_{OL}(s) = \frac{\omega_n^2}{s(s + 2\zeta\omega_n)} = \frac{(2\pi)^2}{s\left[s + 2\left(1/\sqrt{2}\right)(2\pi)\right]} \approx \frac{39.5}{s^2 + 8.9s} . \tag{2.75}$$

If we use (2.75) with (2.61) we get

$$\begin{bmatrix} x_1' \\ x_2' \end{bmatrix} = \begin{bmatrix} 0 & 1 \\ 0 & -8.9 \end{bmatrix} \begin{bmatrix} x_1 \\ x_2 \end{bmatrix} + \begin{bmatrix} 0 \\ 1 \end{bmatrix} e, \; x_p = \begin{bmatrix} 39.5 & 0 \end{bmatrix} \begin{bmatrix} x_1 \\ x_2 \end{bmatrix} \tag{2.76}$$

for the state variable representation.

Using (2.75) with (2.62) yields

$$G_{OL}(z) = \frac{0.0247\left(z^2 + 2z + 1\right)}{0.778z^2 - 2z + 1.223} \tag{2.77}$$

for the $z$-transfer function. In (2.77), we used an update period of $T = 0.05$ seconds.

### 2.4.1.3.3.2 Example Plots

If we use (2.76) in the state variable approach of Section 2.4.1.3.1 or (2.77) in the $z$-transform approach of Section 2.4.1.3.2, we obtain the plots of Figures 2.16, 2.17, and 2.18 for the cases where the input is a step, a ramp, and a quadratic. The top graph contains plots of the input ($x_a(t)$ — dotted line) and the tracker output ($x_p(t)$ — solid line). The bottom graph contains plots of the error ($e(t) = x_a(t) - x_p(t)$). It will be noted that the tracker follows a step input with zero steady state error. For the ramp input, there is a constant steady state error, and for the quadratic input, the error increases with time. These are the expected behaviors because $G_{OL}(s)$ has a single pole at the origin of the $s$ plane, which means the closed loop servo is Type 1. The settling time is about one second, which is consistent with the one-sided, closed loop, bandwidth of 1 Hz. There is a small over shoot in the step response and the error plots, which is consistent with the selection of $\zeta = 1/\sqrt{2}$.

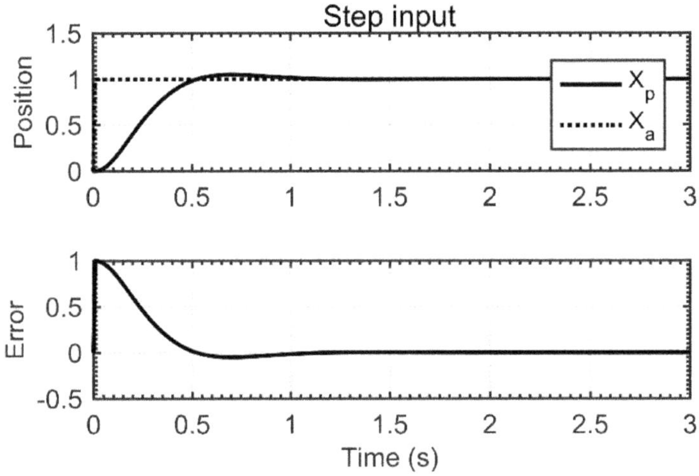


**Figure 2.16** Simulation output for step input.

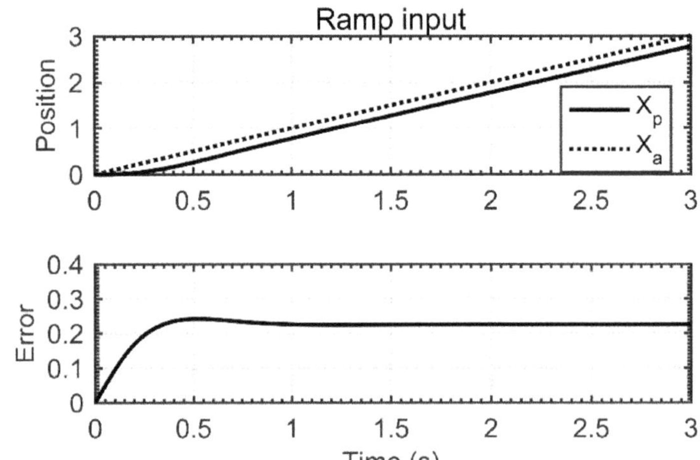


**Figure 2.17** Simulation output for ramp input.

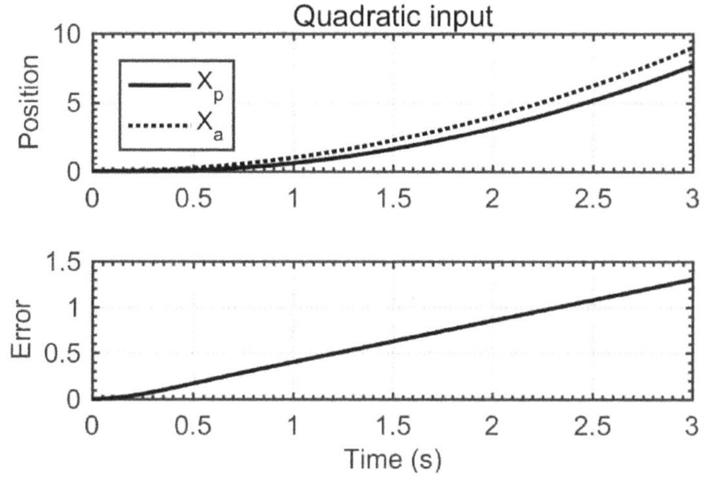


**Figure 2.18** Simulation output for quadratic input.

## 2.4.2 Digital Servo Modeling

If $G_{OL}(z)$ is given, the easiest way to model a digital servo is to use the $z$-transform approach of Section 2.4.1.3.2. As we will see in Chapter 3, α-β, α-β-γ, and Kalman filters are expressed in a discrete time, state variable representation. The general form of this notation is

$$X(n+1) = FX(n) + Ge(n) \text{ and } y(n) = H^T X(n) \tag{2.78}$$

where $X(n)$ is an $k$-element *state vector*, $F$ is an $k\times k$ *state transition matrix*, $G$ is an $k\times r$ *input distribution matrix*, $e(n)$ is an $r$ element *input vector*, $y(n)$ is a $s$ element *output vector,* and $H^T$ is an $s\times k$ element *output distribution matrix*.

To model such filters, the state variable approach of Section 2.4.1.3.1 can be used, with the computation in the first block of the iterative loop replace by

$$\text{Xnew} = \text{FXold} + \text{Ge(n)}$$

and the computation in the second-to-last block replaced by

$$\text{xp} = \text{H}^\text{T}\text{Xold.}$$

Since we sometimes characterize digital filters in terms of their transfer functions, we now discuss a method of deriving the $z$-transfer function from a discrete-time, state variable representation.

### 2.4.2.1 Deriving the $z$-Transfer Function from a State Variable Representation

If we take the $z$-transform of (2.78), we get

$$zX(z) = FX(z) + GE(z) \tag{2.79}$$

and

$$Y(z) = H^T X(z) \tag{2.80}$$

where we assume zero initial conditions (i.e., $X(n)|_{n=0} = 0$).

From (2.79) we get

$$(zI - F)X(z) = GE(z). \tag{2.81}$$

Solving for $X(z)$ yields

$$X(z) = (zI - F)^{-1} GE(z). \tag{2.82}$$

In (2.82), $I$ is an identity matrix that is the same size as the $F$ matrix, and the $-1$ superscript denotes the matrix inverse.

Substituting (2.82) in to (2.80) results in

$$Y(z) = H^T (zI - F)^{-1} GE(z). \tag{2.83}$$

From (2.83) we get the $z$-transfer function as

$$G_F(z) = \frac{Y(z)}{E(z)} = H^T (zI - F)^{-1} G. \tag{2.84}$$

## 2.5 EXERCISES

1. Generate Table 2.1.
2. Sketch the root locus of a unity feedback servo with the open loop transfer function

$$G_{OL}(s) = \frac{K_{OL}(s+4)}{s(s+3)(s+6)} \tag{2.85}$$

3. Show that E(z) in the block diagram of Figure 2.9 is

$$E(z) = \frac{1}{1 + G_{OL}(z)} X_a(z) \tag{2.86}$$

4. Generate Table 2.3.

5. Sketch the root locus of a discrete time, unity feedback servo with the open loop transfer function

$$G_{OL}(z) = \frac{K_{OL}(2z-1)}{(z-1)^2} \tag{2.87}$$

6. Repeat exercise 5 for the open loop transfer function

$$G_{OL}(z) = \frac{K_{OL}\left(z^2 - 0.7z + 0.1\right)}{(z-1)^3}. \tag{2.88}$$

7. Derive (2.51) and (2.56).
8. Implement the state variable approach of Section 2.4.1.3.3.1 and reproduce the plots of Figure 2.16, Figure 2.17, and Figure 2.18.
9. Implement the $z$-transform approach of Section 2.4.1.3.3.2 and reproduce the plots of Figure 2.16, Figure 2.17, and Figure 2.18.
10. Derive a transfer function for a digital filter with the state variable representation

$$\begin{aligned} X(n+1) &= FX(n) + Ge(n) \\ x_p(n) &= H^T X(n) \end{aligned} \tag{2.89}$$

where

$$F = \begin{bmatrix} 1 & T \\ 0 & 1 \end{bmatrix}, G = \begin{bmatrix} \alpha+\beta \\ \beta/T \end{bmatrix}, H^T = \begin{bmatrix} 1 & 0 \end{bmatrix} \tag{2.90}$$

Derive a transfer function for a digital filter with the state variable representation

$$\begin{aligned} X(n+1) &= FX(n) + Ge(n) \\ x_p(n) &= H^T X(n) \end{aligned} \tag{2.91}$$

where

$$F = \begin{bmatrix} 1 & T & T^2/2 \\ 0 & 1 & T \\ 0 & 0 & 1 \end{bmatrix}, G = \begin{bmatrix} \alpha+\beta+\gamma \\ (\beta+2\gamma)/T \\ 2\gamma/T^2 \end{bmatrix}, H^T = \begin{bmatrix} 1 & 0 & 0 \end{bmatrix} \tag{2.92}$$

**References**

[1] C. H. Houpis and S. N. Sheldon, Linear Control System Analysis and Design with MATLAB, Sixth ed., Boca Raton, FL: CRC Press, 2013.

[2] W. R. Evans, "Graphical Analysis of Control Systems," *Transactions of the American Institute of Electrical Engineers,* vol. 67, no. 1, pp. 547-551, January 1948.

[3] R. C. Dorf, *Modern Control Systems*, 5th ed., NY: Addison-Wesley Publishing Company, 1989.

[4] B. C. Kuo, *Digital Control Systems*, Fort Worth, TX: Holt, Rinehart and Winston, Inc., 1980.

[5] M. I. Skolnik, Ed., *Radar Handbook*, 3rd ed., New York: McGraw-Hill, 2008.

[6] H. M. James, N. B. Nichols and R. S. Phillips, Eds., *Theory of Servomechanisms*, New York: McGraw-Hill Book Company, Inc., 1947.

[7] E. Brookner, Ed., *Practical Phased Array Antenna Systems*, Norwood, MA: Artech House, 1991.

[8] R. J. Mailloux, *Phased Array Antenna Handbook*, 2nd ed., Norwood, MA: Artech House, 2005.

[9] J. Sklansky, "Optimizing the dynamic parameters of a track-while-scan system," *RCA Review,* vol. 18, no. 2, pp. 163-185, June 1957.

[10] B. K. Bhagavan and R. J. Polge, "Performance of the g-h Filter for Tracking Maneuvering Targets," *IEEE Transactions on Aerospace and Electronic Systems,* vols. AES-10, no. 6, pp. 864-866, Nov. 1974.

[11] H. R. Simpson, "Performance measures and optimization condition for a third-order sampled-data tracker," *IEEE Transactions on Automatic Control,* vol. 8, no. 2, pp. 182-183, April 1963.

[12] D. T. a. T. Singh, "Optimal design of α-β-(γ) filters," *American Control Conference, 2000. Proceedings of the 2000*, Chicago, IL.

[13] T. Benedict and G. Bordner, "Synthesis of an optimal set of radar track-while-scan smoothing equations," *IRE Transactions on Automatic Control,* vol. 7, no. 4, pp. 27-32, 1962.

[14] R. J. Polge and B. K. Bhagavan, "A Study of the G-H-K Tracking Filter," UAH Research Report No. 176, MICOM Report No. RE-CR-76-1, DTIC ADA021317, 1975.

[15] S. Neal, "Parametric relations for the α-β-γ filter Predictor," *IEEE Transactions on Automatic Control,* vol. 12, no. 3, pp. 315-317, June 1967.

[16] R. E. Kalman, "A new approach to linear filtering and prediction problems," *Transactions of the ASME, Journal of Basic Engineering, Series D,* vol. 82, no. 1, pp. 35-45, March 1960.

[17] R. E. Kalman and R. S. Bucy, "New results in linear filtering and prediction theory," *Transactions of the ASME, Journal of Basic Engineering,* vol. 83, no. 3, pp. 95-107, December 1961.

[18] A. S. Locke, *Guidance: Principles Of Guided Missile Design*, vol. 1, New York: D. Van Nostrand Company, 1955.

[19] P. A. McCollum and B. F. Brown, *Laplace Transform Tables and Theories*, New York, NY: Holt, Rinehart and Winston, 1965.

[20] W. H. Beyer, Ed., *Standard Mathematical Tables*, 26th ed., Boca Raton, FL: CRC Press, 1981.

[21] R. G. Lyons, *Understanding Digital Signal Processing*, 3rd ed., New York: Prentice Hall, 2011.

[22] J. R. Dormand and P. J. Prince, "A family of embedded Runge-Kutta formulae," *J. Comp. Appl. Math.*, vol. 6, pp. 19-26, 1980.

[23] L. F. Shampinel and M. W. Reichelt, "The MATLAB ODE Suite," *SIAM Journal on Scientific Computing,* vol. 18, pp. 1-22, 1997.

[24] C. Runge, "Ueber die numerische Auflösung von Differentialgleichungen," *Math. Ann.,* no. 46, pp. 167-178, 1895.

[25] W. Kutta, "Beitrag zur naherungsweisen Integration von Differentialgleichungen," *Math. und Phys.,* vol. 46, pp. 435-453, 1901.

[26] N. S. Bakhalov, *Numerical methods: analysis, algebra, ordinary differential equations*, MIR, 1977.

[27] "Encyclopedia of Mathematics," Springer, [Online]. Available: www.encyclopediaofmath.org.

[28] R. L. Burden and J. D. Faires, *Numerical Analysis*, 4th ed., Boston: PWS-Kent, 1989.

[29] D. Kincaid and W. Cheney, *Numerical Analysis*, Pacific Grove: Brooks/Cole, 1991.

[30] N. S. Asaithambi, *Numerical Analysis, Theory and Practice*, Fort Worth: Saunders College Publishing, 1995.

[31] D. G. Zill and W. S.Wright, *Advanced Engineering Mathematics*, 4th ed., Jones & Bartlett Publishers, 2011.

[32] D. G. Schultz and J. L. Melsa, *State Functions and Linear Control Systems*, New York, NY: McGraw-Hill, 1967.

[33] G. M. Swisher, *Introduction to Linear Systems Analysis*, Matrix Publishers Inc., 1976.

[34] G. R. Cooper and C. D. McGillem, *Continuous & Discrete Systems*, 3rd ed., Philadelphia: Saunders College Publishing, 1991.

[35] A. V. Oppenheim, A. S. Willisky and I. T. Young, *Signals and Systems*, Englewood Cliffs: Prentice-Hall, Inc., 1983.

[36] R. E. Zeimer, W. H. Tranter and D. R. Fannin, *Signals and Systems: Continuous and Discrete*, 3rd ed., MacMillan Publishing Company, 1993.

[37] A. V. Oppenheim and R. W. Schafer, *Discrete-Time Signal Processing*, Englewood Cliffs: Prentice-Hall, 1989.

[38] J. L. Melsa and D. G. Schultz, *Linear Control Systems*, New York, NY: McGraw-Hill, 1969.

[39] C. E. Shannon, "A Mathematical Theory of Communication," *The Bell System Technical Journal,* vol. 27, no. 3, pp. 379-423, July 1948.

[40] C. E. Shannon, "Communication in the Presence of Noise," *Proceedings of the IRE,* vol. 37, no. 1, pp. 10-21, January 1949.

[41] H. Nyquist, "Certain Topics in Telegraph Transmission Theory," *Transactions of the American Institute of Electrical Engineers,* vol. 47, no. 2, pp. 617-644, April 1928.

[42] V. A. Kotelnikov, "On the Capacity of the 'Ether' and Cables in Electrical," in *Proc. 1st All-Union Conf. Technological Reconstruction of the Commun. Sector and Low-Current Eng.*, Moscow, 1933.

# Chapter 3

## Track Filters

### 3.1 INTRODUCTION

In this chapter, we discuss track filter design. When most people reference track filters, they think of α-β, α-β-γ, or Kalman filters [1–16]. However, a track filter can consist of any transfer function that provides a desired closed loop bandwidth, transient behavior, and noise reduction. Analog closed loop trackers are usually this "transfer function" type, while digital closed loop trackers can be either the transfer function type, or α-β, α-β-γ, or Kalman filters.

### 3.2 KALMAN, α-β, AND α-β-γ TRACK FILTERS

#### 3.2.1 Background

The α-β and α-β-γ track filters, and the linear Kalman filter take the same form, consisting of a *prediction* stage and an *update* or *smoothing* stage [1–15]. The prediction stage is of the form

$$X_p(n) = F(n-1)X_s(n-1) \tag{3.1}$$

and the smoothing stage has the form

$$X_s(n) = X_p(n) + K(n)\left[y(n) - y_p(n)\right] \tag{3.2}$$

where

$X_p(n)$ is the *predicted* state (vector).

$X_s(n)$ is the *smoothed* state (vector).

$F(n-1)$ is the *state transition* matrix.

$K(n)$ is the *weighting* or *gain* matrix.

$y(n)$ is the measurement.

$y_p(n)$ is the predicted measurement.

The predicted and smoothed state vectors are of the form

$$\begin{bmatrix} position \\ velocity \end{bmatrix} = \begin{bmatrix} p \\ p' \end{bmatrix} = \begin{bmatrix} p \\ dp/dt \end{bmatrix} \tag{3.3}$$

for the α-β filter and

$$\begin{bmatrix} position \\ velocity \\ acceleration \end{bmatrix} = \begin{bmatrix} p \\ p' \\ p'' \end{bmatrix} = \begin{bmatrix} p \\ dp/dt \\ d^2p/d^2t \end{bmatrix} \tag{3.4}$$

for the α-β-γ filter. They can be either form in the Kalman filter, depending on the order of the filter (second or third). In a range tracker, position would be range, velocity would be range-rate, and acceleration would be range acceleration. In an angle tracker, position would be angle while velocity and acceleration would be successive derivatives of angle.

The state transition matrix [17–20], $F(n-1)$, is of the form

$$F(n-1) = \begin{bmatrix} 1 & T_{n-1} \\ 0 & 1 \end{bmatrix} \tag{3.5}$$

for a α-β and second order Kalman filter and

$$F(n-1) = \begin{bmatrix} 1 & T_{n-1} & T_{n-1}^2/2 \\ 0 & 1 & T_{n-1} \\ 0 & 0 & 1 \end{bmatrix} \tag{3.6}$$

for a α-β-γ and third order Kalman filter[4]. The parameter, $T_{n\text{-}1}$, is the *update period* of the closed loop tracker. It is the time between the previous measurement, $y(n-1)$, and the current measurement, $y(n)$. It is shown as a function of *stage*, $n$, because the time between measurements may not be constant. For the case where the update period is a constant value of $T$, the *track rate*, $f_s$, is the reciprocal of $T$.

For both the α-β and α-β-γ filters, it is assumed the measurement, $y(n)$, is position (e.g., range or angle). Similarly, the predicted measurement, $y_p(n)$, is the predicted positon, which is

---

[4] The Kalman filters considered in this book are linear Kalman filters where the states are in the same coordinate system as the measurements. That is, the states of a Kalman filter used in a range tracker are range, range-rate and possibly range acceleration. The measurement is range. In another variation of the Kalman filter, known as the *extended Kalman filter*, the states could be in a Cartesian coordinate system while the measurements are range and angles. That type of Kalman filter is not considered in this book.

the first element of $X_p(n)$, the predicted state vector. The same applies to the Kalman filter that uses only one measurement. In equation form, consistent with Chapter 2, $y_p(n)$ is given by

$$y_p(n) = H^T X_p(n) \tag{3.7}$$

where

$$H^T = \begin{bmatrix} 1 & 0 \end{bmatrix} \tag{3.8}$$

for the α-β filter and second order Kalman filter, and

$$H^T = \begin{bmatrix} 1 & 0 & 0 \end{bmatrix} \tag{3.9}$$

for the α-β-γ and third order Kalman filter.

The weighting matrix, $K(n)$, is of the form

$$K(n) = \begin{bmatrix} \alpha \\ \beta/T \end{bmatrix} \tag{3.10}$$

for the α-β filter and

$$K(n) = \begin{bmatrix} \alpha \\ \beta/T \\ 2\gamma/T^2 \end{bmatrix} \tag{3.11}$$

for the α-β-γ filter. For a Kalman filter that uses only one measurement, $K(n)$ is a 2×1 or 3×1 matrix, depending on the filter order.

In some range trackers, the Kalman filter can use both range and range-rate measurements [2, 3, 9, 12], where the range-rate measurement is derived from target Doppler frequency [21]. In that case, $K(n)$ is a 2×2 or 3×2 matrix depending on the order of the Kalman filter (2 or 3, respectively).

During operation, the prediction equation, (3.1), *propagates*, via the state transition matrix, the smoothed state from the previous track cycle (i.e., stage $n-1$) to the time of the next measurement (i.e., stage $n$). This propagation yields the predicted state. The predicted state is then used to compute the predicted measurement. The predicted measurement is used to prepare the radar for making the measurement (at stage $n$) by steering the beam to the predicted target angle or positioning the range gate in time. The filter then forms the *error*, or *residual*, between the actual and predicted measurements, weights it by $K(n)$, and uses the weighted error to, hopefully, improve the predicted state to produce the new smoothed state.

In applications where the output of the tracker is sent to an ancillary system, the *smoothed* position is used. That is, the first element of $X_s(n)$. The smoothed position incorporates the latest measurement, $y(n)$.

## 3.3 THE PREDICTION EQUATION

The equation used in the prediction stage of the α-β, α-β-γ, and Kalman filters is based on a model that is supposed to represent the dynamics of the system that gives rise to the position, velocity, and accelerations states. For the radar tracking problem, such a model should, ostensibly, account for all of the dynamics of the target and the environment in which the target operates. However, such a model would be very complicated in that it would need to account for different target types and operating conditions, such as the environment, maneuvers, and pilot actions. Development of such a model is not mathematically feasible. In lieu of such a model, designers of α-β, α-β-γ, and Kalman track filters have commonly adopted a simpler approach based on the Taylor series expansion. In particular, they decide what states to include in the filter (i.e., position, velocity, acceleration) and represent each of them by a Taylor series expansion. In developing a Taylor series [12, 22] expansion model, designers take care to preserve the relation between states. For example, they preserve the fact that velocity is the derivative of position, and acceleration is the derivative of velocity. We will illustrate the procedure by an example.

We assume the states to be included in the filter are position, $p(t)$, velocity, $p'(\mathrm{t}) = \mathrm{d}p(t)/\mathrm{d}t$, and acceleration $p''(t) = \mathrm{d}p'(t)/\mathrm{d}t = \mathrm{d}^2p(t)/\mathrm{d}t^2$. We are interested in the state at $t+\Delta t$; that is $p(t+\Delta t)$, $p'(t+\Delta t)$, and $p''(t+\Delta t)$. If we expand each of these into a Taylor series about $t$, we get

$$\begin{aligned}
p(t+\Delta t) &= p(\tau)\Big|_{\tau=t} + \frac{\partial p(\tau)}{\partial \tau}\Big|_{\tau=t}\Delta t + \frac{\partial^2 p(\tau)}{\partial \tau^2}\Big|_{\tau=t}\frac{\Delta t^2}{2!} + \cdots \\
&\quad + \frac{\partial^m p(\tau)}{\partial \tau^m}\Big|_{\tau=t}\frac{\Delta t^m}{m!} + \cdots \\
p'(t+\Delta t) &= p'(\tau)\Big|_{\tau=t} + \frac{\partial p'(\tau)}{\partial \tau}\Big|_{\tau=t}\Delta t + \frac{\partial^2 p'(\tau)}{\partial \tau^2}\Big|_{\tau=t}\frac{\Delta t^2}{2!} + \cdots \\
&\quad + \frac{\partial^m p'(\tau)}{\partial \tau^m}\Big|_{\tau=t}\frac{\Delta t^m}{m!} + \cdots \\
p''(t+\Delta t) &= p''(\tau)\Big|_{\tau=t} + \frac{\partial p''(\tau)}{\partial \tau}\Big|_{\tau=t}\Delta t + \frac{\partial^2 p''(\tau)}{\partial \tau^2}\Big|_{\tau=t}\frac{\Delta t^2}{2!} + \cdots \\
&\quad + \frac{\partial^m p''(\tau)}{\partial \tau^m}\Big|_{\tau=t}\frac{\Delta t^m}{m!} + \cdots
\end{aligned} \tag{3.12}$$

If we use the associations, $p'(\tau) = \partial p(\tau)/\partial\tau$ and $p''(\tau) = \partial p'(\tau)/\partial\tau = \partial^2 p(\tau)/\partial\tau^2$, we can rewrite (3.12) as

$$\begin{aligned} p(t+\Delta t) &= p(t) + p'(t)\Delta t + p''(t)\frac{\Delta t^2}{2} + \cdots + p^{(m)}(t)\frac{\Delta t^m}{m!} + \cdots \\ p'(t+\Delta t) &= p'(t) + p''(t)\Delta t + p'''(t)\frac{\Delta t^2}{2} + \cdots + p'^{(m)}(t)\frac{\Delta t^m}{m!} + \cdots \\ p''(t+\Delta t) &= p''(t) + p'''(t)\Delta t + p^{(4)}(t)\frac{\Delta t^2}{2} + \cdots + p''^{(m)}(t)\frac{\Delta t^m}{m!} + \cdots \end{aligned} \quad . \tag{3.13}$$

In (3.13), we are interested in $p(t)$, $p'(t)$, and $p''(t)$ but not the higher derivatives of $p(t)$. Thus, we gather all of the higher order derivative terms into variables we denote as $HOT(t)$, $HOT'(t)$, and $HOT''(t)$ where $HOT$ stands for *higher order terms*. This gives,

$$\begin{aligned} p(t+\Delta t) &= p(t) + p'(t)\Delta t + p''(t)\frac{\Delta t^2}{2} + HOT(t) \\ p'(t+\Delta t) &= p'(t) + p''(t)\Delta t + HOT'(t) \\ p''(t+\Delta t) &= p''(t) + HOT''(t) \end{aligned} \tag{3.14}$$

We next let $\Delta t = T$, which is assumed to be fixed for now (recall that, generally, the update period between stage $n-1$ and $n$ is $T_{n-1}$). We further let $t = (n-1)T$ so that $t+\Delta t = nT$. These substitutions yield:

$$\begin{aligned} p(nT) &= p\left[(n-1)T\right] + p'\left[(n-1)T\right]T + p''\left[(n-1)T\right]\frac{T^2}{2} + HOT\left[(n-1)T\right] \\ p'(nT) &= p'\left[(n-1)T\right] + p''\left[(n-1)T\right]T + HOT'\left[(n-1)T\right] \\ p''(nT) &= p''\left[(n-1)T\right] + HOT''\left[(n-1)T\right] \end{aligned} \tag{3.15}$$

By dropping $T$ from the arguments of the variables, equation (3.15) can be written in matrix form as:

$$\begin{bmatrix} p(n) \\ p'(n) \\ p''(n) \end{bmatrix} = \begin{bmatrix} 1 & T & T^2/2 \\ 0 & 1 & T \\ 0 & 0 & 1 \end{bmatrix} \begin{bmatrix} p(n-1) \\ p'(n-1) \\ p''(n-1) \end{bmatrix} + \begin{bmatrix} HOT(n-1) \\ HOT'(n-1) \\ HOT''(n-1) \end{bmatrix} \tag{3.16}$$

or

$$X(n) = F(n-1)X(n-1) + \left[HOT\right]_{n-1}. \tag{3.17}$$

We note (3.17) is the form of the prediction equation, (3.1), except for the inclusion of the *HOT* terms and the subscripts on the *X* vector in (3.1).

The *HOT* terms pose a problem because they represent higher order derivatives of $p(t)$, about which we have no information. In the α-β and α-β-γ track filters, the *HOT* terms are set to zero. They are also set to zero in the Kalman filter, however, designers recognize that the resulting system dynamics model is inaccurate. They attempt to account for this by characterizing the *HOT* terms as *model uncertainties* and include that fact in the *tuning* process associated with the Kalman filter.

By setting the *HOT* terms to zero, we are saying the acceleration term of (3.16) is constant. Thus, (3.16) represents a *constant acceleration* system model.

If we were to derive an equation like (3.16) for a second order system, it would take the form:

$$\begin{bmatrix} p(n) \\ p'(n) \end{bmatrix} = \begin{bmatrix} 1 & T \\ 0 & 1 \end{bmatrix} \begin{bmatrix} p(n-1) \\ p'(n-1) \end{bmatrix} + \begin{bmatrix} HOT(n-1) \\ HOT'(n-1) \end{bmatrix} \tag{3.18}$$

If we again set the *HOT* terms to zero, the velocity term is constant. With this, we say (3.18) represents a *constant velocity* system model.

In the next section, we discuss methods for determining $\alpha$, $\beta$, and $\gamma$ for the α-β and α-β-γ track filters. Two methods for each are discussed. The first is based on minimizing mean squared error, or equivalently, maximizing the noise reduction offered by the filters. This method is termed the Benedict-Bordner [4] method for the α-β filter and the Polge-Bhagavan [8] method for the α-β-γ filter. The second method is based on control theory [17, 18, 20, 23, 24] and uses the criteria of achieving a given closed loop bandwidth and transient behavior (see Chapter 2).

Before presenting the filter design approaches, we examine the structure of the α-β and α-β-γ filters, and the single input Kalman filter, in the context of the closed loop tracker discussions of Chapter 1. This will allow us to characterize the properties of the filters using the concepts of Chapter 2. Specifically, we will characterize the system type and transient behavior of the filter. We also present equations that are used to characterize the noise reduction properties of the α-β and α-β-γ filters. This development sets the stage for developing the Benedict-Bordner and Polge-Bhagavan design techniques.

### 3.3.1 Closed Loop Tracker Structure

Figure 3.1 contains a block diagram of (3.1) and (3.2) in the form of the closed loop servo diagram of Figure 2.1. The $K_R$ scaling constant of Figure 2.1 has been omitted because it is unity in this case. Also, the controlled element block has been omitted because a controlled element, per. se., is not part of the closed loop servo for the α-β, α-β-γ, and Kalman filters. The filter transfer function of Figure 2.1, $G_F(z)$, consists of the elements in the dashed box. $G_F(z)$ is the open loop transfer function discussed in Chapter 2.

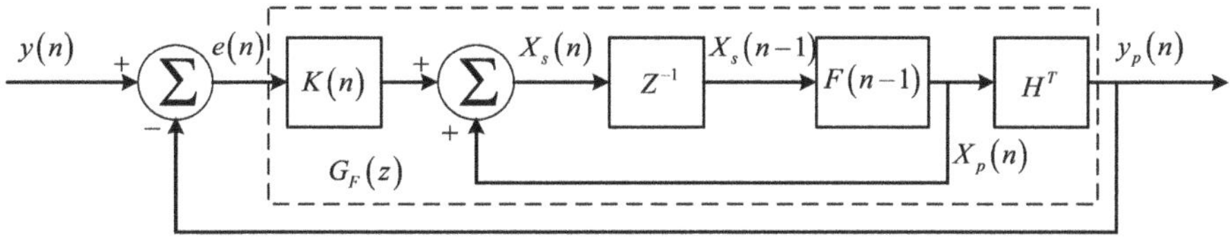


**Figure 3.1** Block diagram for α-β, α-β-γ, and simple Kalman filter.

To derive an equation for $G_F(z)$, we begin by writing (3.1) and (3.2) in a form consistent with Figure 3.1. Specifically, we replace $y(n) - y_p(n)$ with $e(n)$. This yields

$$X_p(n) = F(n-1)X_s(n-1) \tag{3.19}$$

and

$$X_s(n) = X_p(n) + K(n)e(n) \tag{3.20}$$

We next combine (3.19) and (3.20) in a fashion that hides the smoothed state, $X_s(n)$. This yields

$$\begin{aligned} X_p(n) &= F(n-1)\left[X_p(n-1) + K(n-1)e(n-1)\right] \\ &= F(n-1)X_p(n-1) + F(n-1)K(n-1)e(n-1) \end{aligned}. \tag{3.21}$$

From Figure 3.1, we have that the output, $y_p(n)$, is (see (3.7))

$$y_p(n) = H^T X_p(n). \tag{3.22}$$

To derive the transfer function from (3.21) and (3.22), we need to drop the dependency of $F$ and $K$ on $n-1$ and write (3.21) as[5]

$$X_p(n) = FX_p(n-1) + FKe(n-1). \tag{3.23}$$

Equations (3.22) and (3.23) are in the forms discussed in Section 2.4.2.1. Thus, we can use the technique discussed in that section to write the open loop transfer function, $G_F(z)$ as

$$G_F(z) = \frac{Y_p(z)}{E(z)} = H^T(zI - F)^{-1} FK. \tag{3.24}$$

For the α-β filter, $G_F(z)$ is (see Problem 10, Chapter 2)

[5] We drop the dependency of $F$ and $K$ on $n-1$ because transfer functions, in general, are only defined for linear, time-invariant systems.

$$G_F(z) = \frac{(\alpha+\beta)z - \alpha}{(z-1)^2} \tag{3.25}$$

and for the α-β-γ filter, it is

$$G_F(z) = \frac{(\alpha+\beta+\gamma)z^2 + (\gamma-\beta-2\alpha)z + \alpha}{(z-1)^3}. \tag{3.26}$$

For the α-β filter, $G_F(z)$ has two poles at $z = 1$. This means the α-β filter is Type 2 and will be able to follow a ramp input (a constant velocity input) with zero steady-state error. This is consistent with an earlier statement where we said the prediction model for a α-β filter is a constant velocity model.

$G_F(z)$ for the α-β-γ filter has three poles at $z = 1$, and thus the α-β-γ filter is a Type 3 servo. This means the α-β-γ filter will track a quadratic input (a constant acceleration input) with zero steady-state error. Again, this is consistent with a previous statement that the prediction model for a α-β-γ filter is a constant acceleration model.

The $G_F(z)$ for second and third order, single input Kalman filters will have the same form as (3.25) and (3.26), respectively, except that the numerator coefficients will be different. That means second and third order Kalman filters are also Type 2 and Type 3 servos. It must be noted that this statement only applies to the case where the Kalman filter is operating in steady state, where the Kalman gain becomes constant. Since Kalman filters are not usually operated in steady state, the concept of a transfer function for such filters is somewhat meaningless.

### 3.3.2 Filter Stability and Variance Reduction

Now that we have the general forms of the α-β and α-β-γ filter we want to address general requirements on $\alpha$, $\beta$, and $\gamma$ as relates to filter stability. We also introduce the concept of a *variance ratio* since this is one of the criteria used to derive the Benedict-Bordner α-β filter and the Polge-Bhagavan α-β-γ filter. Simply stated, the variance ratio is the ratio of the variance on the smoothed position state (the first element of $X_s(n)$) to the variance on $y(n)$. If possible, we would like the α-β and α-β-γ filters to provide good dynamic behavior, while reducing the variance on the measurements. This is what the Benedict-Bordner and Polge-Bhagavan strive filters achieve.

#### 3.3.2.1 Stability Triangle

A primary requirement on any tracker is that it be stable and have a well behaved response. The stability requirement dictates that all of the poles of the closed loop transfer function of the tracker be inside of the unit circle of the $z$-plane.

To have a well behaved response, all the poles must lie in the right half of the unit circle. If any of the poles lie in the left half of the unit circle, the tracker response could oscillate from sample to sample or demonstrate some other type of undesired behavior [20, 25].

The requirements of stability and well behaved transient response lead to a concept termed the *stability triangle* [26; 27; 28; 29, p. 262][6]. For α-β filters, it defines values of $\alpha$ and $\beta$ for which the filter will be stable, and regions where its poles will be in the right half of the unit circle of the $z$-plane.

A plot of the stability triangle for the α-β filter is illustrated in Figure 3.2. If $\alpha$ and $\beta$ are such that they lie in the larger triangle bounded by the α and β axes and the line

$$\beta = 4 - 2\alpha \tag{3.27}$$

the tracker will be stable. If they lie in the polygon bounded by the α and β axes, and the lines

$$\beta = 2 - \alpha \text{ and } \alpha = 1 \tag{3.28}$$

the poles of the closed loop transfer function will lie in the right half of the unit circle of the $z$-plane.

As will be discussed shortly, if $\alpha$ and $\beta$ lie in the triangle bounded by the α and β axes and the lines

$$\beta = 2\alpha \text{ and } \alpha = 1 \tag{3.29}$$

the variance ratio will be less than one. That is, the variance on the smoothed position state will be less than the variance on the measurement.

Since the α-β-γ tracker has three parameters, we use a pair of stability triangles to define the values of $\alpha$, $\beta$, and $\gamma$ for which the tracker will be stable. These triangles are illustrated in Figure 3.3. The boundaries indicated in Figure 3.3 are defined by the equations

$$0 < \alpha < 2,\ \gamma > 0,\ \gamma < 2\alpha,\ \beta > 0 \text{ and } \beta < 4 - \alpha\,. \tag{3.30}$$

As discussed in Appendix 3A, there is another restriction given by

$$\gamma < \frac{\beta\alpha}{2 - \alpha}\,. \tag{3.31}$$

The two triangles of Figure 3.3 provide separate $\alpha$-$\beta$ and $\alpha$-$\gamma$ regions, while (3.31) effectively ties the two regions together.

[6] The various equations presented in this section are derived in Appendix 3A.

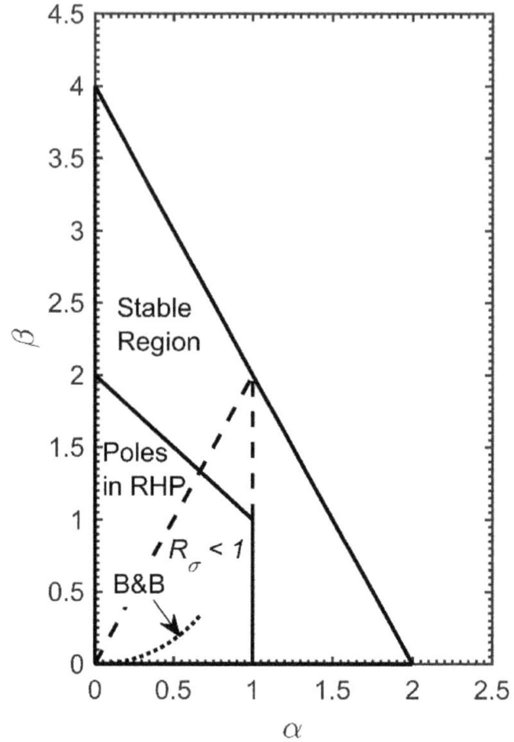


**Figure 3.2** Stability triangle for an α-β tracker.

The additional requirements on $\alpha$, $\beta$, and $\gamma$ for the closed loop poles to be in the right half of the unit circle are

$$\alpha < 1, \tag{3.32}$$

$$3 - \alpha - \beta - \gamma > 0, \tag{3.33}$$

$$3 - 2\alpha - \beta + \gamma > 0 \tag{3.34}$$

and

$$(3 - \alpha - \beta - \gamma)(3 - 2\alpha - \beta + \gamma) - (1 - \alpha) > 0. \tag{3.35}$$

Unfortunately, we have not found a way of graphing these additional requirements. Having said that, (3.32) is easy to represent on Figure 3.3 and restricts the regions to lie to the left of the dashed lines

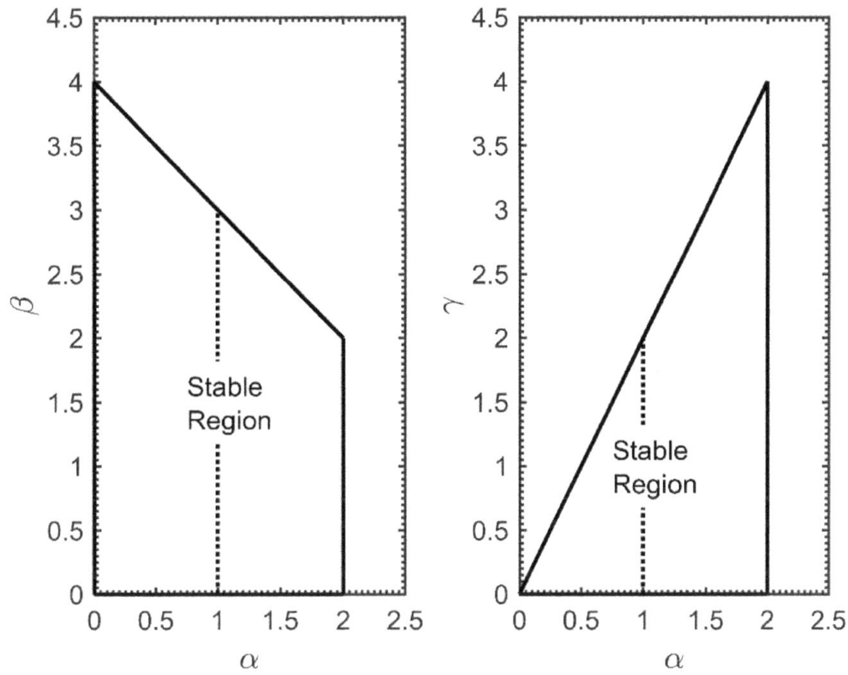


**Figure 3.3** Stability triangle for an α-β-γ tracker.

### 3.3.2.2 Variance Ratio

A key aspect of the Benedict-Bordner and Polge-Bhagavan filter design methods is the idea of reducing the variance on the measurements. As will be discussed in the next two sections, their optimization criterion is based on minimizing the variance on the position element of the smoothed state, $X_s(n)$, while providing good transient behavior.

In the process of deriving their formulation, Benedict and Bordner define a variance ratio as[7]

$$R_\sigma = \frac{\sigma_p^2}{\sigma_y^2} = \frac{2\alpha^2 + \beta(2-3\alpha)}{\alpha(4-\beta-2\alpha)}. \tag{3.36}$$

Ideally, if the filter is to reduce the variance, $\alpha$ and $\beta$ should be chosen to make $R_\sigma$ as small as possible. Using that criterion alone, it would be possible to design a filter such that $R_\sigma$ was zero, by choosing $\alpha = 0$ and $\beta = 0$, which would be nonsensical. That is why Benedict and Bordner, and Polge and Bhagavan, use a constrained minimization, where the constraint is based on transient behavior.

---

[7] As a note, Benedict and Bordner define $R_\sigma$ as a variance *reduction* ratio. Strictly speaking, $1/R_\sigma$ is a variance reduction ratio.

If we include the requirement $R_\sigma < 1$ to the requirement that the closed loop poles of the α-β filter lie in the right half plane, we find $\alpha$ and $\beta$ must lie inside of the triangle defined by

$$0<\beta<2-\alpha,\ 0<\alpha<1, \text{ and } \beta<2\alpha . \tag{3.37}$$

This is the region inside of the triangle bounded by the α and β axes, and the dashed lines of Figure 3.2. In general, $R_\sigma = 1$ on the dashed line boundaries, and drops below 1 as $\alpha$ and $\beta$ move toward the origin. As an interesting side note, the plot of $\beta$ versus $\alpha$, for the Benedict-Bordner is the curved line of Figure 3.2.

The variance ratio for an α-β-γ is (see Appendix 3A)

$$R_\sigma = \frac{\sigma_p^2}{\sigma_y^2} = \frac{\beta^2(2-3\alpha)+2\alpha^2(\beta+\gamma)+\alpha\gamma(\beta-4)}{(\gamma(\alpha-2)+\alpha\beta)(4-\beta-2\alpha)} . \tag{3.38}$$

Unfortunately, as with the stability triangle, (3.38) does not provide insight into how to limit $\alpha$, $\beta$, and $\gamma$ so at to assure $R_\sigma$ will be less than one. It is interesting to note (3.38) and (3.36) have two terms in common.

### 3.3.2.3 Noise Bandwidth and Variance Ratio

Suppose we have an ideal low-pass, digital filter with the frequency response

$$|G(f)|^2 = |G(z)|^2_{z=e^{j2\pi fT}} = \text{rect}\left[\frac{f}{B_n}\right]. \tag{3.39}$$

where

$$\text{rect}\left[\frac{f}{B_n}\right] = \begin{cases} 1 & |f| \le B_n/2 \\ 0 & |f| > B_n/2 \end{cases}. \tag{3.40}$$

Suppose further we input to the filter wide-sense stationary (WSS), zero-mean, white, discrete-time (sampled data) noise with a variance, and spectral density, of $\sigma_y^2$. From the theory of discrete time systems and signals [21], the variance, and power, of the noise at the output is

$$\begin{aligned}\sigma_o^2 &= T\int_{-1/2T}^{1/2T} \sigma_y^2 |G(f)|^2 df = T\int_{-1/2T}^{1/2T} \sigma_y^2 \text{rect}\left[\frac{f}{B_n}\right] df \\ &= T\int_{-B_n/2}^{B_n/2} \sigma_y^2 df = TB_n\sigma_y^2\end{aligned} . \tag{3.41}$$

We can solve (3.41) for $B_n$ as

$$B_n = \frac{1}{T}\left(\frac{\sigma_o^2}{\sigma_y^2}\right). \tag{3.42}$$

We define $B_n$ as *effective noise bandwidth* and use it to define the effective noise bandwidth of the σ-β and α-β-γ filters as [9]

$$B_n = \frac{1}{T}R_\sigma. \tag{3.43}$$

Since $R_\sigma$ is related to α, β, and γ, so is the effective noise bandwidth.

As defined by (3.43), $B_n$ is only loosely related to the two-sided, half-power bandwidth defined earlier. In fact, while $B_n$ for the ideal filter can never be larger than $1/T$, the effective noise bandwidth of the α-β and α-β-γ filters can be much larger than $1/T$. This is because $R_\sigma$ can be greater than one. That is, α-β and α-α-γ filters can amplify noise if not designed properly.

## 3.4 BENEDICT-BORDNER METHOD FOR α-β FILTER DESIGN

Benedict and Bordner base their design approach on two performance measures. One is the variance on $y_s(n)$, the filter output when the input, $y(n)$, is zero-mean, unit-variance, white noise. The second measure is the square of the error between the filter output and a unit ramp, when the input is a unit ramp. They term the first measure a *noise reduction* measure and the second a *transient performance* measure. They discuss a couple of methods for minimizing the two performance measures and provide details on one approach where they seek to minimize the noise reduction measure subject to the constraint that the α–β filter follow a ramp input as closely as possible. We will outline their approach here and discuss it in some detail in the Appendix 3B.

We start by defining the impulse response of the (closed loop) α-β filter as $g(n)$. If the input to the filter is $y(n)$, we can write the filter output as

$$y_s(n) = g(n) * y(n) = \sum_{k=0}^{\infty} g(k)y(n-k). \tag{3.44}$$

The summation in (3.44) is restricted to go from 0 to ∞ because realizability (i.e., filter must be causal) requires that $g(n) = 0$ for $n < 0$ [30, 31]. We note that the α-β filter is typically asymptotically stable [24]. This means

$$\lim_{n\to\infty} g(n) = 0. \tag{3.45}$$

If we assume $y(n)$ is zero-mean, unit-variance, white noise [32], we denote as $\eta(n)$ and have

$$E\{y_{s\eta}(n)\} = E\left\{\sum_{k=0}^{\infty} g(k)\eta(n-k)\right\} = \sum_{k=0}^{\infty} g(k)E\{\eta(n-k)\} = 0 \tag{3.46}$$

and

$$\begin{aligned}\sigma_{ys\eta}^2 &= E\left\{\left(y_{p\eta}(n) - E\{y_{s\eta}(n)\}\right)^2\right\} = E\{y_{s\eta}^2(n)\} \\ &= E\left\{\left(\sum_{k=0}^{\infty} g(k)\eta(n-k)\right)\left(\sum_{l=0}^{\infty} g(l)\eta(n-l)\right)\right\}. \\ &= \sum_{k=0}^{\infty}\sum_{l=0}^{\infty} g(k)g(l)E\{\eta(n-k)\eta(n-l)\}\end{aligned} \tag{3.47}$$

Since $\eta(n)$ is white and unit variance,

$$E\{\eta(n-k)\eta(n-l)\} = \sigma_\eta^2 \delta(k-l) = \delta(k-l) \tag{3.48}$$

where $\delta(n)$ is the Kronecker delta[8] (which is the discrete time counterpart to the Dirac delta function) defined by [33–35]

$$\delta(n) = \begin{cases} 1 & n = 0 \\ 0 & n \neq 0 \end{cases}. \tag{3.49}$$

Using (3.48) in (3.47) yields

$$\sigma_{ys\eta}^2 = \sum_{k=0}^{\infty}\sum_{l=0}^{\infty} g(k)g(l)\delta(k-l) = \sum_{k=0}^{\infty} g^2(k). \tag{3.50}$$

Since the α-β filter is asymptotically stable, the sum of (3.50) converges.

To formulate the transient performance measure, we let $y(n) = nU(n)$, that is, apply a unit ramp to the filter. In this expression, $U(n)$ is the unit step function. The filter output will be

$$r(n) = \sum_{k=0}^{\infty} g(k)y(n-k) = \left[\sum_{k=0}^{\infty} g(k)(n-k)\right]U(n) \tag{3.51}$$

With this, the transient performance measure is

---

[8] Named after German mathematician Leopold Kronecker (1823–1891), though there seems to be some dispute as to this delta function's inventor

$$D = \sum_{n=0}^{\infty} \left[ n - r(n) \right]^2 \tag{3.52}$$

where the symbol $D$ is used because the measure is the mean squared distance between the actual filter output, $r(n)$, and a ramp, $nU(n)$. Since the α-β filter is a Type 2 servo, it will follow a ramp with zero steady state error (see Chapter 2). This means $D$ will also converge.

Benedict and Bordner define the optimization problem as minimizing $\sigma^2_{ys\eta}$ subject to the constraint that $r(n)$ approach the input ramp, $y(n) = nU(n)$. They use the method of Lagrange to solve the problem. To that end, they define the overall performance measure as

$$J(g(n), \lambda) = \sigma^2_{ys\eta} + \lambda D \tag{3.53}$$

where $\lambda$ is the Lagrange multiplier [24, 36, 37].

To facilitate the solution, they use an interim step that allows them to write the impulse response, $g(n)$, as a finite series. Specifically, they note they can write the $z$-transform of $r(n)$ in terms of $G(z)$ as

$$R(z) = G(z) \mathcal{Z}\left[ nU(n) \right] = G(z) \frac{z}{(z-1)^2}. \tag{3.54}$$

Solving for $G(z)$ yields

$$G(z) = \frac{(z-1)^2}{z} R(z) = \left( z - 2 + z^{-1} \right) R(z) \tag{3.55}$$

and taking the inverse $z$-transform of (3.55) gives

$$g(n) = r(n+1) - 2r(n) + r(n-1). \tag{3.56}$$

Using (3.56) in (3.50), combining the result with (3.52), and substituting the combination in (3.53), yields

$$\begin{aligned} J(g(n), \lambda) &= J(r(n), \lambda) \\ &= \sum_{n=0}^{\infty} \left[ r(n+1) - 2r(n) + r(n-1) \right]^2 + \lambda \left[ n - r(n) \right]^2. \end{aligned} \tag{3.57}$$

The problem now becomes that of finding the $r(n)$ that minimizes the performance measure defined by (3.57).

Benedict and Bordner use a calculus of variations approach to solving the minimization problem [37]. Specifically, they assume the optimal solution is $r(n)$ then perturb it to find alternate solutions as

$$r_{alt}(n) = r(n) + \varepsilon h(n) \tag{3.58}$$

where $h(n)$ is an arbitrary, but realizable, function. The realizability constraint says that $h(n) = 0$ for $n \le 0$. They then use the calculus of variation approach of forming

$$\left.\frac{\partial J\left[r(n) + \varepsilon h(n), \lambda\right]}{\partial \varepsilon}\right|_{\varepsilon=0} = 0 \tag{3.59}$$

to establish the properties of $r(n)$. Finally, they use (3.55), and a considerable amount of manipulation, to arrive at a relation between $\alpha$ and $\beta$. Specifically,

$$\beta = \frac{\alpha^2}{2-\alpha}. \tag{3.60}$$

They do not specify a value for $\alpha$. Instead they indicate its value is determined by the bandwidth of the tracker. We will investigate this later. The details of the derivation of (3.60) are contained in the appendix.

### 3.5 POLGE-BHAGAVAN METHOD FOR α-β-γ FILTER DESIGN

Polge and Bhagavan took an approach similar to the Benedict-Bordner approach to derive relations between $\alpha$, $\beta$, and $\gamma$ for the α-β-γ filter. A key difference, aside from filter order, is that Polge and Bhagavan used the constraint that $r(n)$ should approach a weighted average of a quadratic input and a ramp input. Specifically, the input they used to form $r(n)$ is

$$y(n) = \frac{\left(n^2 + n\right)}{2} U(n). \tag{3.61}$$

As a note, since the α-β-γ track filter is a Type 3 servo, $r(n)$ will approach the $y(n)$ of (3.61) as $n \to \infty$. A detailed discussion of their development was published in a University of Alabama in Huntsville (UAH) research report [8]. Thus, it will not be repeated here. The relations between $\alpha$, $\beta$, and $\gamma$ they derived are

$$\beta = \frac{\alpha^2\left(\alpha - 2 + \sqrt{\alpha^2 - 16\alpha + 16}\right)}{4(1-\alpha)} \tag{3.62}$$

and

$$\gamma = \frac{\beta(2-\alpha) - \alpha^2}{\alpha}. \tag{3.63}$$

As with Benedict and Bordner, they did not specify $\alpha$. As will be shown, $\alpha$ is determined by the desired filter bandwidth.

## 3.6 CALCULATION OF α FOR THE BENEDICT-BORDNER AND POLGE-BHAGAVAN FILTERS

As indicated earlier, $\alpha$ of the Benedict-Bordner and Polge-Bhagavan filters is determined from the desired bandwidth [4, 8]. Once $\alpha$ is determined, the value of $\beta$ given by (3.60) results in the optimal $\alpha$-$\beta$ filter as defined by Benedict and Bordner. The values of $\beta$ and $\gamma$ given by (3.62) and (3.63) results in the optimal $\alpha$-$\beta$-$\gamma$ filter as defined by Polge and Bhahavan.

We specify the one-sided, half-power bandwidth of the $\alpha$-$\beta$ and $\alpha$-$\beta$-$\gamma$ filters as the value of $f = f_c$ that satisfies the relation

$$\left|G_{CL}(z)\right|^2\Big|_{z=e^{j2\pi f_c T}} = \frac{1}{2}\left|G_{CL}(z)\right|^2\Big|_{z=1}. \tag{3.64}$$

For both the $\alpha$-$\beta$ and $\alpha$-$\beta$-$\gamma$ filters,

$$\left|G_{CL}(z)\right|^2\Big|_{z=1} = 1 \tag{3.65}$$

so that the condition of (3.64) becomes

$$\left|G_{CL}(z)\right|^2\Big|_{z=e^{j2\pi f_c T}} = \frac{1}{2}. \tag{3.66}$$

As a note, the two-sided, half power bandwidth, or simply the half power bandwidth, of the filters is $f_B = 2f_c$. This definition derives from the fact that the frequency response is symmetric about $f = 0$. Thus, the filter has half power points at $f_c$ and $-f_c$. This means the separation between half power points is $f_B = 2f_c$.

To find the value of $\alpha$ that satisfies (3.66) for a given $f_c$, we would evaluate (3.66) as a function of $\alpha$, using the relations between $\beta$, $\gamma$, and $\alpha$ given by (3.60), (3.62), and (3.63). Unfortunately, such an approach is, at best, very tedious. In lieu of this, we propose a technique where we find an empirical relation between $\alpha$ and $f_c$. To develop the empirical relation, we compute the frequency response of the filters for a range of $\alpha$ values, determine the frequency where (3.66) is satisfied, and generate a table of $f_c$ and associated $\alpha$ values. We then fit a polynomial to the $f_c$ and $\alpha$ values.

To bound the range of $\alpha$ values that must be considered, we imposed the requirement that the poles of $G_{CL}(z)$ lie in the right half of the unit circle. This requirement derives from two observations: (1) If $G_{CL}(z)$ has poles outside of the unit circle, the filter will be unstable. (2) If $G_{CL}(z)$ has poles in the left half of the unit circle, the time response of the filter could oscillate from sample to sample, or exhibit otherwise unexpected and undesired behavior. The restriction that the poles of $G_{CL}(z)$ lie in the right half of the unit circle leads to the requirement $0 < (1 - \alpha) < 1$, which translates to the requirement $0 < \alpha < 1$.

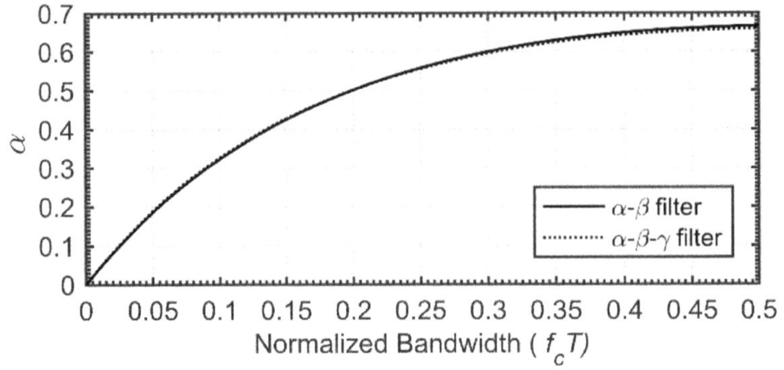


**Figure 3.4** Relation between α and normalized bandwidth for α-β and α-β-γ filters.

Figure 3.4 contains plots of $\alpha$ versus normalized, one-sided (half power) bandwidth for the Benedict-Bordner α-β filter and the Polge-Bhagavan α-β-γ filter. The normalized, one-sided bandwidth is defined as $f_cT$, or $f_c/f_s$, where $T$ is the track update period, and $f_s = 1/T$ is the track rate. It is interesting to note that the curves are almost identical. Thus, the relation between $\alpha$ and $f_c$ appears to be virtually the same for each filter type. We also note $f_cT$ extends to only 0.5. This stems from the fact that the theoretical, maximum, two-sided bandwidth is limited to $f_B = 2f_c = f_s = 1/T$.

The curves of Figure 3.4 can be approximated by the polynomial

$$\alpha = \sum_{n=1}^{5} c_n \left(f_c T\right)^n \tag{3.67}$$

using the coefficient values in Table 3.1.

## 3.7 RESPONSES OF THE OPTIMAL α-β AND α-β-γ FILTERS

Figure 3.5 contains example frequency and step responses for α-β and α-β-γ filters that are based on the optimal design techniques discussed by Benedict-Bordner and Polge-Bhagavan.

The responses were generated for filters with a one-sided bandwidth of 1 Hz and an update (track) rate of 10 Hz ($f_s = 10$ Hz, $T = 0.1$ s). As indicated by the plots, the optimal α-β filter exhibits a slightly underdamped response [23], and the optimal α-β-γ filter exhibits a response that is considerably underdamped.

**Table 3.1**
Coefficients Used to Generate Figure 3.4

| Coefficient | α-β filter | α-β-γ filter |
|---|---|---|
| c1 | 4.302 | 4.4193 |
| c2 | -12.9834 | -14.1397 |
| c3 | 25.6727 | 29.6192 |
| c4 | -31.8607 | -38.1654 |
| c5 | 17.358 | 21.3524 |

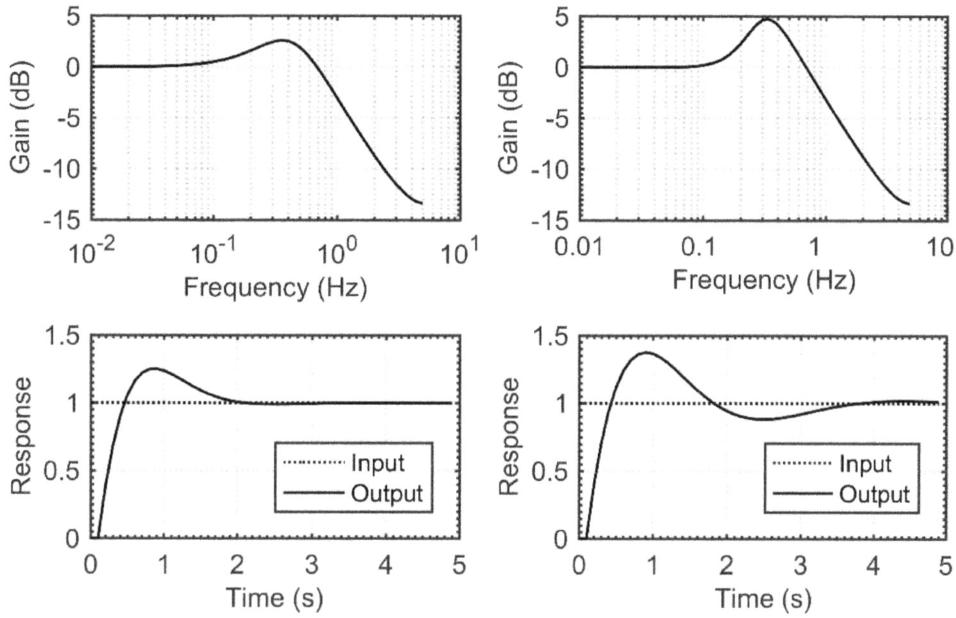


**Figure 3.5** Responses of the optimal α-β (left) and α-β-γ (right) filters – $f_c$ = 1 Hz, $T$ = 0.1 s.

## 3.8 CONTROL THEORY APPROACH

The optimal design approaches discussed in Sections 3.3 and 3.4 are somewhat limiting in that once $\alpha$ is selected to achieve a desired, closed loop bandwidth, the transient response of the filters is fixed because $\beta$ and $\gamma$ are determined from $\alpha$. In this section, we investigate the design of α-β and α-β-γ filters based on control theory concepts, with the hope such an approach will provide more design flexibility.

### 3.8.1 Type 1 Servo

Before we discuss the α-β and α-β-γ filters, we digress and discuss a unity feedback, Type 1 servo. Doing so provides us a third tracker type alternative, and also establishes some terminology and techniques we will use in the α-β and α-β-γ tracker designs.

The Type 1 tracker we consider has an open loop transfer function of the form

$$G_{OL}(z) = \frac{a}{(z-1)(z-b)}. \tag{3.68}$$

The pole as $z = 1$ makes the tracker a Type 1 servo, and the pole at $z = b$, along with the parameter $a$, allows us to set the bandwidth and transient behavior of the tracker.

The closed loop transfer function of the tracker is

$$G_{CL}(z) = \frac{G_{OL}(z)}{1+G_{OL}(z)} = \frac{a}{z^2 - (1+b)z + a + b} = \frac{a}{(z-z_1)(z-z_2)}. \tag{3.69}$$

If we assume the poles of $G_{CL}(z)$ are located at

$$z_1, z_2 = e^{-\zeta\omega_n T \pm j\omega_n T\sqrt{1-\zeta^2}} \tag{3.70}$$

we can write $G_{CL}(z)$ as

$$G_{CL}(z) = \frac{a}{z^2 - 2e^{-\zeta\omega_n T}\cos\left(\omega_n T\sqrt{1-\zeta^2}\right)z + e^{-2\zeta\omega_n T}} \tag{3.71}$$

from which we get

$$b = 2e^{-\zeta\omega_n T}\cos\left(\omega_n T\sqrt{1-\zeta^2}\right) - 1 \text{ and } a = e^{-2\zeta\omega_n T} - b. \tag{3.72}$$

The denominator of $G_{CL}(z)$ is a standard quadratic form analogous to the standard quadratic form $s^2+2\zeta\omega_n s+\omega_n^2$ of continuous-time servos [23, 38, 39, 40]. As such, $\zeta$ and $\omega_n$ set the one-sided bandwidth and transient behavior of the servo. Specifically, they set the damping and closed loop bandwidth, as discussed in Chapter 2. The properties also apply to the $\zeta$ and $\omega_n$ of (3.71). Normally, $\zeta$, is selected based on a desired transient behavior, and $\omega_n$ is determined from

$$\omega_n = 2\pi f_c\sqrt{\sqrt{\left(2\zeta^2-1\right)^2+1}+2\zeta^2-1} \tag{3.73}$$

where, as before, $f_c$ is the one-sided bandwidth.

Figure 3.6 contains frequency and step responses of the Type 1 servo for a one sided bandwidth of $f_c = 1$ Hz, a sample period of $T = 0.1$ s (a track rate of $f_s = 10$ Hz) and $\zeta$ values of $1/\sqrt{2}$, 0.5 and 0.3. Recall from Chapter 2 that $\zeta = 1$ results in a critically damped response, $\zeta > 1$ results in an overdamped response, and $\zeta < 1$ results in an underdamped response. Thus, all of the plots represent an underdamped response. It is interesting to note the frequency and step responses for $\zeta = 0.5$ are similar to the responses for the optimal α-β filter of Figure 3.3. Similarly, the frequency and step responses for $\zeta = 0.3$ are similar to the responses for the optimal α-β-γ filter.

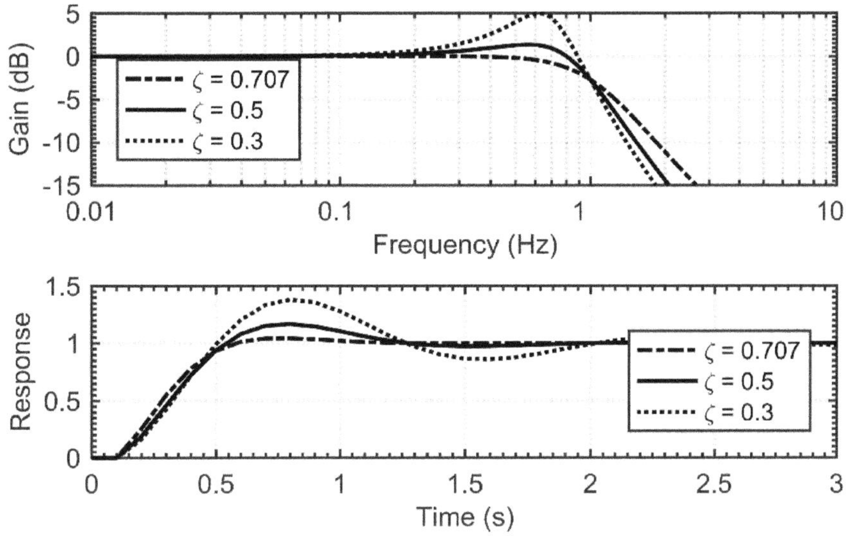


**Figure 3.6** Frequency and step responses of a 1-Hz, one-sided bandwidth, Type 1 servo.

We can put the Type 1 servo in the form of (3.1) and (3.2) by using

$$F = \begin{bmatrix} 1 & T \\ 0 & b \end{bmatrix} \text{ and } K = \begin{bmatrix} -a/b \\ a/(bT) \end{bmatrix}. \tag{3.74}$$

We might term the result a "α filter" where we use the single variable, $\alpha$, to remind us that it is a Type 1 servo and not a Type 2 servo like the α-β filter or a Type 3 servo like the α-β-γ filter.

### 3.8.2 α-β Tracker

We can think of the α-β tracker as a minimum order, Type 2 servo. We say minimum order because such a servo has an open loop transfer function with two poles at $z = 1$ and one zero. The two poles at $z = 1$ are required by the Type 2 specification, and the zero is required to make the servo stable.

The statement about the zero can be explained by examining the root locus of the servo. The left diagram of Figure 3.7 is a sketch of a root locus for the case where the open loop transfer function has only a double pole at $z = 1$. The right diagram is a sketch of a root locus for the case where the open loop transfer function has a double pole at $z = 1$ and a zero at some other value of $z$. For the double-pole-only case, the root locus is outside of the unit circle for all open loop gains greater than zero. However, when the open loop transfer function includes a zero, the locus remains mostly inside of the unit circle. Based on this discussion, we write the open loop transfer function of a second order, Type 2 servo as

$$G_{OL}(z) = \frac{az - b}{(z-1)^2} \tag{3.75}$$

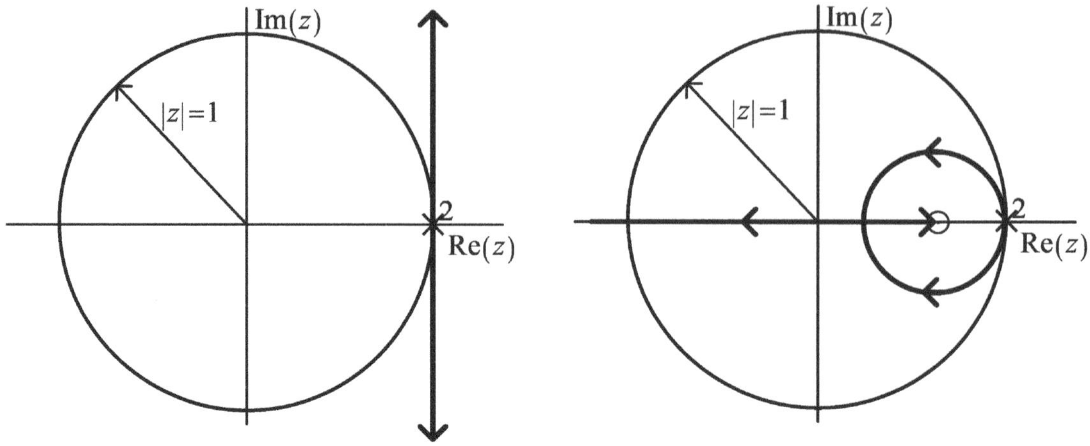


**Figure 3.7** Example root loci for a Type 2 servo.

The form of (3.75) gives us two parameters to control the behavior of the resulting Type 2 servo. Based on the discussion of Section 3.6.1, we expect the two parameters will give us some control over both the closed loop bandwidth and the transient behavior.

The closed loop transfer function of the servo is

$$G_{CL}(z) = \frac{G_{OL}(z)}{1+G_{OL}(z)} = \frac{az-b}{z^2-(2-a)z+1-b} \tag{3.76}$$

or, writing the denominator in the standard quadratic form,

$$G_{CL}(z) = \frac{az-b}{z^2-2e^{-\zeta\omega_n T}\cos\left(\omega_n T\sqrt{1-\zeta^2}\right)z+e^{-2\zeta\omega_n T}}. \tag{3.77}$$

With this we have

$$b = 1-e^{-2\zeta\omega_n T} \text{ and } a = 2-2e^{-\zeta\omega_n T}\cos\left(\omega_n T\sqrt{1-\zeta^2}\right). \tag{3.78}$$

As with the Type 1 servo of Section 3.6.1, $\zeta$ and $\omega_n$ control the transient behavior and one-sided bandwidth of the closed loop tracker. However, because of the numerator of $G_{CL}(z)$, the relation between $\zeta$ and $\omega_n$, and transient behavior of $f_c$ is not exactly the same as with the Type 1 servo. We will see this via examples. However, we first need to relate $\zeta$ and $\omega_n$ to $f_c$.

To determine a relation between $\zeta$, $\omega_n$, and $f_c$, we consider an analog equivalent of this Type 2 digital servo. For the analog servo, we let

$$G_{OL}(s) = \frac{2\zeta\omega_n s + \omega_n^2}{s^2}. \tag{3.79}$$

The resulting closed loop transfer function is

$$G_{CL}(s) = \frac{G_{OL}(s)}{1 + G_{OL}(s)} = \frac{2\zeta\omega_n s + \omega_n^2}{s^2 + 2\zeta\omega_n s + \omega_n^2}. \tag{3.80}$$

The one-sided bandwidth of the closed loop servo is the value of $\omega_c = 2\pi f_c$ where $|G_{CL}(s)|^2 = \frac{1}{2}|G_{CL}(0)|^2$. Since $G_{CL}(0) = 1$, the one-sided bandwidth is the $f_c$ where[9]

$$\begin{aligned} |G_{CL}(j\omega_c)|^2 &= \left| \frac{j2\zeta\omega_n\omega_c + \omega_n^2}{-\omega_c^2 + j2\zeta\omega_n\omega_c + \omega_n^2} \right|^2 \\ &= \frac{4\zeta^2\omega_n^2\omega_c^2 + \omega_n^4}{\omega_c^4 + 2\left(2\zeta^2 + 1\right)\omega_n^2\omega_c^2 + \omega_n^4}. \\ &= \frac{1}{2} \end{aligned} \tag{3.81}$$

After considerable manipulation, (3.81) leads to (see Problem 7)

$$\omega_n^4 + 2\left(2\zeta^2 + 1\right)\omega_n^2\omega_c^2 - \omega_c^4 \tag{3.82}$$

which can be solved to give

$$\omega_n^2 = \omega_c^2\left(-\left(1 + 2\zeta^2\right) \pm \sqrt{\left(1 + 2\zeta^2\right)^2 + 1}\right) \tag{3.83}$$

If we use the negative sign on the square root, we note that $\omega_n^2 < 0$, which would lead to a non-real value for $\omega_n$. Thus, we choose the positive sign and write

$$\omega_n = \omega_c\sqrt{\sqrt{\left(1 + 2\zeta^2\right)^2 + 1} - \left(1 + 2\zeta^2\right)}. \tag{3.84}$$

Once we have $\zeta$ and $\omega_n$, we can find $a$ and $b$ of the digital servo from (3.78).

---

[9] We owe this development to Dr. Craig Newborn of Dynetics, Inc.

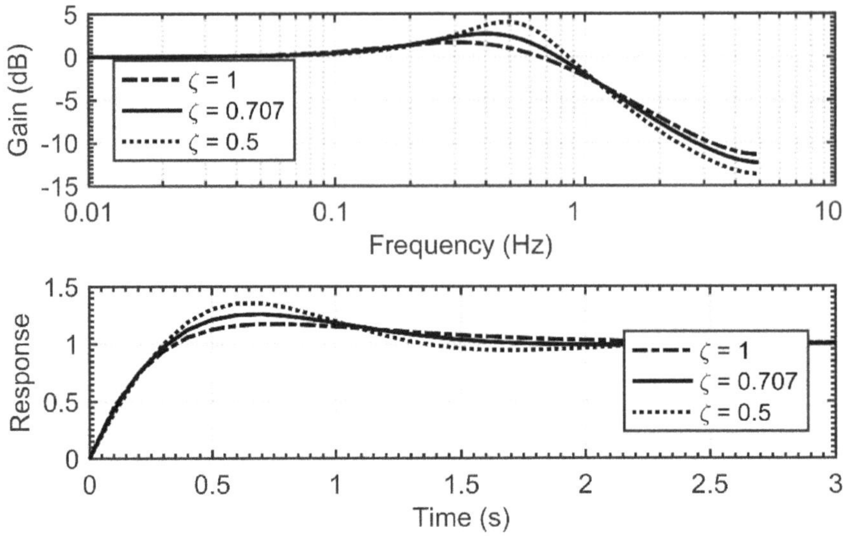


**Figure 3.8** Frequency and step responses for a general Type 2 tracker.

Figure 3.8 contains plots of closed loop frequency and step responses for $f_c = 1$ Hz, $T = 0.1$ s, and several values of $\zeta$. For the standard second order servo, we expect $\zeta = 1$ to correspond to the critically damped case (see Chapter 2 and Section 3.6.1). However, because of the open loop zero, the response for $\zeta = 1$ is slightly under damped. Similarly, the damping for the other values of $\zeta$ is less than we would get for a standard second order servo. In spite of this, we note that this method of designing Type 2 servos (or α-β filters) gives us more flexibility in controlling the transient behavior of the tracker.

We note that this method provides no guarantee the resulting $\alpha$ and $\beta$ will yield a variance ratio less than one. That would need to be investigated using (3.36). If the variance ratio is not less than one, it may be necessary to adjust $\zeta$ in an attempt to reduce the variance ratio.

As a final step, we can relate $\alpha$ and $\beta$ of the α-β filter to $a$ and $b$ by relating (3.75) to (3.25). That is, by writing

$$G_{OL}(z) = \frac{(\alpha+\beta)z - \alpha}{(z-1)^2} = \frac{az - b}{(z-1)^2}, \tag{3.85}$$

we have

$$\alpha = b \text{ and } \beta = a - b. \tag{3.86}$$

### 3.8.3 α-β-γ Tracker

The α-β-γ tracker is a minimum order, Type 3 servo. Its open loop transfer function has three poles at $z = 1$ and two zeros. As with the α-β tracker, the zeros are needed to "draw" the root locus of the tracker to the inside of the unit circle. Two zeros are needed because of the three poles at $z = 1$. This can be seen by sketching the root locus with none, one, and two zeros (see Problem 10). With no or only one zero, branches of the root locus are always outside of the unit circle.

We can write the open loop transfer function of a generic, minimum order, Type 3 servo as

$$G_{OL}(z) = \frac{az^2 - bz + c}{(z-1)^3}. \tag{3.87}$$

The associated closed loop transfer function is

$$G_{CL}(z) = \frac{G_{OL}(z)}{1 + G_{OL}(z)} = \frac{az^2 - bz + c}{z^3 - (3-a)z^2 + (3-b)z - 1 + c}. \tag{3.88}$$

We can also write the $G_{CL}(z)$ with a denominator that is the product of the standard quadratic form and a single real pole, or the product of three poles. The former would be

$$G_{CL}(z) = \frac{az^2 - bz + c}{\left[z^2 - 2e^{-\zeta\omega_n T}\cos\left(\omega_n T\sqrt{1-\zeta^2}\right)z + e^{-2\zeta\omega T}\right](z - z_3)} \tag{3.89}$$

and the latter would be

$$G_{CL}(z) = \frac{az^2 - bz + c}{(z - z_1)(z - z_2)(z - z_3)}. \tag{3.90}$$

Because of the extra closed loop pole at $z = z_3$, relating $\zeta$, $\omega_n$, and $z_3$ or $z_1$, $z_2$, and $z_3$, to bandwidth and transient behavior poses significant challenges. Instead of attempting to consider the general case, we examine two special cases. One is the "critically damped"[10] case discussed in the literature [1, 8, 11]. For that case, all three closed loop poles are set to the same value of $z_0$. For the second case, we set the two open loop zeros to the same value. The first special case will provide control of the one-sided bandwidth but not the transient behavior, because $z_0$ is the only parameter that can be varied. Once $z_0$ is determined, *a*, *b*, and *c* will be determined by the requirement that the three poles of (3.88) be equal to $z_0$.

---

[10] The term critically damped is a carryover from the standard second order transfer function. For that case, $\zeta = 1$ is defined as the critically damped case. With $\zeta = 1$, its two poles are located at the same value of z.

Two parameters will be available for the second case—the location of the pair of open loop zeros and the loop gain. Because of that, we hope to have some control over both the one sided bandwidth and the transient behavior.

3.8.3.1 Critically Damped Case

For the critically damped case we have

$$G_{CL}(z) = \frac{az^2 - bz + c}{(z - z_0)^3} = \frac{az^2 - bz + c}{z^3 - 3z_0 z^2 + 3z_0^2 z - z_0^3} \tag{3.91}$$

which, with (3.88), leads to

$$a = 3(1 - z_0),\ b = 3\left(1 - z_0^2\right) \text{ and } c = 1 - z_0^3. \tag{3.92}$$

As we did with the previous trackers, we use an equivalent analog, Type 3 servo to determine $z_0$, the closed loop pole location.

The open loop transfer function for the equivalent Type 3, analog servo is

$$G_{OL}(s) = \frac{ds^2 + es + f}{s^3} \tag{3.93}$$

and the closed loop transfer function is

$$G_{CL}(s) = \frac{G_{OL}(s)}{1 + G_{OL}(s)} = \frac{ds^2 + es + f}{s^3 + ds^2 + es + f} = \frac{ds^2 + es + f}{(s + p)^3} \tag{3.94}$$

where the denominator of the last term comes from setting the three poles of $G_{CL}(s)$ to the same value of $s = -p$.

We now want to relate $p$ to $f_c$. This means we want to find the value of $f_c$ where

$$\left|G_{CL}(j2\pi f_c)\right|^2 = \frac{1}{2}\left|G_{CL}(0)\right|^2. \tag{3.95}$$

As before, $G_{CL}(0) = 1$, so the problem becomes one of finding the $p$ such that

$$\left|G_{CL}(j2\pi f_c)\right|^2 = \left|\frac{\left(p^3 - 3p\omega_c^2\right) + j3p^2\omega_c}{(j\omega_c + p)^3}\right|^2 = \frac{1}{2}. \tag{3.96}$$

In (3.96), we used $\omega_c = j2\pi f_c$.

Considerable, tedious, algebraic manipulation of (3.96) leads to the rather simple result of (see Problem 12)

$$p = 2\pi f_c \left( \sqrt{\left(7-\sqrt{113}\right)^{1/3} + \left(7+\sqrt{113}\right)^{1/3} - 1} \right) = 1.6115 f_c . \tag{3.97}$$

If we make use of the relation $z = e^{sT}$, we have

$$z_0 = e^{-1.6115 f_c T} . \tag{3.98}$$

Figure 3.9 contains plots of the frequency and step response for the critically damped, minimum order, Type 3 servo with $f_c = 1$ Hz and an update period of $T = 0.1$ s ($f_s = 10$ Hz). Even though we term this the "critically damped" case, the response is still under damped because of the double zero. When we compare the response of Figure 3.9 to the right side of Figure 3.3, we note the response of the critically damped, minimum order, Type 3 servo is considerably more damped than the optimal α-β-γ filter.

As a final step, we need to relate $\alpha$, $\beta$, and $\gamma$ of the α-β-γ filter to $z_0$. To do so, we find the closed loop transfer function of the α-β-γ filter by using (3.26) in the equation

$$G_{CL}(z) = \frac{G_{OL}(z)}{1+G_{OL}(z)} = \frac{G_F(z)}{1+G_F(z)} . \tag{3.99}$$

This leads to

$$G_{CL}(z) = \frac{(\alpha+\beta+\gamma)z^2 + (\gamma-\beta-2\alpha)z + \alpha}{z^3 - (3-\alpha-\beta-\gamma)z^2 + (3-2\alpha-\beta+\gamma)z + \alpha - 1} . \tag{3.100}$$

Finally, equating denominator coefficients of (3.100) and (3.91) leads to

$$3-\alpha-\beta-\gamma = 3z_0,\ 3-2\alpha-\beta+\gamma = 3z_0^2 \text{ and } \alpha - 1 = -z_0^3 . \tag{3.101}$$

or

$$\alpha = 1 - z_0^3,\ \beta = \frac{3}{2}(z_0-1)^2(z_0+1) \text{ and } \gamma = \frac{1}{2}(1-z_0)^3 . \tag{3.102}$$

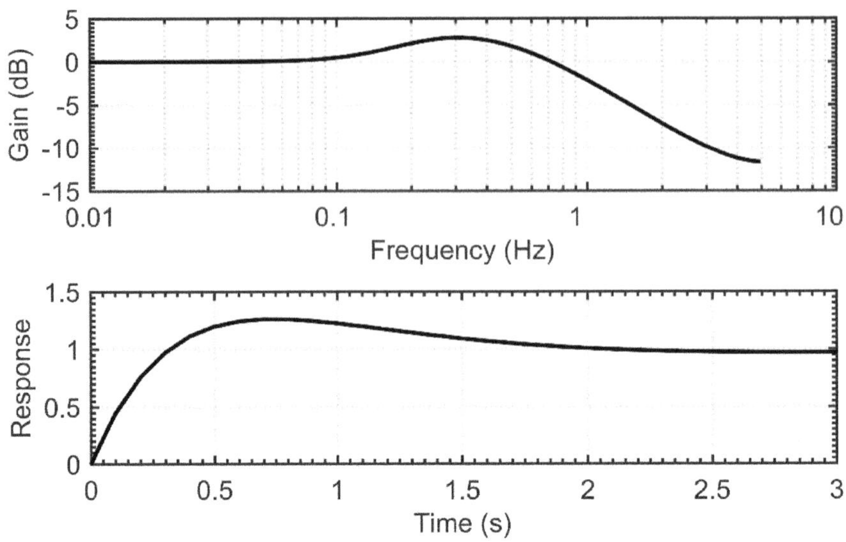


**Figure 3.9** Frequency and step responses the "critically damped" Type 3 tracker.

### 3.8.3.2 Type 3 Servo with Equal Open Loop Zeros

For the case where we have two, equal, open loop zeros, the open loop transfer function is

$$G_{OL}(z) = \frac{K(z - z_0)^2}{(z-1)^3} \tag{3.103}$$

and the associated closed loop transfer function is

$$G_{CL}(z) = \frac{G_{OL}(z)}{1 + G_{OL}(z)} = \frac{K(z - z_0)^2}{z^3 - (3-K)z^2 + (3 - 2Kz_0)z - 1 + Kz_0^2}. \tag{3.104}$$

In this case, we have two parameters, $K$ and $z_0$, we can vary to control the bandwidth and transient behavior of the system. As before, we will set one parameter, $z_0$, and choose the other, $K$, to provide a desired one-sided bandwidth. In this case, we work directly with the $z$-transfer function rather than the equivalent analog transfer function. This saves the step of specifying the analog equivalent and translating the results back to the digital servo. It has the added feature that it allows direct placement of the pair of zeros.

As before, our criterion is that we want to select $K$ and $z_0$ so that

$$\left|G_{CL}(z)\right|^2\Big|_{e^{j2\pi f_c T}} = \frac{1}{2}\left|G_{CL}(z)\right|^2\Big|_{z=1}. \tag{3.105}$$

From (3.104), we note that $|G_{CL}(1)|^2 = 1$ so that (3.105) reduces to

$$\left|G_{CL}\left(e^{j2\pi f_c T}\right)\right|^2 = \frac{1}{2}. \quad (3.106)$$

Substituting (3.104) into (3.106) and solving for $K$ yields

$$K = \frac{AB' + A'B + \sqrt{(AB' + A'B)^2 + 4AA'BB'}}{2AA'} \quad (3.107)$$

where $A = (z_c - z_0)^2$, $A' = (1/z_c - z_0)^2$, $B = (z_c - 1)^3$, $B' = (1/z_c - 1)^3$ , and $z_c = e^{j2\pi f_c T}$. Although not obvious from the form of (3.107), $K$ is real.

With this approach, we use $z_0$ to obtain a desired transient behavior and compute $K$ to give the desired one-sided bandwidth. Experimentation indicates that $z_0$ should be greater than about 0.8 but less than 1.0. The exact value will depend on the closed loop bandwidth and the amount of damping desired. If the closed loop bandwidth is such that $f_c T$ is in the neighborhood of 0.1 to 0.3, values of $z_0$ close to 0.8 would be appropriate. If $f_c T$ is significantly less than 0.1 (i.e., the one sided bandwidth is small relative to the update rate, $f_s = 1/T$), $z_0$ should be chosen close to, but never equal to[11], 1.0. The amount of damping increases as $z_0$ gets larger.

Figure 3.10 contains plots of the frequency and step response of a Type 3 servo with dual open loop (and closed loop) zeros at three values of $z_0$. The one-sided bandwidth is $f_c = 1$ Hz, and the update period is $T = 0.1$ s ($f_s = 10$ Hz). As hoped, the ability to choose $z_0$ and $K$ affords some control over both the bandwidth and transient behavior. As a note, Figure 3.8 supports the statement that the damping increases as the location of the two zeros approaches 1.0.

As a final step, we can relate $\alpha$, $\beta$, and $\gamma$ of the α-β-γ filter to $K$ and $z_0$ by equation $G_{OL}(z)$ of (3.103) to $G_F(z)$ of (3.26), which is the open loop transfer function, $G_{OL}(z)$, of the α-β-γ filter. That is, by writing

$$\frac{(\alpha+\beta+\gamma)z^2 + (\alpha-\beta-2\gamma)z + \alpha}{(z-1)^3} = \frac{K(z-z_0)^2}{(z-1)^3} = \frac{Kz^2 - 2Kz_0 z + Kz_0^2}{(z-1)^3}, \quad (3.108)$$

we have

$$\alpha+\beta+\gamma = K,\ \gamma-\beta-2\alpha = -2Kz_0 \text{ and } \gamma = Kz_0^2. \quad (3.109)$$

which leads to

$$\alpha = Kz_0^2,\ \beta = \frac{K}{2}\left(1+2z_0-3z_0^2\right) \text{ and } \gamma = \frac{K}{2}(1-z_0)^2. \quad (3.110)$$

---

[11] Setting the $z_0$ to 1.0 would reduce the Type 3 servo to a Type 1 servo (see Problem 16).

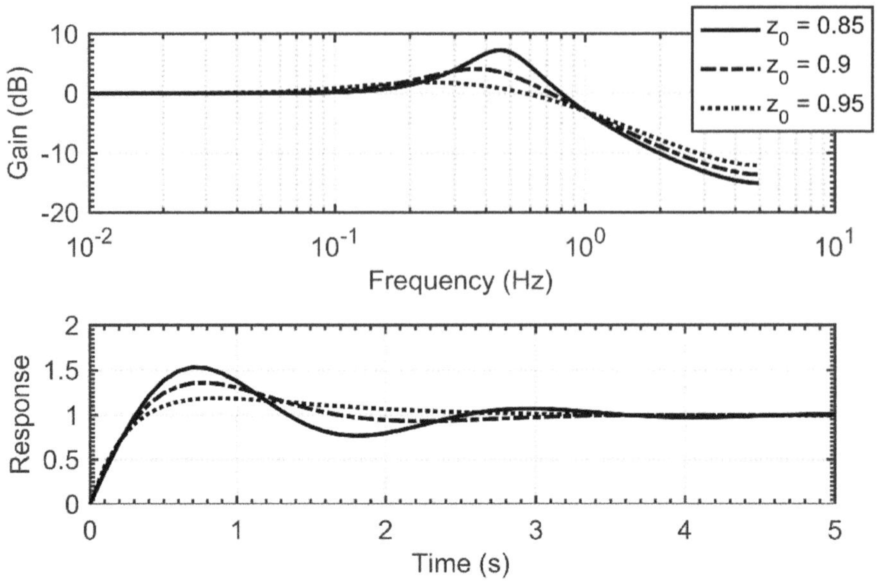


**Figure 3.10** Example frequency and step responses the equal zeros, Type 3 tracker.

As indicated earlier, the design methods of this section provide no guarantee the resulting $\alpha$, $\beta$, and $\gamma$ will yield a variance ratio less than one. That would need to be investigated using 3.38. If the variance ratio is not less than one, it may be necessary to adjust the filter design parameters in an attempt to reduce the variance ratio.

## 3.9 LINEAR KALMAN FILTER

The linear Kalman filter is an optimal filter that minimizes the mean squared error between the actual state of a system, $X(n)$, and the smoothed state, $X_s(n)$, produced by the filter. The statement that the filter is optimal is based on several conditions, some of which are:

- The prediction equation of (3.1) is a perfect model of the actual system, except for an input disturbance, $W(n)$, to the actual system.
- The input disturbance is zero-mean, white noise with a known covariance of $Q(n) = E\{W(n)W^T(n)\}$.
- The noise on the measurements, $V(n)$, is also zero-mean and white with a known covariance of $R(n) = E\{V(n)V^T(n)\}$.
- The measurement noise, system disturbance, and initial state (which is assumed to be a zero-mean random variable with a known covariance) are mutually uncorrelated.

Only the third condition can be reasonably supported. As discussed in Section 3.2.2, the prediction equation of (3.1) is not generally a perfect system model. Also, we treat the *HOT* terms of Section 3.2.2 as the disturbances. However, since they are the higher order derivatives of position (a deterministic quantity), they are not white noise terms. Finally, we usually ignore the fourth condition because of the fact that the first two conditions are not satisfied. In spite of its limitations, the linear Kalman filter, and the non-linear or *extended* Kalman filter, are widely used in radar trackers, especially open loop trackers.

The (linear) Kalman filter is defined by the equations [12, 41]

$$\begin{aligned} X_p(n) &= F(n-1)X_s(n-1) \\ y_p(n) &= H^T X_p(n) \\ e(n) &= y(n) - y_p(n) \\ P_p(n) &= F(n-1)P_s(n-1)F^T(n-1) + Q(n-1). \\ K(n) &= P_p(n)H^T\left[HP_p(n)H^T + R(n)\right]^{-1} \\ X_s(n) &= X_p(n) + K(n)e(n) \\ P_s(n) &= \left[I - K(n)H\right]P_p(n) \end{aligned} \tag{3.111}$$

The 1st, 2nd, and 6th equations are those discussed in Section 3.2 for the α-β and α-β-γ filters. The 4th, 5th,and 7th are needed to compute the gain matrix, $K(n)$.

In (3.111),

$$P_p(n) = E\left\{\left[X(n) - X_p(n)\right]\left[X(n) - X_p(n)\right]^T\right\} \tag{3.112}$$

is the *predicted* covariance matrix and

$$P_s(n) = E\left\{\left[X(n) - X_s(n)\right]\left[X(n) - X_s(n)\right]^T\right\} \tag{3.113}$$

is the *smoothed* covariance matrix. If the Kalman filter is properly tuned, the diagonal terms of these matrices provide an idea of how well the filter is performing. That is, they provide good estimates of the mean squared errors between $X_p(n)$ and $X(n)$, and between $X_s(n)$ and $X(n)$. The operative term is "properly tuned"; and therein lies the problematic aspect of Kalman filters. Specifically, Kalman filters are typically tuned by trial and error. Tuning is accomplished by manipulation of the $Q(n)$ matrix. Generally, designers use properties of the *HOT*s to select an initial $Q(n)$ then adjust the elements to achieve an acceptable level of performance, such that $P_p(n)$ and $P_s(n)$ are reasonable predictors of performance. For complicated and varied target and environment conditions, this can be a lengthy process.

An attractive feature of Kalman filters, aside from the potential usefulness of $P_p(n)$ and $P_s(n)$, is that they are dynamic. Thus, things like target maneuvers and changes in measurement noise can be accommodated by changing $Q(n)$ and $R(n)$. Another attractive feature of Kalman filters is, they can accommodate independent position *and* velocity measurements. This is not possible with the α-β, α-β-γ, and other filters discussed in this chapter.

Kalman filters must be initialized by selecting values for $X_s(0)$ and $P_s(0)$. $X_s(0)$ is usually set to $[p(0)\ 0\ 0]^T$ (for a third order filter) where $p(0)$ is the initial positon estimate. $p(0)$ is usually available from a search radar, or the search and acquisition function of the track radar. $P_s(0)$ is usually set to

$$P_s(0) = \text{diag}\left[\sigma_p^2(0),\ \sigma_{p'}^2(0),\ \sigma_{p''}^2(0)\right] \tag{3.114}$$

where the $\sigma^2(0)$ are estimates of the variances on the elements of $X(0)–X_s(0)$. These are sometimes available from the search radar or search function. At other times, they are estimated based on the parameters of the search radar, or waveform(s) used by the search function. As a note, it is also generally necessary to define $X_s(0)$ for the other track filters discussed in this chapter.

## 3.10 EXAMPLE

Now that we have methods for designing α filters (Type 1 servos), α-β filters (Type 2 servos), and α-β-γ filters (Type 3 servos), we want to discuss how to integrate them into closed loop trackers and compare their performance via an example.

Figure 3.11 contains the block diagram of Figure 2.1 with the $K_R$ replaced by a nonlinearity, the $G_F(z)$ replaced by Figure 3.1, and $G_C(z)$ replaced by a gain of unity. The nonlinearity is termed a discriminator curve and is indicative of the nonlinear behavior exhibited by a sampling-gate range tracker for an unmodulated pulse (see Chapter 4). The discriminator curve has been normalized to provide a slope $K_R = 1$ in the linear region. The linear region extends from $-\Delta R/2$ to $\Delta R/2$ where $\Delta R$ is the range resolution of the pulse. Beyond $\pm\Delta R/2$, the slope goes to zero, and beyond $\pm 1.5\Delta R$, the discriminator output drops to zero.

The controlled element is represented by $G_C(z) = 1/K_R = 1$ because the discriminator was normalized to have a slope of $K_R = 1$, and because it is assumed the controlled element has no dynamics.

To test the trackers, we use the scenario of Figure 3.12. In this scenario, the target is flying a circular path with a radial dimension of 10 km. The tangential velocity of the target is constant at approximately 175 m/s, or about Mach 0.5 [42].[12] The constant radial velocity of 175 m/s and radius of 10 km means the target is experiencing about 1/3 g of centripetal acceleration [43]. For convenience, we assume the target is at an altitude of 0 m.

We assume tracking starts when the target is at the right of the circle and continues for about 200s, or until the target is near the left of the circle.

[12] Named after Austrian physicist Ernst Mach, Mach number is the ratio of airspeed to the speed of sound. Note that the speed of sound is 340 m/s at sea level, but varies with temperature and air density.

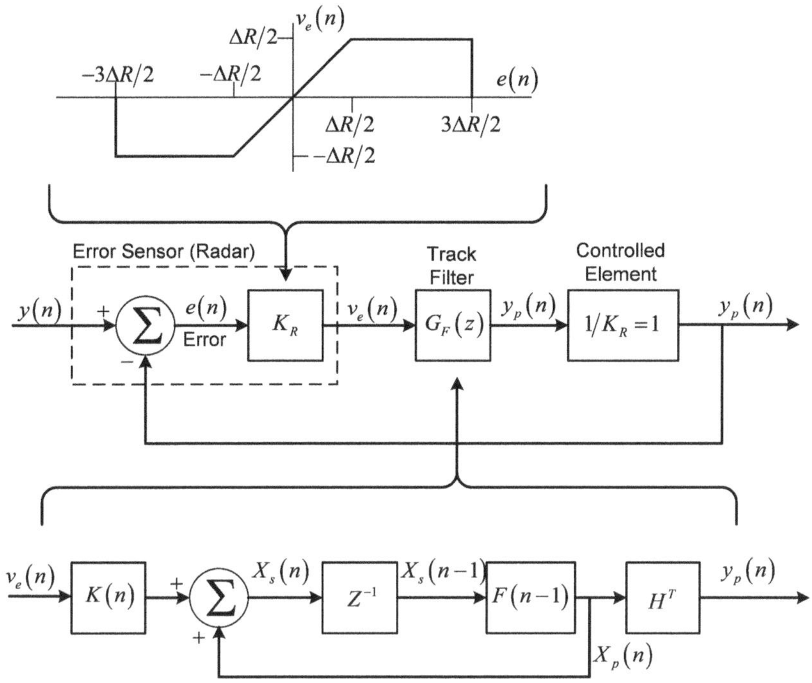


**Figure 3.11** Example closed loop range tracker.

All trackers are designed to have a one-sided bandwidth of $f_c = 1$ Hz and a track update period of $T = 0.1$ s ($f_s = 10$ Hz track rate). The α filter is designed to have a $\zeta$ of about 0.707. As a result, from (3.72) and (3.73), we have

$$a = 0.2532 \text{ and } b = 0.1581. \tag{3.115}$$

We use the Benedict and Bordner form of the α–β filter and the Polge-Bhagavan form of the α-β-γ filter. From (3.67) and (3.60), we have

$$\alpha = 0.3230 \text{ and } \beta = 0.0622 \tag{3.116}$$

for the Benedict and Bordner form, and

$$\alpha = 0.3265,\ \beta = 0.0643 \text{ and } \gamma = 0.00316 \tag{3.117}$$

for the Polge-Bhagavan form. We assume we have a perfect initial range estimate but no range rate or range acceleration estimate. Because of this, we initialize $X_p$ to $[60 \text{ km } 0]^T$ for the α and α-β filters and $[60 \text{ km } 0\ 0]^T$ for the α-β-γ filter. We assume the radar is using a 1 μs, unmodulated pulse so that $\Delta R = 150$ m.

Figure 3.13 contains plots of actual ($R_{act}(n)$) and tracked ($R_p(n)$) range, and track error ($R_e(n) = R_{act}(n) - R_p(n)$) for the three trackers. The plots of $R_{act}(n)$ and $R_p(n)$ appear as one curve because the range error for all three trackers is small relative to $R_{act}(n)$ and $R_p(n)$.

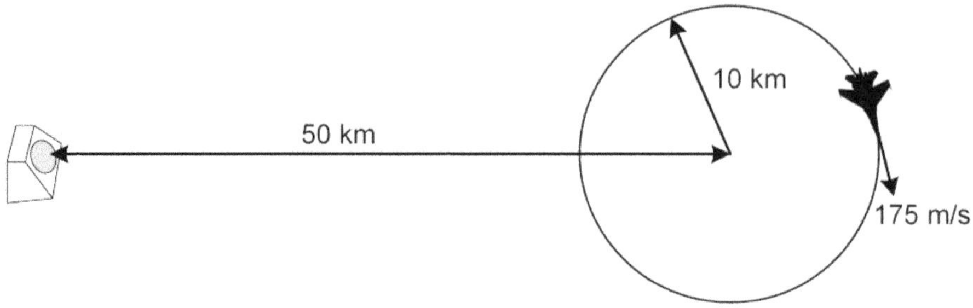


**Figure 3.12** Experiment scenario.

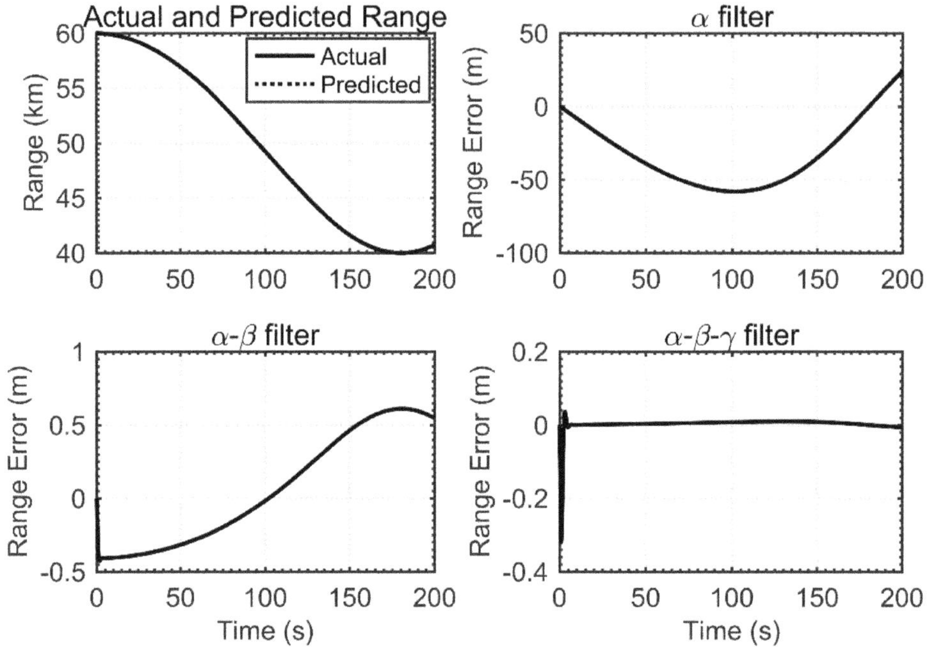


**Figure 3.13** Simulation outputs for the three tracker types and a 1-$\mu$s pulse.

The plots of range error are interesting and demonstrate the effect of system type on track error. The tracker with the α filter is a Type 1 servo and thus can track a constant range (i.e., stationary) target with zero error. However, if the range rate and range acceleration are not zero, as is the case in this example, the tracker can experience large errors. This is illustrated in the upper right plot of Figure 3.13.

The tracker with the α-β filter is a Type 2 servo and can thus track a constant rate target (zero acceleration) with zero steady state error. In this example, the acceleration is not zero but is small. Thus, we anticipate the tracker will experience a smaller error than the tracker that uses the α filter. This is illustrated in the lower left figure.

The tracker with the α-β-γ filter is a Type 3 servo, which means it can track a constant acceleration target with zero steady state error. In this example, the acceleration is not constant, but it changes fairly slowly (low acceleration rate, also known as jerk). As a result, we anticipate the track error will be smaller than with the other two trackers. This is illustrated in the lower right figure.

Figure 3.13 applied to a radar that used a 1 $\mu$s pulse for tracking. The results for the case where the radar uses a 0.5 $\mu$s pulse are quite different. This is illustrated in Figure 3.14. The left pair of plots applies to the tracker that uses the α filter, and the right pair of plots applies to the tracker that uses the α-β filter. As can be seen, the tracker with the α filter loses track while the tracker with the α-β filter does not. The reason for this can be explained with the help of Figure 3.13. For the tracker with the α filter, the error peaked at a value of –60m. This is well in excess of the ±37.5m linear region of the discriminator curve for a 0.5 $\mu$s pulse. This means the output of the discriminator will saturate at ±37.5m and eventually go to zero when the error exceeds ±112.5m. Once the discriminator output goes to zero, the predicted range from the tracker will remain fixed, as shown by the dashed curve of the upper left plot. Since the actual range continues to decrease, the range error magnitude becomes larger. This phenomenon doesn't occur with the tracker that uses the α-β filter because the error never exceeds the linear region of the discriminator curve (for this example).

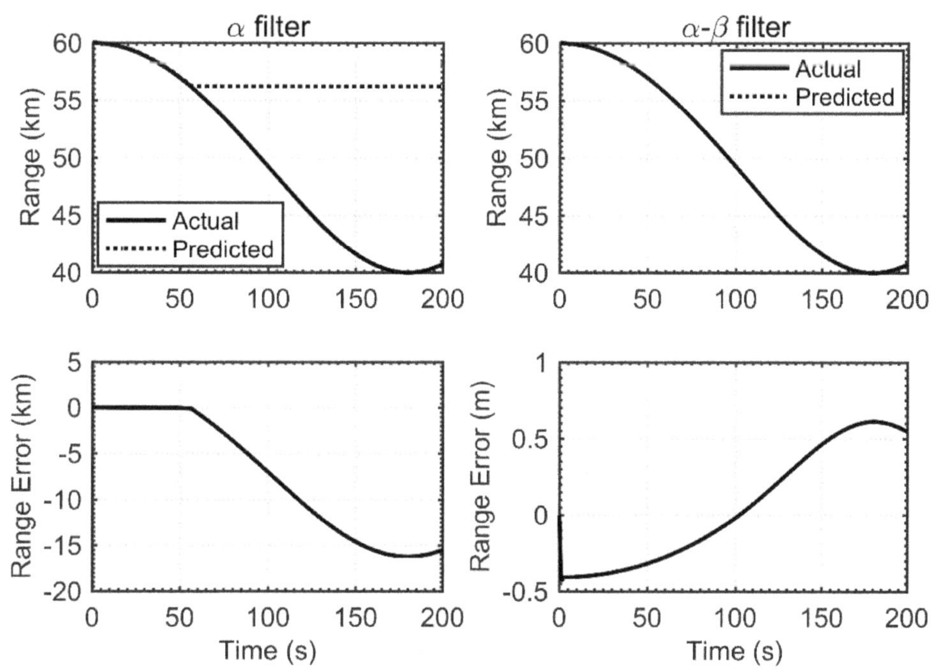


**Figure 3.14** Simulation output for the optimal α-β and α-β-γ trackers with a 0.5-$\mu$s pulse.

## 3.11 EXERCISES

1. Derive (3.24).
2. Derive (3.25) and (3.26).
3. Use (3.67) to find α and β for the optimum α-β filter with a normalized, one-sided bandwidth of $f_cT = 0.2$. Use this to generate plots like the left two of Figure 3.3.
4. Use (3.67) to find α, β, and γ for the optimum α-β-γ filter with a normalized, one-sided bandwidth of $f_cT = 0.2$. Use this to generate plots like the right two of Figure 3.3.
5. Derive (3.73).
6. Reproduce the plots of Figure 3.4.
7. Derive (3.82) and (3.84).
8. Reproduce Figure 3.6.
9. Derive (3.86).
10. Sketch the root loci for a Type 3, digital servo for the cases where the open loop transfer function has none, one, and two zeros. What can you say about the minimum number of zeros needed to make the servo stable?
11. Derive (3.92) from (3.91) and (3.88).
12. Derive (3.97).
13. Reproduce Figure 3.7.
14. Derive (3.101).
15. Derive (3.104).
16. Show that the Type 3 servo of Section 3.6.3.2 reduced to a Type 1 servo is $z_0 = 0$.
17. Derive (3.107).
18. Reproduce Figure 3.8.
19. Derive (3.109) and (3.110).
20. Implement the range tracker of Figure 3.9 and reproduce Figure 3.11.
21. Reproduce Figure 3.12.
22. Derive (3B.26).
23. Use (3B.26) and (3B.29) to derive (3B.30).
24. Plot the absolute value of $z_1$, $z_2$, $z_3$, and $z_4$ of the appendix versus λ as λ varies from 0 to 1. Does your plot verify the assertion in the sentence below (3B.20)?

### References

[1] J. Sklansky, "Optimizing the dynamic parameters of a track-while-scan system," *RCA Review,* vol. 18, no. 2, pp. 163-185, June 1957.

[2] R. E. Kalman, "A new approach to linear filtering and prediction problems," *Transactions of the ASME, Journal of Basic Engineering, Series D,* vol. 82, no. 1, pp. 35-45, March 1960.

[3] R. E. Kalman and R. S. Bucy, "New results in linear filtering and prediction theory," *Transactions of the ASME, Journal of Basic Engineering,* vol. 83, no. 3, pp. 95-107, December 1961.

[4] T. Benedict and G. Bordner, "Synthesis of an optimal set of radar track-while-scan smoothing equations," *IRE Transactions on Automatic Control,* vol. 7, no. 4, pp. 27-32, 1962.

[5] H. R. Simpson, "Performance measures and optimization condition for a third-order sampled-data tracker," *IEEE Transactions on Automatic Control,* vol. 8, no. 2, pp. 182-183, April 1963.

[6] S. Neal, "Parametric relations for the α-β-γ filter Predictor," *IEEE Transactions on Automatic Control,* vol. 12, no. 3, pp. 315-317, June 1967.

[7] B. K. Bhagavan and R. J. Polge, "Performance of the g-h Filter for Tracking Maneuvering Targets," *IEEE Transactions on Aerospace and Electronic Systems,* Vols. AES-10, no. 6, pp. 864-866, Nov. 1974.

[8] R. J. Polge and B. K. Bhagavan, "A Study of the G-H-K Tracking Filter," UAH Research Report No. 176, MICOM Report No. RE-CR-76-1, DTIC ADA021317, 1975.

[9] S. S. Blackman and R. Popoli, Design and Analysis of Modern Tracking Systems, Norwood, MA: Artech House, 1999.

[10] D. Tenne and T. Singh, "Optimal design of α-β-(γ) filters," in *American Control Conference, 2000. Proceedings of the 2000*, Chicago, IL.

[11] D. Tenne and T. Singh, "Characterizing performance of α-β-γ filters," *IEEE Transactions on Aerospace and Electronic Systems,* vol. 38, no. 3, pp. 1072-1087, July 2002.

[12] M. C. Budge Jr., "EE 706 - Kalman Filtering Techniques," 2010. [Online]. Available: http://www.ece.uah.edu/courses/material/EE706-Merv/index.htm. [Accessed 28 February 2017].

[13] 徐振来 (Xu Zhenlai), 相控阵雷达数据处理 (Phased Array Radar Data Processing), 北京: 国防工业出版社 (Beijing: National Defense Industry Press), 2009.

[14] 蔡慶宇 (Cai Qingyu) and 张伯彦 (Zhang Boyan), 相控阵雷达数据处理及其仿真技术 (Data Processing and Simulation Technology of Phased Array Radar), 北京: 国防工业出版社 (Beijing: National Defense Industry Press), 1997.

[15] 何友 (He You), 修建娟 (Jian Juan) and 关欣 (Guan Xin), 雷达数据处理及应用 (Radar Data Processing and Application), 北京: 电子工业出版社 (Beijing: Publishing House of Electronics Industry), 2013.

[16] В. С. Верба (V. S. Verba), Авиационные комплексы радиолокационного дозора и наведения. Состояниеи тенденции развития (Aviation Radar Patrol and Guidance Complexes. State of trending developments), Москва: Радиотехника (Moscow: Radio Engineering), 2008.

[17] D. G. Schultz and J. L. Melsa, *State Functions and Linear Control Systems*, New York, NY: McGraw-Hill, 1967.

[18] J. L. Melsa and D. G. Schultz, *Linear Control Systems*, New York, NY: McGraw-Hill, 1969.

[19] S. C. Gupta, *Transform and State Variable Methods in Linear Systems*, Huntington, New York: Robert E. Kreiger Publishing Co., Inc., 1971.

[20] R. C. Dorf and R. H. Bishop, *Modern Control Systems*, 13th ed., Pearson Education, Inc., 2016.

[21] M. C. Budge Jr. and S. R. German, *Basic Radar Analysis*, Norwood, MA: Artech House, 2015.

[22] E. Kreyszig, *Advanced Engineering Mathematics*, 7th ed., New York: John Wiley & Sons, 1993.

[23] C. H. Houpis and S. N. Sheldon, *Linear Control System Analysis and Design with MATLAB*, 6th ed., Boca Raton, FL: CRC Press, 2013.

[24] G. M. Siouris, *An Engineering Approach to Optimal Control and Estimation Theory*, New York: Wiley-Interscience, 1996.

[25] B. C. Kuo, *Digital Control Systems*, Fort Worth, TX: Holt, Rinehart and Winston, Inc., 1980.

[26] C. F. Asquith, "Weight Selection in First-Order Linear Filters," U. S. Army Missile Command, Redstone Arsenal, Alabama, Rep. No. RG-TR-69-12, 1969. Available from DTIC as AD859332.

[27] W.-S. Lu and T. Hinamoto, "Optimal design of IIR digital filters with robust stability using conic-quadratic-programming updates," *IEEE Transactions on Signal Processing,* vol. 51, no. 6, pp. 1581-1592, 2003.

[28] A. Antoniou, *Digital Signal Processing: Signals, Systems, and Filters*, New York: McGraw-Hill, 2005.

[29] S. K. Mitra, *Digital Signal Processing: a Computer-Based Approach*, 2nd ed., NY: McGraw-Hill, 2001.

[30] R. E. Zeimer, W. H. Tranter and D. R. Fannin, *Signals & Systems Continuous and Discrete*, 4th ed., Upper Saddle River, NJ: Prentice Hall, 1984.

[31] H. Urkowitz, *Signal Theory and Random Processes*, Dedham, MA: Artech House, 1983.

[32] A. Papoulis, *Probability, Random Vairables, and Stochastic Processes*, 3rd ed., New York: McGraw-Hill, Inc., 1991.

[33] L. P. Eisenhart, *Riemannian Geometry*, Princeton, New Jersey: Princeton University Press, 1997.

[34] I. Grattan-Guinness, Ed., *Companion Encyclopedia of the History and Philosophy of the Mathematical Sciences*, New York: Routledge, 1993.

[35] J. A. Cadzow and H. R. Martens, *Discrete-Time and Computer Control Systems*, F. F. Kuo, Ed., Englewood Cliffs, New Jersey: Prentice hall, 1970.

[36] J. L. Lagrange, *Mecanique Analytique*, Paris: Courcier, 1811.

[37] G. M. Ewing, *Calculus of Variations with Applications*, New York: Dover Publications, Inc., 2016.

[38] W. R. Ahrendt, *Servomechanism Practice*, New York: McGraw-Hill Book Company, Inc., 1954.

[39] S. A. Davis, *Outline of Servomechanisms*, New York: Unitech, 1966.

[40] H. M. James, N. B. Nichols and R. S. Phillips, Eds., *Theory of Servomechanisms*, New York: McGraw-Hill Book Company, Inc., 1947.

[41] И. Н. Синицын (I. N. Sinitsyn), Фильтры Калмана и Пугачева (Kalman and Pugachev Filters), Москва: Логос (Moscow: Logos), 2006.

[42] M. Williamson, *Dictionary of Space Technology*, Cambridge University Press, 2010.

[43] H. D. Young and R. A. Freedman, *Sears and Zemansky's University Physics*, 10th ed., Addison-Wesley, 1999.

[44] Y. Bar-Shalom, X.-R. Li and T. Kirubarajan, *Estimation with Applications to Tracking and Navigation*, New York: John Wiley & Sons, Inc., 2001.

[45] M. S. Grewal and A. P. Andrews, *Kalman Filtering Theory and Practice Using MATLAB*, 2nd ed., NY: Wiley-Interscience, 2001.

[46] A. V. Oppenheim and R. W. Schafer, *Discrete-Time Signal Processing*, Englewood Cliffs: Prentice-Hall, 1989.

[47] R. G. Lyons, *Understanding Digital Signal Processing*, 3rd ed., New York: Prentice Hall, 2011.

[48] S. J. Orfanidis, *Introduction to Signal Processing*, Englewood Cliffs, New Jersey: Prentice Hall, 1996.

[49] J. G. Proakis and D. G. Manolakis, *Digital Signal Processing*, 4th ed., Upper Saddle River, New Jersey: Pearson Prentice Hall, 2007.

[50] E. O. Brigham, *The Fast Fourier Transmform and Its Applications*, Englewood Cliffs, NJ: Preintice-Hall, Inc., 1988.

[51] G. R. Cooper and C. D. McGillem, *Continuous & Discrete Systems*, 3rd ed., Philadelphia: Saunders College Publishing, 1991.

[52] А. Б. Сергиенко (A. B. Sergienko), Цифровая обработка сигналов (*Digital Signal Processing*), Москва: ЗАО Издательский дом «Питер» (Moscow: ZAO Peter Publishing House), 2002.

## APPENDIX 3A: STABILITY TRIANGLE AND VARIANCE RATIO

In this appendix, we present the development that led to the stability triangles of Figures 3.2 and 3.3. We also discuss derivations of the variance ratios for the α-β and α-β-γ trackers.

### 3A.1 Stability Triangle

The stability triangle is a graphic that is used to find the range of values of $\alpha$, $\beta$, and $\gamma$ that result in stable trackers. To develop the equations for the triangles, we start with the closed loop transfer function of the trackers.

### 3A.2 Stability Triangle—α-β Tracker

The closed loop transfer function of the α-β tracker is

$$G_{CL}(z) = \frac{G_F(z)}{1+G_F(z)} = \frac{(\alpha+\beta)z+\alpha}{z^2-(2-\alpha-\beta)z+1-\alpha} \tag{3A.1}$$

where we used (3.25) for $G_F(z)$. We can also write $G_{CL}(z)$ as

$$G_{CL}(z) = \frac{az+b}{(z-z_1)(z-z_2)} = \frac{az+b}{z^2-(z_1+z_2)z+z_1z_2}. \tag{3A.2}$$

Working directly with (3A.1) is not straight forward in terms of finding the values of $\alpha$ and $\beta$ for which the tracker is stable. Because of this, we transform $G_{CL}(z)$ to the $s$-domain using the bilinear transform [29, p. 430]. Specifically, we use

$$z = \frac{1+s}{1-s}. \tag{3A.3}$$

With this we get

$$G_{CL}(s) = \frac{(\alpha+\beta)\frac{1+s}{1-s}+\alpha}{\left(\frac{1+s}{1-s}\right)^2-(2-\alpha-\beta)\frac{1+s}{1-s}+1-\alpha} = \frac{as^2+bs+c}{(4-2\alpha-\beta)s^2+2\alpha s+\beta}. \tag{3A.4}$$

If we invoke the Routh-Hurwitz stability criterion [20, 23] we have the requirement that

$$4-2\alpha-\beta>0,\ \alpha>0 \text{ and } \beta>0 \tag{3A.5}$$

or

$$4-2\alpha-\beta<0,\ \alpha<0 \text{ and } \beta<0. \tag{3A.6}$$

To eliminate the second condition, we consider the term $z_1z_2$ in (3A.2). For the filter to be stable, we require that $|z_1| < 1$ and $|z_2| < 1$, which implies $|z_1z_2| < 1$. From the relation between (3A.1) and (3A.2), this means

$$|1-\alpha|<1 \text{ or } -1<1-\alpha<1 \text{ or } 0<\alpha<2. \tag{3A.7}$$

From this we conclude (3A.5) is the appropriate inequality to use. This leads to the further conclusion that, for the tracker to be stable, we require

$$0<\alpha<2,\ \beta>0 \text{ and } \beta<4-2\alpha. \tag{3A.8}$$

This gives rise to the larger triangle of Figure 3.2.

The polygon of Figure 3.2 defines the bounds on $\alpha$ and $\beta$ for the poles of $G_{CL}(z)$ to be in the right half of the $z$-plane. To derive the equations for the bounding lines of the polygon, we require $\mathrm{Re}(z_1) > 0$ and $\mathrm{Re}(z_2) > 0$. This leads to the conditions

$$0<z_1z_2<1 \text{ and } \mathrm{Re}(z_1+z_2)>0 \tag{3A.9}$$

which subsequently lead to the conditions

$$0<1-\alpha<1 \text{ and } 2-\alpha-\beta>0. \tag{3A.10}$$

These, along with (3A.8) lead to,

$$0<\alpha<1,\ \beta>0 \text{ and } \beta<2-\alpha, \tag{3A.11}$$

as the boundaries of the polygon of Figure 3.2.

### 3A.3 Stability Triangle—α-β-γ Tracker

We can write the closed loop transfer function for the α-β-γ tracker as

$$\begin{aligned} G_{CL}(z) &= \frac{G_F(z)}{1+G_F(z)} \\ &= \frac{(\alpha+\beta+\gamma)z^2+(\gamma-\beta-2\alpha)z+\alpha}{z^3-(3-\alpha-\beta-\gamma)z^2+(3+\gamma-\beta-2\alpha)z+\alpha-1}, \end{aligned} \tag{3A.12}$$

where we used (3.26) for $G_F(z)$. We can also write $G_{CL}(z)$ as

$$\begin{aligned} G_{CL}(z) &= \frac{az^2+bz+c}{(z-z_1)(z-z_2)(z-z_3)} \\ &= \frac{az^2+bz+c}{z^3-(z_1+z_2+z_3)z^2+(z_1z_2+z_2z_3+z_1z_3)z-z_1z_2z_3}. \end{aligned} \tag{3A.13}$$

Using (3A.3) in (3A.12) we get the equivalent $s$-domain transfer function as

$$G_{CL}(s) = \frac{ds^2+es+f}{(8-2\beta-4\alpha)s^3+(4\alpha-2\gamma)s^2+2\beta s+2\gamma}. \tag{3A.14}$$

If we again invoke the Routh-Hurwitz criterion, we have the requirement that

$$4-\beta-\alpha>0,\ 2\alpha-\gamma>0,\ \beta>0 \text{ and } \gamma>0 \tag{3A.15}$$

or

$$4-\beta-\alpha<0,\ 2\alpha-\gamma<0,\ \beta<0 \text{ and } \gamma<0, \tag{3A.16}$$

and the additional requirement that

$$2\beta(4\alpha-2\gamma)>2\gamma(8-2\beta-4\alpha) \tag{3A.17}$$

We again use the conditions $|z_1| < 1$, $|z_2| < 1$ and $|z_3| < 1$, along with (3A.14) to arrive at $|z_1z_2z_3| < 1$ and

$$|\alpha-1|<1 \text{ or } 0<\alpha<2\,. \tag{3A.18}$$

We use this in (3A.16) to conclude $\gamma > 2\alpha > 0$ and $\gamma < 0$, which is contradictory. Thus, we conclude (3A.15) defines the proper set of conditions. This leads to the conditions

$$0<\alpha<2,\ \gamma>0,\ \gamma<2\alpha,\ \beta>0 \text{ and } \beta<4-\alpha\,. \tag{3A.19}$$

Equation (3A.17) leads to the additional restriction

$$\gamma<\frac{\beta\alpha}{2-\alpha}\,. \tag{3A.20}$$

Equation (3A.19) leads to the two stability triangles of Figure 3.2, and (3A.20) leads to the additional restriction noted on the figure.

We can determine the additional requirements for the poles of $G_{CL}(z)$ to also lie in the right half of the unit circle by applying the Routh-Hurwitz criterion to (3A.12), after replacing $z$ by $-z$. This leads to the additional conditions

$$\alpha<1, \tag{3A.21}$$

$$3-\alpha-\beta-\gamma>0, \tag{3A.22}$$

$$3-2\alpha-\beta+\gamma>0 \tag{3A.23}$$

and

$$(3-\alpha-\beta-\gamma)(3-2\alpha-\beta+\gamma)-(1-\alpha)>0\,. \tag{3A.24}$$

Unfortunately, except for (3A.21), how these would be applied to further restrict the values of $\alpha$, $\beta$, and $\gamma$ is not obvious. However, the first condition is useful in that it places an upper bound of one on $\alpha$.

### 3A.4 Variance Ratio

To determine the variance ratio, we use a method we term *covariance propagation*. This method appears in Kalman filter theory and error analysis, and is used to predict the variance on the states of a linear system, when the input to the system is zero-mean, white noise [12; 44, p. 191; 45, p. 88]. There are forms for continuous time systems and discrete time systems. Since we are interested in discrete time systems, we use that form.

We consider a linear system that we represent by the vector difference equation

$$X(n)=A(n-1)X(n-1)+B(n)u(n) \tag{3A.25}$$

where $u(n)$ is wide-sense stationary, zero-mean, white noise with a variance of $\sigma_u^2$. $u(n)$ is uncorrelated with $X(0)$. We can write the *covariance matrix*, $P(n)$, of the state, $X(n)$, as

$$P(n)=A(n-1)P(n-1)A^T(n-1)+B(n)B^T(n)\sigma_u^2. \tag{3A.26}$$

In (3A.25) and (3A.26), we assume $X(n)$, $A(n)$, $B(n)$, and $u(n)$ are real.

Given $P(n)$ is defined as

$$P(n)=E\{X(n)X^T(n)\}, \tag{3A.27}$$

we recognize the diagonal elements of $P(n)$ are the variances on the various elements of $X(n)$, and the off diagonal terms as the covariances between the elements of $X(n)$.

For our purposes, we are interested in the steady state value of $P(n)$, assuming $A(n)$ and $B(n)$ are constant matrices. In particular, we are interested in the first diagonal element of $P = P(\infty)$, since this is the variance on the position state of the α-β and α-β-γ filters. To find the first diagonal element, $P_{11}$, we need to solve the matrix equation,

$$P=APA^T+BB^T\sigma_u^2. \tag{3A.28}$$

For our particular problem, we are interested in the variance of the smoothed position state of the filter. Thus, we combine (3.1) and (3.2) to produce

$$X_s(n)=\left(I-K(n)H^T\right)F(n-1)X_s(n-1)+K(n)y(n). \tag{3A.29}$$

In (3A.29), we assume $y(n)$ is wide-sense stationary, zero-mean, white noise with a variance of $\sigma_y^2$. To put (3A.29) in the form of (3A.25), we use

$$A(n-1)=\left(I-K(n)H^T\right)F(n-1) \text{ and } B(n)=K(n), \tag{3A.30}$$

or, if we use the steady state values of the matrices,

$$A=\left(I-KH^T\right)F \text{ and } B=K. \tag{3A.31}$$

For the next step, we use these in (3A.28) to solve (with help from Maple or the Matlab symbolic tool box, or a large amount of tedious labor) for $P_{11} = \sigma_p^2$, the steady-state variance on the position element $X_s(n)$. Finally, we use this to find the variance ratio as

$$R_\sigma = \frac{\sigma_p^2}{\sigma_y^2} = \frac{2\alpha^2 + \beta(2-3\alpha)}{\alpha(4-\beta-2\alpha)} \tag{3A.32}$$

for the α-β filter and

$$R_\sigma = \frac{\sigma_p^2}{\sigma_y^2} = \frac{2\alpha^2(\beta+\gamma)+\beta^2(2-3\alpha)+\alpha\gamma(\beta-4)}{(\gamma\alpha-2\gamma+\alpha\beta)(4-\beta-2\alpha)}. \tag{3A.33}$$

for the α-β-γ filter.

## APPENDIX 3B: DERIVATION OF (3.60)—BENEDICT AND BORDNER α-β RELATION

The Benedict-Bordner approach seeks to choose the $r(n)$ that minimize the cost function

$$\begin{aligned} J(g(n),\lambda) &= \sigma_{ys\eta}^2 + \lambda D \\ &= \sum_{n=0}^{\infty}\left[r(n+1)-2r(n)+r(n-1)\right]^2 + \lambda\left[n-r(n)\right]^2 \end{aligned} \tag{3B.1}$$

where $r(n)$ is related to $g(n)$ by

$$g(n)=r(n+1)-2r(n)+r(n-1). \tag{3B.2}$$

Alternately, $G(z)$ and $R(z)$ are related by

$$G(z)=\frac{(z-1)^2}{z}R(z)=\left(z-2+z^{-1}\right)R(z). \tag{3B.3}$$

Benedict and Bordner use a calculus of variations approach to solving the problem of minimizing $J(g(n),\lambda)$. Specifically, they assume the optimal solution is $r(n)$, then perturb it to find alternate solutions as

$$r_{alt}(n)=r(n)+\varepsilon h(n). \tag{3B.4}$$

Through this procedure, they determine the form of the optimal solution.

In (3B.4), $h(n)$ is chosen to be an arbitrary, but realizable, function that satisfies $h(n)=0$ for $n \le 0$. The caveat "realizable" means $h(n)$ is finite (or *bounded*) for all $n > 0$. The specific calculus of variations approach they use is to form

$$\left.\frac{\partial J(r(n)+\varepsilon h(n),\lambda)}{\partial \varepsilon}\right|_{\varepsilon=0}=0 \tag{3B.5}$$

and use it to derive an equation for $r(n)$. The following is an implementation of that approach.

Using (3.53) in (3B.5) results in

$$\begin{aligned}\frac{\partial J}{\partial \varepsilon}=\sum_{n=0}^{\infty}\Bigg\{&\begin{bmatrix}(r(n+1)+\varepsilon h(n+1))-2(r(n)+\varepsilon h(n))\\+(r(n-1)+\varepsilon h(n-1))\end{bmatrix}\\&\times[h(n+1)-2h(n)+h(n-1)]\\&+\lambda[n-(r(n)+\varepsilon h(n))]h(n)\Bigg\}\end{aligned} \tag{3B.6}$$

and, from (3B.5),

$$\begin{aligned}\left.\frac{\partial J}{\partial \varepsilon}\right|_{\varepsilon=0}&=\sum_{n=0}^{\infty}\{[r(n+1)-2r(n)+r(n-1)][h(n+1)-2h(n)+h(n-1)]\\&+\lambda[n-r(n)]h(n)\}=0\end{aligned}. \tag{3B.7}$$

Grouping terms with similar $h(n)$ indices yields

$$\begin{aligned}\left.\frac{\partial J}{\partial \varepsilon}\right|_{\varepsilon=0}&=\sum_{n=0}^{\infty}h(n+1)[r(n+1)-2r(n)+r(n-1)]\\&-\sum_{n=0}^{\infty}h(n)[2r(n+1)-4r(n)+2r(n-1)-\lambda n+\lambda r(n)].\\&+\sum_{n=0}^{\infty}h(n-1)[r(n+1)-2r(n-1)+r(n-2)]=0\end{aligned} \tag{3B.8}$$

In the first summation, let $m=n+1$, and in the third summation, let $k=n-1$ to obtain

$$\begin{aligned}\left.\frac{\partial J}{\partial \varepsilon}\right|_{\varepsilon=0}&=\sum_{m=-1}^{\infty}h(m)[r(m)-2r(m-1)+r(m-2)]\\&-\sum_{n=0}^{\infty}h(n)[2r(n+1)-4r(n)+2r(n-1)-\lambda n+\lambda r(n)].\\&+\sum_{k=1}^{\infty}h(k)[r(k+2)-2r(k+1)+r(k)]=0\end{aligned} \tag{3B.9}$$

Since $h(n) = 0$ for $n \leq 0$, $h(m)|_{m=-1} = 0$, and we can set the lower limit of the first summation to 0. Using similar reasoning, we have $h(k)|_{k=0} = 0$, and we can set the lower limit of the third summation to zero. The result of these operations is

$$\begin{aligned}\left.\frac{\partial J}{\partial \varepsilon}\right|_{\varepsilon=0} &= \sum_{m=0}^{\infty} h(m)\left[r(m) - 2r(m-1) + r(m-2)\right] \\ &-\sum_{n=0}^{\infty} h(n)\left[2r(n+1) - 4r(n) + 2r(n-1) - \lambda n + \lambda r(n)\right] \\ &+\sum_{k=0}^{\infty} h(k)\left[r(k+2) - 2r(k+1) + r(k)\right] \\ &= 0\end{aligned} \quad (3B.10)$$

If we let $m = n$ in the first summation and $k = n$ in the third summation, we get

$$\begin{aligned}\left.\frac{\partial J}{\partial \varepsilon}\right|_{\varepsilon=0} &= \sum_{n=0}^{\infty} h(n)\left[r(n) - 2r(n-1) + r(n-2)\right] \\ &-\sum_{n=0}^{\infty} h(n)\left[2r(n+1) - 4r(n) + 2r(n-1) - \lambda n + \lambda r(n)\right] \\ &+\sum_{n=0}^{\infty} h(n)\left[r(n+2) - 2r(n+1) + r(n)\right] \\ &= 0\end{aligned} \quad (3B.11)$$

Since the three summations are over the same limits, we can combine them into a single summation and get

$$\begin{aligned}\left.\frac{\partial J}{\partial \varepsilon}\right|_{\varepsilon=0} &= \sum_{n=0}^{\infty} h(n)\left[r(n) - 2r(n-1) + r(n-2)\right. \\ &\quad -2r(n+1) + 4r(n) - 2r(n-1) - \lambda n + \lambda r(n) \\ &\quad \left. +r(n+2) - 2r(n+1) + r(n)\right] \\ &= 0\end{aligned} \quad (3B.12)$$

or

$$\begin{aligned}\left.\frac{\partial J}{\partial \varepsilon}\right|_{\varepsilon=0} &= \sum_{n=0}^{\infty} h(n)\left[r(n+2) - 4r(n+1) + 6r(n)\right. \\ &\quad \left. - 4r(n-1) + r(n-2) - \lambda n + \lambda r(n)\right] = 0\end{aligned} \quad (3B.13)$$

We define

$$\begin{aligned} f(n) = r(n+2) - 4r(n+1) + 6r(n) \\ -4r(n-1) + r(n-2) - \lambda\left[n - r(n)\right] \end{aligned} \quad (3B.14)$$

and write (3B.13) as

$$\left.\frac{\partial J}{\partial \varepsilon}\right|_{\varepsilon=0} = \sum_{n=0}^{\infty} h(n) f(n) = 0 . \quad (3B.15)$$

Since $h(n)$ is arbitrary for $n > 0$, the only way (3B.15) can be satisfied is if

$$f(n) = 0 \ \forall n > 0 . \quad (3B.16)$$

From $z$-transform theory, the condition of (3B.16) means the region of convergence (ROC) of the $z$-transform, $F(z)$, of $f(n)$ is $|z| < a$ ($a$ is a positive number) [46–49].

We will also impose the requirement that the Fourier transform of $f(n)$ exist [49–52]. This imposes the further restriction that the ROC not include $|z| = 1$. From this, we conclude that the ROC for $F(z)$ is $|z| < 1$, or the inside of the unit circle.

If we form the $z$-transform of $f(n)$ we have, from (3B.14),

$$\begin{aligned} F(z) &= \left(z^2 - 4z + 6 - 4z^{-1} + z^{-2}\right) R(z) - \lambda\left(\frac{z}{(z-1)^2} - R(z)\right) \\ &= \frac{1}{z^2}\left(z^4 - 4z^3 + 6z^2 - 4z + 1\right) R(z) - \lambda\left(\frac{z}{(z-1)^2} - R(z)\right) \\ &= \frac{(z-1)^4}{z^2} R(z) - \lambda\left(\frac{z}{(z-1)^2} - R(z)\right) \\ &= \frac{(z-1)^4 - \lambda z^2}{z^2} R(z) - \lambda \frac{z}{(z-1)^2} \end{aligned} . \quad (3B.17)$$

In (3B.17), $R(z)$ is the $z$-transform of $r(n)$, and $z/(z-1)^2$ is the $z$-transform of the unit ramp, $nU(n)$ (see Chapter 2).

We need to represent $F(z)$ in terms of $G(z)$, the $z$-transform of the α-β filter. To that end, we substitute (3.55) into (3B.17) to get

$$F(z)=\left[\frac{(z-1)^4-\lambda z^2}{z^2}\right]\frac{zG(z)}{(z-1)^2}-\lambda\frac{z}{(z-1)^2}$$
$$=\frac{\left[(z-1)^4-\lambda z^2\right]G(z)-\lambda z^2}{z(z-1)^2} \qquad (3B.18)$$

Rearranging (3B.18) yields

$$z(z-1)^2F(z)+\lambda z^2=\left[(z-1)^4-\lambda z^2\right]G(z). \qquad (3B.19)$$

From the ROC discussions, $F(z)$ has no poles inside of the unit circle (i.e., no poles in the ROC). Thus, $z(z-1)^2F(z)+\lambda z^2$ has no poles inside of the unit circle. This means $[(z-1)^4-\lambda z^2]G(z)$ has no poles inside of the unit circle. However, since $G(z)=G_{CL}(z)$ is the closed loop $z$-transform of the α–β filter, it will have two poles inside of the unit circle. For this and (3B.19) to be satisfied, these two poles must be equal to two of the roots of $(z-1)^4-\lambda z^2$.

At this point, Benedict and Bordner take the approach of finding the roots of $(z-1)^4-\lambda z^2$. They then show that two of the roots lie inside of the unit circle, and two lie outside of the unit circle. Their equations for the four roots are

$$z_{1,2}=\frac{2\pm j\lambda-\sqrt{-\lambda^2\pm 4j\lambda}}{2}$$
$$z_{3,4}=\frac{2\pm j\lambda+\sqrt{-\lambda^2\pm 4j\lambda}}{2} \qquad (3B.20)$$

where $z_{1,2}$ lie inside of the unit circle, and $z_{3,4}$ lie outside of the unit circle, for all $\lambda\geq 0$. They then proceed to argue that this can be used to determine the relation between $\alpha$ and $\beta$.

Since the Benedict-Bordner approach to deriving the relation between $\alpha$ and $\beta$ based on $z_{1,2}$ is difficult to follow, we use the alternate approach suggested by Polge and Bhagavan. Specifically, we write

$$(z-1)^4-\lambda z^2=z^4-4z^3+(6-\lambda)z^3-4z+1$$
$$=(z-a)(z-b)(z-c)(z-d) \qquad (3B.21)$$
$$=z^4-Az^3+Bz^2-Cz+D$$

which leads to

$$\begin{aligned} A &= 4 = a+b+c+d \\ B &= 6-\lambda = ab+ac+ad+bc+bd+cd \\ C &= 4 = abc+abd+acd+bcd \\ D &= 1 = abcd \end{aligned} . \tag{3B.22}$$

We assume the two poles inside of the unit circle are $a$ and $b$. The method proposed by Polge and Bhagavan doesn't lead to specific values for $a$ and $b$. Instead, it leads to an equation that relates $a$ to $b$. That relation is then used to find the relation between $\alpha$ and $\beta$.

To determine the relation between $a$ and $b$, we want to eliminate $c$ and $d$ from (3B.22). We do this by manipulating the equations for $A$, $B$, and $D$. We use the $A$ equation to obtain

$$a+b=4-(c+d) \tag{3B.23}$$

and the $D$ equation to obtain

$$cd=\frac{1}{ab}. \tag{3B.24}$$

We then rewrite the $C$ equation as

$$4=ab(c+d)+(a+b)cd \tag{3B.25}$$

and use (3B.24) and (3B.23) to arrive at the relation

$$ab(4-a-b)+\frac{a+b}{ab}=4. \tag{3B.26}$$

From (3B.21) we have

$$G(z)=\frac{N(z)}{(z-a)(z-b)}=\frac{N(z)}{z^2-(a+b)z+ab}. \tag{3B.27}$$

For the α-β filter we have

$$G(z)=G_{CL}(z)=\frac{G_{OL}(z)}{1+G_{OL}(z)}$$
$$=\frac{\dfrac{(\alpha+\beta)z-\alpha}{(z-1)^2}}{\dfrac{(\alpha+\beta)z-\alpha}{(z-1)^2}+1}$$
$$=\frac{(\alpha+\beta)z-\alpha}{z^2-(2-\alpha-\beta)z+1-\alpha} \quad \text{(3B.28)}$$

With this and (3B.27) we get

$$a+b=2-\alpha-\beta \text{ and } ab=1-\alpha. \quad \text{(3B.29)}$$

Finally, we can use (3B.29) in (3B.26) to arrive at the final Benedict and Bordner answer of

$$\beta=\frac{\alpha^2}{2-\alpha}. \quad \text{(3B.30)}$$

As did Benedict and Bordner, we found the relation between $\alpha$ and $\beta$. If we wanted to compute a specific value of $\alpha$, we would select a $\lambda > 0$ and find numeric values for $a$ and $b$. From that, we could use (3B.29) to find $\alpha$ then $\beta$. However, such an approach would be of little use to us because, per the discussions in the chapter, we want to find $\alpha$ as a function of desired one-sided bandwidth.

# Chapter 4

# Closed Loop Range Tracking

## 4.1 INTRODUCTION

According to Barton and others, [1–5; 6, p. 411], the optimum range estimator, which we term a range tracker, consists of a matched filter, a differentiator, and a zero-crossing detector. The differentiator forms the derivative of the envelope of the matched filter output, and the zero-crossing detector determines the time at which the derivative goes through zero. This time, relative to the leading edge of the transmit pulse is the range delay, $\tau_R$, to the target, plus a bias due to delays in the amplifiers, matched filter, and other components of the receiver. The premise of using the derivative and zero crossing detector is that the envelope of the matched filter output will peak at $\tau_R$ (plus the aforementioned delays). This peak can be found by determining the time at which the derivative of the envelope of the matched filter output changes sign.

Figure 4.1 contains one, approximate, implementation of the optimum range tracker. We call it a *closed loop tracker* for reasons we will explain later. The input to the tracker is the pulse returned from the target. It occurs at some time, $\tau_R$, relative to the transmit pulse, which is defined as $\tau = 0$. The returned pulse is processed through the receiver and matched filter, whose output is subsequently sent to the block labeled "env det". This block is the aforementioned envelope detector. In radars where the matched filter is implemented at some intermediate frequency (IF), it consists of a rectifier followed by a low-pass filter [2; 7; 8, p. 376]. In modern radars where the matched filter is implemented at base band [9; 10, p. 150] using complex signals, it forms the magnitude of the complex signal out of the matched filter.

The differentiator is approximated by the pair of switches and the block that follows them. In an analog radar, the switches, which we term *range gates* represent ideal, impulse samplers (and hold networks, which are not shown) that sample the detector output at two times denoted $\tau_E$ and $\tau_L$. In radars that use a digital matched filter and detector, the switches are software samplers that record the detector output at $\tau_E$ and $\tau_L$. They are also considered impulse samplers. If the tracker is working properly, $\tau_E$ occurs shortly before the peak of the detector output, and $\tau_L$ occurs shortly after the peak of the detector output.

The block following the switches approximates the derivative as a first order difference (i.e., $V_L - V_E$). The division by $V_L + V_E$, and the multiplication by $S_\tau$ normalize the output of the differentiator so it will be compatible with the signal expected by the track filter. We term the combination of the detector, range gates and differentiator, a *range discriminator*.

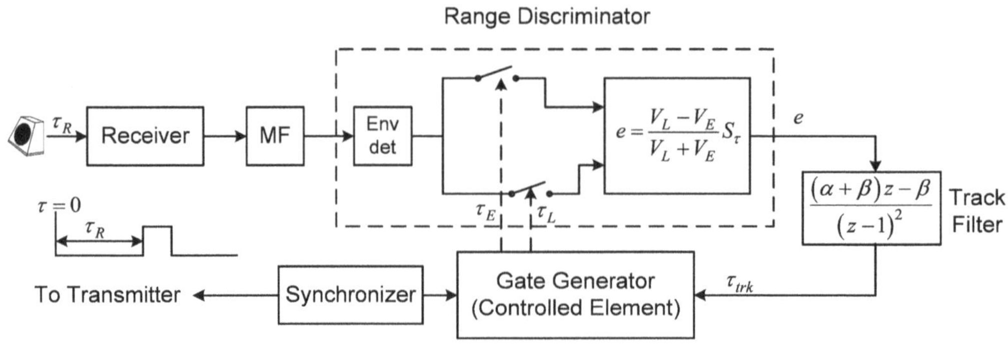


**Figure 4.1** Closed loop range tracker block diagram.

The track filter and gate generator make up the zero-crossing detector. Basically, the track filter attempts to maintain the signal out of the range discriminator at zero by adjusting the timing on the range gates, via the gate generator. In this way, it is finding the range delay where the derivative of the envelope detector output is zero. The fact that there is feedback, from the discriminator output to the range gates, makes the range tracker a closed loop tracker, or a closed loop servo.

Since the range gates are impulse samplers, we term the range discriminator of Figure 4.1 a *sampling gate discriminator*. This discriminator uses only two samples to approximate a derivative. We will later discuss another derivative approximation that is based on the use of several samples. We will term it a *summing gate discriminator*. It is somewhat analogous to an integrating gate discriminator used in analog radars [6, p. 414; 11, p. 545]. We also discuss a technique we term *direct range measurement*. With this method, we do not use a range discriminator. Instead, we use a weighted interpolation to estimate range from samplers of the detector output. This would be analogous to the range measurement technique used in search radars [12, p. 56]. However, instead of performing the interpolation over a sliding range window, it is performed over a range window centered on the predicted range from the track filter.

Following the discussion of range discriminators and direct range measurement, we present several examples that illustrate how to build or model closed loop range trackers.

## 4.2 SAMPLING GATE RANGE DISCRIMINATOR

We will analyze the operation of the sampling gate discriminator with the help of Figure 4.2. The triangle in Figure 4.2 represents the output of the envelope detector for an unmodulated pulse with a width of $\tau_p$. The time $\tau_R$ notionally represents the range delay to the target. $\tau_R$ is the input to the tracker ($y$) in the block diagrams of Chapters 2 and 3. $\tau_{trk}$ is the range delay provided by the tracker; it is $y_p$ in the block diagrams of Chapters 2 and 3. The range delay error is

$$\Delta\tau = \tau_R - \tau_{trk} . \tag{4.1}$$

$\tau_E$ and $\tau_L$ are the sample times of the early and late gates (see Figure 4.1). The times $\tau_E$ and $\tau_L$ normally occur at equal time separations before and after the range delay provided by the tracker ($\tau_{trk}$). $\tau_{trk}$ is also termed the *track point* or the *predicted range delay*. The normal values of $\tau_E$ and $\tau_L$ are $\tau_E = \tau_{trk} - \tau_{res}/2$ and $\tau_L = \tau_{trk} + \tau_{res}/2$, but we will consider different locations to see how they affect the error signal, $e$. We specifically consider $\tau_E = \tau_{trk} - q\tau_{res}$ and $\tau_L = \tau_{trk} + q\tau_{res}$, as indicated in Figure 4.2. In these equations, $\tau_{res}$ is the range resolution of the waveform.[13] The parameter, $q$ can be between 0 and 1.

In some implementations, $q$ can be different for $\tau_E$ and $\tau_L$. If different values of $q$ are used, the tracker will have a bias error. Such a bias might be useful to remove biases due to other sources. In this book, we consider a single value for $q$.

The parameters used in the differentiator are $V_E$ and $V_L$, the output of the envelope detector at $\tau_E$ and $\tau_L$. In radars that use analog matched filters, they are voltages. In radars that use digital matched filters, they are numbers (digital words) that represent voltages. As a note, the radar could measure $V_E^2$ and $V_L^2$, the early and late gate "powers". We assume $V_E$ and $V_L$ are positive.

In the example depicted by Figure 4.2, the track point, $\tau_{trk}$, is later than the actual target range delay, thus $V_E > V_L$. Had the track point been earlier than the actual target range delay, we would have the opposite situation, i.e., $V_E < V_L$. If $\tau_{trk} = \tau_R$, $V_E = V_L$. This leads to

$$\Delta V = V_L - V_E \begin{cases} <0 & \text{if} \quad \tau_{trk} > \tau_R \quad \text{or} \quad \Delta\tau = \tau_R - \tau_{trk} < 0 \\ =0 & \text{if} \quad \tau_{trk} = \tau_R \quad \text{or} \quad \Delta\tau = \tau_R - \tau_{trk} = 0 \\ >0 & \text{if} \quad \tau_{trk} < \tau_R \quad \text{or} \quad \Delta\tau = \tau_R \quad \tau_{trk} > 0 \end{cases} \tag{4.2}$$

Equation (4.2) holds as long as $\tau_E$ and $\tau_L$ are on the left and right sides, of the triangle (respectively). We will generalize this shortly. Equation (4.2) can be written as

$$\Delta V = V_L - V_E = K_R\left(\tau_R - \tau_{trk}\right) = K_R \Delta\tau \tag{4.3}$$

which is the form required by the tracker. The important property illustrated by (4.3) is, the error signal, $\Delta V$, is directly proportional to the sign and magnitude of $\Delta\tau$.

We now want to consider $K_R$. Working under the current assumption that $\tau_E$ and $\tau_L$ are in the left and right halves of the triangle, $V_E$ and $V_L$ are

$$\begin{aligned} V_E &= \frac{V_R}{\tau_p}\left[\tau_E - \left(\tau_R - \tau_p\right)\right] \\ V_L &= -\frac{V_R}{\tau_p}\left[\tau_L - \left(\tau_R + \tau_p\right)\right] \end{aligned} \tag{4.4}$$

---

[13] For an unmodulated pulse $\tau_{res} = \tau_p$, the transmitted pulse width. For phase-modulated pulses $\tau_{res}$ is approximately $1/B$, where $B$ is the waveform bandwidth.

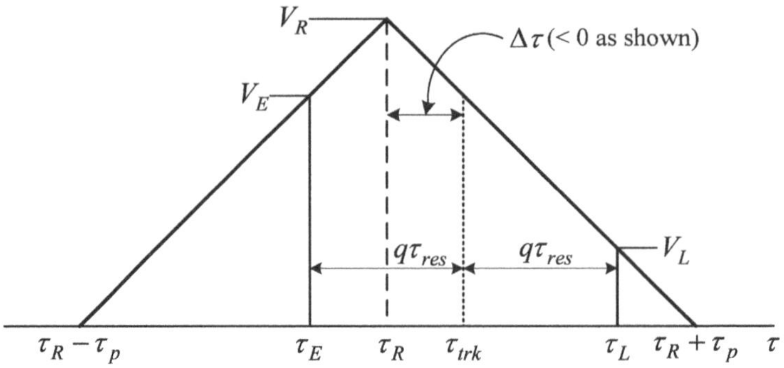


**Figure 4.2** Envelope of matched filter output for an unmodulated pulse.

For an unmodulated pulse $\tau_{res} = \tau_p$ and, thus,

$$\tau_E = \tau_{trk} - q\tau_p \text{ and } \tau_L = \tau_{trk} + q\tau_p . \tag{4.5}$$

Using this in (4.4) gives

$$\begin{aligned} V_E &= \frac{V_R}{\tau_p}\left(\tau_{trk} - q\tau_p - \tau_R + \tau_p\right) = \frac{V_R}{\tau_p}\left[-\Delta\tau + (1-q)\tau_p\right] \\ V_L &= -\frac{V_R}{\tau_p}\left(\tau_{trk} + q\tau_p - \tau_R - \tau_p\right) = -\frac{V_R}{\tau_p}\left[-\Delta\tau - (1-q)\tau_p\right] \end{aligned} \tag{4.6}$$

where we made use of $\Delta\tau = \tau_R - \tau_{trk}$. With (4.6), $\Delta V$ becomes

$$\begin{aligned} \Delta V = V_L - V_E &= -\frac{V_R}{\tau_p}\left[-\Delta\tau - (1-q)\tau_p\right] - \frac{V_R}{\tau_p}\left[-\Delta\tau + (1-q)\tau_p\right] \\ &= -\frac{V_R}{\tau_p}\left[-\Delta\tau - (1-q)\tau_p - \Delta\tau + (1-q)\tau_p\right] \\ &= \frac{2V_R}{\tau_p}\Delta\tau \end{aligned} . \tag{4.7}$$

Comparing this with (4.3), we get

$$K_R = \frac{2V_R}{\tau_p} . \tag{4.8}$$

We immediately note a problem in that $K_R$ depends on $V_R$, which depends on target range and RCS (and the rest of the terms in the radar range equation), and can thus vary. If we assume the controlled element gain (see Figure 3.11) is fixed, which we do, then the fact $K_R$ depends on $V_R$ means the open loop gain of the tracker will vary with signal amplitude. This can have an impact on the transient and noise reduction behavior of the tracker and could even cause the tracker to become unstable (see Chapter 3).

One way of making $K_R$ independent of $V_R$ is to use an automatic gain control (AGC) circuit. An AGC dynamically changes the receiver gain so that, effectively, $V_R$ becomes a known, constant value. AGC normalization is used in older (generally analog) radars that use analog discriminators that cannot explicitly implement multiplication or division. AGC is essentially a way to implement division using analog hardware [6, 13, 14].

Modern radars that utilize digital trackers and digital computers can explicitly implement division. Because of this, these radars often implement normalization using the measured signals, $V_E$ and $V_L$. Specifically, they divide by the average of $V_E$ and $V_L$. That is, the error is formed as

$$\Delta V_{norm} = \frac{V_L - V_E}{V_{avg}} \tag{4.9}$$

where

$$V_{avg} = (V_L + V_E)/2 \, . \tag{4.10}$$

In practice, the factor of 2 is omitted, and $\Delta V_{norm}$ is written as

$$\Delta V_{norm} = \frac{V_L - V_E}{V_L + V_E} \, . \tag{4.11}$$

Using (4.6),

$$\begin{aligned} V_L + V_E &= -\frac{V_R}{\tau_p}\left[-\Delta\tau - (1-q)\tau_p\right] + \frac{V_R}{\tau_p}\left[-\Delta\tau + (1-q)\tau_p\right] \\ &= \frac{2V_R}{\tau_p}(1-q)\tau_p \end{aligned} \tag{4.12}$$

and

$$\Delta V_{norm} = \frac{V_L - V_E}{V_L + V_E} = \frac{2V_R \Delta\tau/\tau_p}{2V_R(1-q)\tau_p/\tau_p} = \frac{\Delta\tau}{(1-q)\tau_p} \, , \tag{4.13}$$

which is close to the desired result in that $\Delta V_{norm}$ is independent of $V_R$. To make the final error signal, $e$ equal to $\Delta\tau$, i.e., $K_R = 1$, we use

$$e = \Delta V_{norm} S_\tau = \frac{V_L - V_E}{V_L + V_E} S_\tau \tag{4.14}$$

with

$$S_\tau = (1-q)\tau_p. \tag{4.15}$$

The development thus far is based on the assumption, $\tau_L$ and $\tau_E$ are on opposite sides of the triangle. Examination of Figure 4.2, reveals this will be the case as long as

$$\tau_R - \tau_p < \tau_E < \tau_R \text{ and } \tau_R < \tau_L < \tau_R + \tau_p. \tag{4.16}$$

With $\tau_L = \tau_{trk} + q\tau_p$, $\tau_E = \tau_{trk} - q\tau_p$, and $\Delta\tau = \tau_R - \tau_{trk}$, the inequalities of (4.16) can be written as

$$-q\tau_p < \Delta\tau < (1-q)\tau_p \text{ and } -(1-q)\tau_p < \Delta\tau < q\tau_p. \tag{4.17}$$

Combining the two inequalities of (4.17) yields

$$-\min([1-q]\tau_p, q\tau_p) < \Delta\tau < \min([1-q]\tau_p, q\tau_p) \tag{4.18}$$

or

$$|\Delta\tau| < \min([1-q]\tau_p, q\tau_p). \tag{4.19}$$

As an aside, (4.19) leads to the restriction $0 < q < 1$. Without this restriction, the argument of the minimum operation could be negative, which would lead to the conclusion $|\Delta\tau|$ could be negative; which is not allowed.

Equations (4.13), (4.14), (4.15), and (4.19) lead to the result

$$e = \Delta\tau \text{ for } |\Delta\tau| < \min([1-q]\tau_p, q\tau_p). \tag{4.20}$$

If $q$ and $\Delta\tau$ are such that $\tau_E$ is on a side (either one) of the triangle of Figure 4.2, but $\tau_L$ is not, $V_E \neq 0$ and $V_L = 0$. This leads to

$$e = \frac{V_L - V_E}{V_L + V_E}(1-q)\tau_p = \frac{0 - V_E}{0 + V_E}(1-q)\tau_p = -(1-q)\tau_p. \tag{4.21}$$

This condition can occur if (see Problem 1)

$$\tau_L > \tau_R + \tau_p \text{ and } \tau_R - \tau_p < \tau_E < \tau_R + \tau_p \tag{4.22}$$

or

$$\Delta\tau < -(1-q)\tau_p \text{ and } -(1+q)\tau_p < \Delta\tau < (1-q)\tau_p \tag{4.23}$$

or

$$-(1+q)\tau_p < \Delta\tau < -(1-q)\tau_p. \quad (4.24)$$

If $q$ and $\Delta\tau$ are such that $\tau_L$ is on a side (either one) of the triangle of Figure 4.2 but $\tau_E$ is not, $V_L \neq 0$ and $V_E = 0$. This leads to

$$e = \frac{V_L - V_E}{V_L + V_E}(1-q)\tau_p = \frac{V_L - 0}{V_L + 0}(1-q)\tau_p = (1-q)\tau_p. \quad (4.25)$$

This condition can occur if (see Problem 1)

$$\tau_E < \tau_R - \tau_p \text{ and } \tau_R - \tau_p < \tau_L < \tau_R + \tau_p \quad (4.26)$$

or

$$\Delta\tau > (1-q)\tau_p \text{ and } (1-q)\tau_p < \Delta\tau < (1+q)\tau_p \quad (4.27)$$

or

$$(1-q)\tau_p < \Delta\tau < (1+q)\tau_p. \quad (4.28)$$

Combining (4.20), (4.21), (4.24), (4.25), and (4.28) leads to (see Problem 2)

$$e = \begin{cases} \Delta\tau & |\Delta\tau| \leq \min\left([1-q]\tau_p, q\tau_p\right) \\ (1-q)\tau_p \operatorname{sgn}(\Delta\tau) & (1-q)\tau_p < |\Delta\tau| \leq (1+q)\tau_p \\ \text{indeterminate} & |\Delta\tau| > (1+q)\tau_p \end{cases}. \quad (4.29)$$

The condition on $|\Delta\tau|$ in the last term means neither $\tau_E$ nor $\tau_L$ are in the horizontal span of the triangle. In this case, $V_E$ and $V_L$ will be zero and $e$ will be the ratio 0/0, which is indeterminate. As a note, the indeterminate condition will seldom occur in practice because of thermal noise. With noise, it is very unlikely both $V_E$ and $V_L$ will be zero. Therefore, $e$ will have some value, albeit a meaningless one in terms of tracking.

As long as $q \geq ½$, the regions indicated in (4.29) are adjacent. However, if $q < ½$, there will be a gap between the central region and the outer regions. When $q < ½$, $\min([1-q]\tau_p, q\tau_p) = q\tau_p$. This means the central region ends at $|\Delta\tau| = q\tau_p$. However, the flat regions don't start until $|\Delta\tau| = (1-q)\tau_p$. This means the region where $q\tau_p < |\Delta\tau| < (1-q)\tau_p$ is not accounted for. A more general form that accounts for this possibility is derived in Appendix 4A. The result from that appendix is

$$e = \begin{cases} \Delta\tau & 0 \le |\Delta\tau| \le \min([1-q]\tau_p, q\tau_p) \\ \dfrac{\text{sgn}(\Delta\tau) q\tau_p}{\tau_p - |\Delta\tau|}(1-q)\tau_p & \min([1-q]\tau_p, q\tau_p) < |\Delta\tau| \le (1-q)\tau_p \\ \text{sgn}(\Delta\tau)(1-q)\tau_p & (1-q)\tau_p < |\Delta\tau| \le (1+q)\tau_p \\ \text{indeterminate} & |\Delta\tau| > (1+q)\tau_p \end{cases} \quad (4.30)$$

Figure 4.3 contains plots of $e$ versus $\Delta\tau$, for $q = 1/4$, $1/2$, and $3/4$, and a pulse width of 1 μs. We term these *discriminator curves*. As planned by the selection of $S_\tau$, the central region of all three discriminator curves has a slope of unity. The region of $\Delta\tau$ beyond the curves is the indeterminate region. For example, $|\Delta\tau|>1.5$ is the indeterminate region for the case where $q = 1/2$.

As $q$ increases, the span of the discriminator curves where $e$ is not indeterminate increases. As $q$ decreases, the span decreases. Because of this, we term the case where $q < 1/2$ the *narrow gate mode,* and the case where $q > 1/2$ the *wide gate mode*. The case where $q = 1/2$ is termed the *normal gate mode*.

The curved portion of the discriminator curve for $q = 1/4$ is caused by the second term of (4.30). If $\Delta\tau$ is such that it lies in one of these regions, the slope of the discriminator curve will vary. This will cause a change in the closed loop gain of the tracker, which could affect the transient and error reduction behavior.

### 4.2.1 LFM Pulse

Given a transmitted LFM pulse of the form,

$$u(t) = e^{j\pi\alpha t^2}\,\text{rect}\left[\frac{t}{\tau_p}\right], \quad (4.31)$$

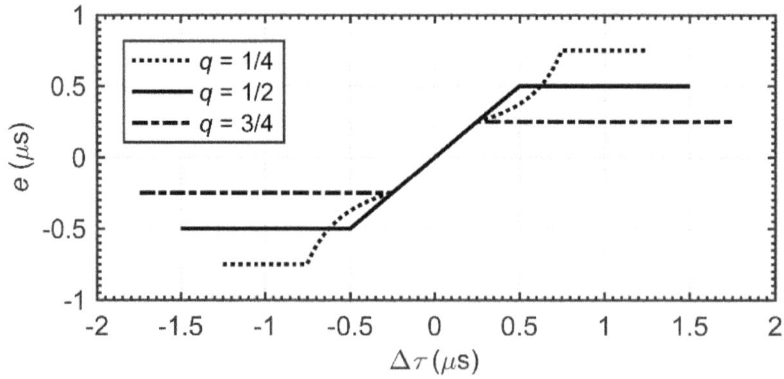


**Figure 4.3** Discriminator curves for a 1-μs, unmodulated pulse and three values of $q$.

the normalized envelope of the matched filter output is given by

$$|MF(\tau)| = \left|\frac{\left(\tau_p - |\tau - \tau_R|\right)\operatorname{sinc}\left(\alpha[\tau - \tau_R]\left[\tau_p - |\tau - \tau_R|\right]\right)}{\tau_p}\right| \operatorname{rect}\left[\frac{\tau - \tau_R}{2\tau_p}\right], \tag{4.32}$$

assuming the matched filter is matched to the transmit pulse. In (4.31) and (4.32), $\alpha$ is the LFM slope and is given by $\alpha = \pm B_{LFM}/\tau_p$ where $B_{LFM}$ is the bandwidth of the LFM pulse, and $\tau_p$ is the transmitted pulse width. The plus sign is used if the frequency increases across the pulse (termed *upchirp*), and the negative sign is used if the frequency decreases across the pulse (termed *downchirp*) [15, 16]. The terms sinc($x$) and rect[$x$] are defined as

$$\operatorname{sinc}(x) = \frac{\sin(\pi x)}{\pi x} \tag{4.33}$$

and

$$\operatorname{rect}[x] = \begin{cases} 1 & |x| \le 1/2 \\ 0 & |x| > 1/2 \end{cases}. \tag{4.34}$$

As before, $\tau_R$ is the range delay to the target.

Figure 4.4 contains a plot of $|MF(\tau)|$ for a $\tau_p = 15$ μs pulse and an LFM bandwidth of $B_{LFM} = 1$ MHz. As a comparison, the figure also contains a plot of the envelope of the matched filter output for a 1-μs, unmodulated pulse (i.e., a plot like Figure 4.2 with $\tau_p = 1$ μs – in that plot). As can be seen, the width of the main lobe of matched filter output for the LFM pulse is about the same as the width of the triangle for the 1 μs, unmodulated pulse. This means both types of pulses will have about the same range resolution; or that $\tau_{res}$ is about 1 μs for both.

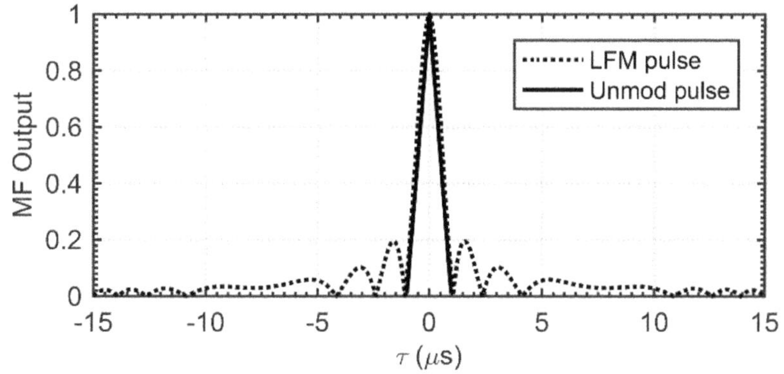


**Figure 4.4** Matched filter output for a 1-MHz, 15-μs LFM pulse (dotted curve) and a 1-μs unmodulated pulse.

Deriving an equation like (4.30) for an LFM pulse is not as easy as for an unmodulated pulse. However, since the span of $|MF(\tau)|$ is typically much larger than the range resolution of the pulse, it is unlikely $\Delta\tau = \tau_R - \tau_{trk}$ will be such that $\tau_E$ and $\tau_L$ will not both be in the span of $|MF(\tau)|$. In terms of Figure 4.4, this means $\tau_E$ and $\tau_L$ will almost always be in the range of –15 μs to 15 μs. Since $\tau_E$ and $\tau_L$ are in the span of $|MF(\tau)|$, the values for $|MF(\tau_E)|$, and $|MF(\tau_L)|$ can be determined from (4.32), without the need to consider indeterminate regions, as was necessary for the unmodulated pulse. This means we can write the equation for the discriminator curve as

$$e = \frac{V_E - V_L}{V_E + V_L} S_\tau = \frac{|MF(\tau_E)| - |MF(\tau_L)|}{|MF(\tau_E)| + |MF(\tau_L)|} S_\tau . \tag{4.35}$$

With $\tau_E = \tau_{trk} - q\tau_{res}$ and $\Delta\tau = \tau_R - \tau_{trk}$ we get, using (4.32),

$$\begin{aligned} |MF(\tau_E)| &= \left| \frac{(\tau_p - |\tau_E - \tau_R|)\,\text{sinc}\left(\alpha[\tau_E - \tau_R][\tau_p - |\tau_E - \tau_R|]\right)}{\tau_p} \right| \\ &= \left| \frac{(\tau_p - |\Delta\tau + q\tau_{res}|)\,\text{sinc}\left(\alpha[\Delta\tau + q\tau_{res}][\tau_p - |\Delta\tau + q\tau_{res}|]\right)}{\tau_p} \right| \end{aligned} \tag{4.36}$$

where we used the fact that $\tau_E$ was in the span of $|MF(\tau)|$. Similarly, with $\tau_L = \tau_{trk} + q\tau_{res}$ and $\Delta\tau = \tau_R - \tau_{trk}$

$$|MF(\tau_L)| = \left| \frac{(\tau_p - |\Delta\tau - q\tau_{res}|)\,\text{sinc}\left(\alpha[\Delta\tau - q\tau_{res}][\tau_p - |\Delta\tau - q\tau_{res}|]\right)}{\tau_p} \right| . \tag{4.37}$$

Recall that $S_\tau$ is chosen so that the slope of the discriminator curve is unity near $\Delta\tau = 0$. To find $S_\tau$ we would need to solve

$$\frac{\partial}{\partial\Delta\tau} \left( \frac{|MF(\tau_E)| - |MF(\tau_L)|}{|MF(\tau_E)| + |MF(\tau_L)|} S_\tau \right) \Bigg|_{\Delta\tau=0} = 1 \tag{4.38}$$

for $S_\tau$, which is quite tedious. In lieu of doing that, we use the development of Section 4.2.1 and speculate that

$$S_\tau = (1-q)\tau_{res} \tag{4.39}$$

might be a reasonable approximation for the LFM discriminator. Recall that $\tau_{res}$ is the range resolution of the matched filter output. For an LFM pulse, $\tau_{res} \approx 1/B_{LFM}$.

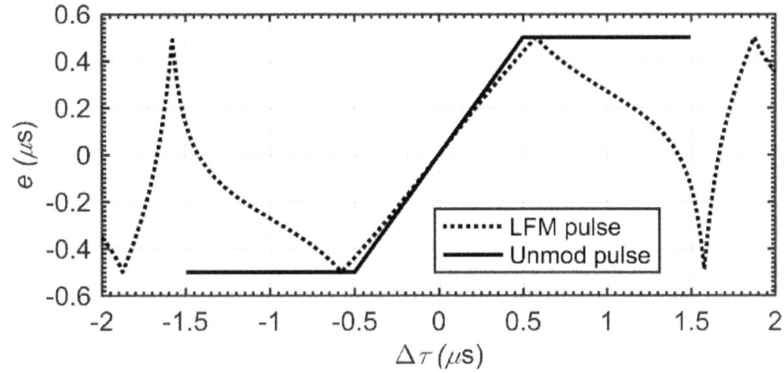


**Figure 4.5** Discriminator curves for a 1-μs unmodulated pulse and a 15-μs, 1-MHz LFM pulse.

Figure 4.5 contains plots of discriminator curves for the aforementioned 15-μs, 1-MHz LFM pulse, and a 1-μs unmodulated pulse. In both cases, $q = ½$. Since the slopes of both discriminator curves are about the same near $\Delta\tau = 0$, the $S_\tau$ defined by (4.39), with $\tau_{res} \approx 1/B_{LFM}$, is a reasonable approximation.

While the discriminator curve for the unmodulated pulse becomes flat and then indeterminate for $|\Delta\tau| > (1+q)\tau_{res}$ (= 1.5 μs for this example), the discriminator curve for the LFM pulse does not. Instead, it appears to oscillate between $\pm(1-q)\tau_{res}$, or ±½ μs for this example. However, that observation is misleading because, then $|\Delta\tau| > (1+q)\tau_{res}$ $\tau_E$ or/and $\tau_L$ will be in the sidelobe region of the matched filter response. This means $V_E$ or/and $V_L$ will be dominated by thermal noise. As a result, the discriminator curve will jump randomly between $\pm(1-q)\tau_{res}$, as was the case when the discriminator curve for the unmodulated pulse was in the indeterminate region.

### 4.2.2 Other Waveforms

Discriminator curves for other types of pulses (e.g., LFM with sidelobe reduction weighting, Barker coded pulses, PRN coded pulses [15]) can be generated by finding their matched filter responses, $|MF(\tau)|$, and using (4.35). As an example, Figure 4.6 contains a plot of the discriminator curve for a 13-bit Barker coded pulse with a chip width of 1 μs. It also contains a discriminator curve for a 1-μs unmodulated pulse. In both cases, $q = ½$. The discriminator curve for the Barker coded pulse is similar to the discriminator curve for the unmodulated pulse, except when $\Delta\tau > (1+q)\tau_{res}$. While the discriminator curve for the unmodulated pulse becomes indeterminate in this region, the discriminator curve for the Barker coded pulse is zero. However, with thermal noise, the behavior of the two discriminators will be about the same because the discriminator output will be dominated by noise when $|\Delta\tau| > (1+q)\tau_{res}$.

## 4.3 SUMMING GATE RANGE DISCRIMINATOR

An examination of Figure 4.2 reveals that, if $\tau_{trk} = \tau_R$ (the radar is tracking the target) and $q = ½$, $V_E$ and $V_L$ will be half of the peak amplitude out of the matched filter. This represents a SNR loss of 6 dB, compared to the SNR at the peak of the matched filter output. If we were to use a smaller value for $q$, the early and late samples would occur closer to the peak, which would result in a smaller SNR loss. However, spacing the early and late samples closer together would reduce the size of the central, linear region of the discriminator curve

A potential way to avoid the SNR loss is to use an approach analogous to the split gate discriminator discussed in [6, p. 417; 8; 11, p. 549]. A block diagram of this configuration, which we will term a *summing gate discriminator*, is illustrated in Figure 4.7. With this configuration, the impulse samplers of the sampling gate discriminator are replaced by switches that close for a finite time. Specifically, the switch labeled early gate closes at $\tau_E$ and stays closed until $\tau_{trk}$. The switch labeled late gate closes at $\tau_{trk}$ and stays closed until $\tau_L$. During the times the switches are closed, the integrators integrate the output of the envelop detector. The integrator outputs are then used for form $e$. The integrator outputs are zeroed, or quenched, after the error signal is sent to the track filter. As indicated in Figure 4.7, $e$ is given by

$$e = \frac{I_L - I_E}{I_L + I_E} S_\tau \quad (4.40)$$

Since the signals into the integrator include the peak of the matched filter output, the summing gate discriminator will experience less of a SNR loss than the sampling gate discriminator (with $q = ½$). There will be some loss if the gate widths, $\tau_{trk} - \tau_E$ and $\tau_L - \tau_{trk}$, become large enough to allow integration over intervals where there is noise but no signal.

In radars that use digital signal processing, the integrators would be replaced by summers. Also, the signal input to the matched filter would need to be sampled at a rate higher than the waveform bandwidth to provide a sufficient number of samples to the summer.

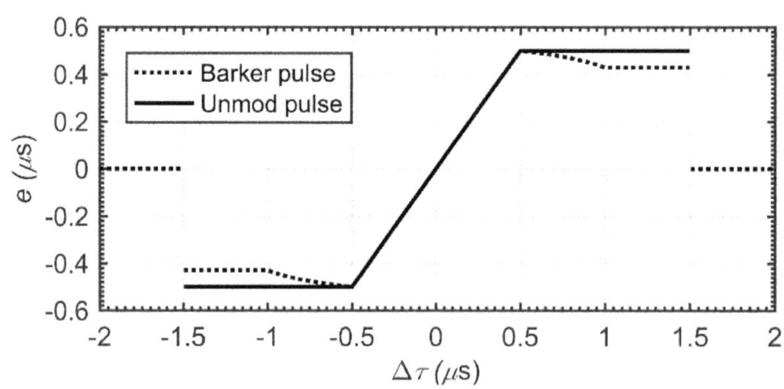


**Figure 4.6** Discriminator curves for a 1-μs unmodulated pulse and a 13-bit Barker coded pulse with a 1-μs chip width.

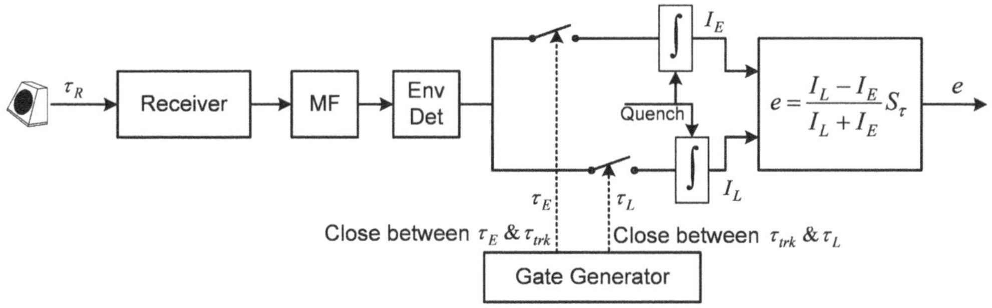


**Figure 4.7** Block diagram of a summing gate discriminators.

### 4.3.1 Unmodulated Pulse

Figure 4.8 contains an illustration of how an integrating gate discriminator works. At $\tau = \tau_E$, the early gate switch closes, and the integrator begins integrating the output of the envelope detector. This continues until $\tau = \tau_{trk}$, at which time the early gate switch opens, and the integration stops. When the early gate switch opens, the integrator output will equal the area under the triangle between $\tau_E$ and $\tau_{trk}$. This area is represented by the vertical lines.

When $\tau = \tau_{trk}$, the late gate switch closes, and the late gate integrator begins integrating the output of the envelope detector. This continues until $\tau = \tau_L$, at which time the late gate switch opens. When the late gate switch opens, the integrator output is the area under the triangle between $\tau_{trk}$ and $\tau_L$, which is represented by the horizontal lines in Figure 4.8. The two integrator outputs are then sent to the error formation algorithm, or circuit, where $e$ is computed.

If $\tau_E$ and $\tau_L$ are symmetrically located about $\tau_{trk}$, and $\tau_{trk} = \tau_R$, $I_E$ and $I_L$ will be equal and $e$ will be zero. If $\tau_{trk} > \tau_R$, as illustrated in Figure 4.8, $I_E$ will be greater than $I_L$, and $e$ will be negative. Since $\Delta\tau = \tau_R - \tau_{trk}$, the condition $\tau_{trk} > \tau_R$ means $\Delta\tau < 0$. Thus, $\Delta\tau$ and $e$ have the same sign, which is what we want. If $\tau_{trk} < \tau_R$, $\Delta\tau > 0$, $I_L > I_E$, and $e > 0$.

A potential advantage of a summing gate discriminator is, the spacing between $\tau_{trk}$ and $\tau_E$ or $\tau_L$ can be greater than the range resolution, $\tau_{res}$. That is, we can have $q > 1$. If that condition is allowed in a sampling gate discriminator, both $\tau_E$ and $\tau_L$ could be outside the span of the main lobe of the matched filter output, even though $\tau_{trk}$ is within the span. Thus, the early and late signals ($V_E$ and $V_L$) of a sampling gate discriminator would contain only noise. With a summing gate discriminator, even if $q > 1$, one or both of the gates may encompass part of the main lobe of the matched filter output. Thus, the discriminator output will provide a valid error signal to the track filter. This property makes summing gate discriminators more tolerant of large range errors, when compared to sampling gate discriminators. Such a feature is useful during acquisition, or during target maneuvers, since the range error could be large in these situations.

If the spacing between $\tau_E$ and $\tau_L$ is large, the integrator outputs could contain a significant amount of noise, which would degrade steady state tracking error. An approach to minimizing the effect of the unwanted noise would be to use a somewhat large separation between $\tau_E$ and $\tau_L$ during acquisition (or when a maneuver is detected) and a small separation during steady state tracking.

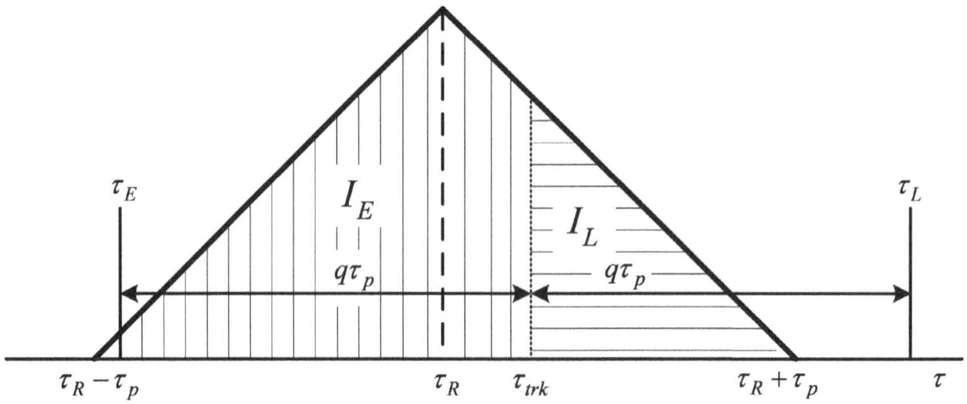


**Figure 4.8** Summing gate operation.

Analytical studies of a summing gate discriminator are difficult because of the need to perform the integration. For an unmodulated pulse, the integration is simpler, but complicated by the fact that the equation for the envelope detector output contains absolute value functions. For an LFM pulse, the detector output contains sinc-like functions and absolute values, which are even more difficult to deal with mathematically. Because of this, we limit the analytical discussion to the special case of an unmodulated pulse with $\tau_E - \tau_L = 2\tau_{res} = 2\tau_p$. This condition equates to the condition $q = 1$.

We consider the case where $\Delta\tau = \tau_R - \tau_{trk} \geq 0$. These conditions are shown graphically in Figure 4.9.

The equations for the early and late gate integrals are

$$\begin{aligned} I_E &= \int_{\tau_E}^{\tau_{trk}} |MF(\tau)| d\tau = \int_{\tau_R-\tau_p}^{\tau_{trk}} |MF_L(\tau)| d\tau \\ &= \int_{\tau_R-\tau_p}^{\tau_{trk}} \left[ \frac{V_R}{\tau_p} \left( \tau_p + (\tau - \tau_R) \right) \right] d\tau = \frac{V_R}{2\tau_p} \left( \tau_p - \Delta\tau \right)^2 \end{aligned} \tag{4.41}$$

and

$$\begin{aligned} I_L &= \int_{\tau_{trk}}^{\tau_L} |MF(\tau)| d\tau = \int_{\tau_{trk}}^{\tau_R} |MF_L(\tau)| d\tau + \int_{\tau_R}^{\tau_L} |MF_R(\tau)| d\tau \\ &= \frac{V_R}{2\tau_p} \left( \tau_p^2 + 2\Delta\tau\tau_p - 2\Delta\tau^2 \right) \end{aligned} \tag{4.42}$$

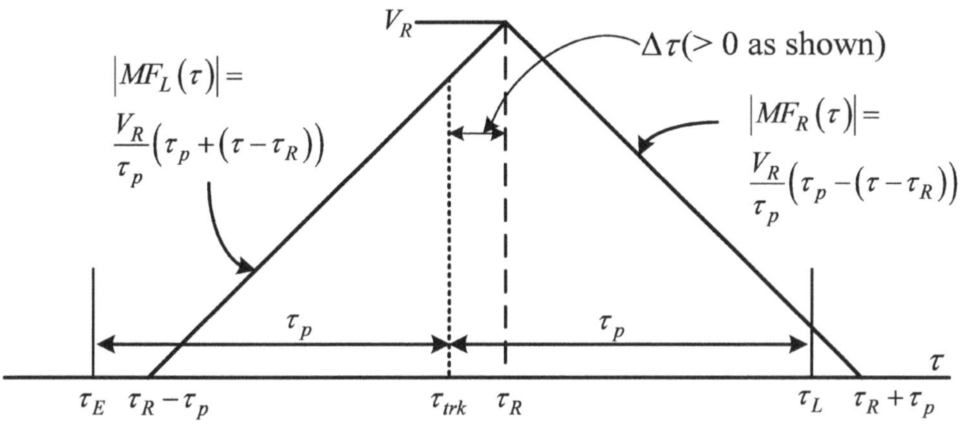


**Figure 4.9** Geometry for an analytical development of an integrating gate discriminator curve.

The derivations of (4.41) and (4.42) are tedious, but straightforward (see Problem 5). From (4.40), the error signal, $e$, is,

$$\begin{aligned} e &= \frac{I_L - I_E}{I_L + I_E} S_\tau \\ &= \frac{4\Delta\tau\tau_p - 3\Delta\tau^2}{2\tau_p^2 - \Delta\tau^2} S_\tau \end{aligned} \tag{4.43}$$

To determine $S_\tau$, we need to compute the derivative of $e$ with respect to $\Delta\tau$ and evaluate it at $\Delta\tau = 0$. This gives

$$\begin{aligned} \left.\frac{\partial e}{\partial \Delta\tau}\right|_{\Delta\tau=0} &= \left.\frac{\partial}{\partial \Delta\tau}\left(\frac{4\Delta\tau\tau_p - 3\Delta\tau^2}{2\tau_p^2 - \Delta\tau^2} S_\tau\right)\right|_{\Delta\tau=0} \\ &= \left.\frac{\left(2\tau_p^2 - \Delta\tau^2\right)\left(4\tau_p - 6\Delta\tau\right) - \left(4\Delta\tau\tau_p - 3\Delta\tau^2\right)\left(-2\Delta\tau\right)}{\left(2\tau_p^2 - \Delta\tau^2\right)^2} S_\tau\right|_{\Delta\tau=0} \\ &= \frac{8\tau_p^3}{4\tau_p^4} S_\tau = \frac{2S_\tau}{\tau_p} \end{aligned} \tag{4.44}$$

Given we want the slope to be one, we have $S_\tau = \tau_p/2$.

The integrals, (4.41) and (4.42), and thus $e$, are valid as long as $\tau_{trk} > \tau_R - \tau_p$, or $\Delta\tau < \tau_p$ (recall that $\Delta\tau \geq 0$). When $\tau_{trk} \leq \tau_R - \tau_p$, or $\Delta\tau \geq \tau_p$, the region between $\tau_E$ and $\tau_{trk}$ will not be in the left side of the triangle. This means $I_E$ will be zero and

$$e = \frac{I_L - I_E}{I_L + I_E} S_\tau = \frac{I_L}{I_L} S_\tau = S_\tau \,. \tag{4.45}$$

When $\tau_L < \tau_R - \tau_p$, or $\Delta\tau > 2\tau_p$, the region between $\tau_E$ and $\tau_L$ will not include the triangle, and both $I_E$ and $I_L$ will be zero. This means $e$ will be 0/0, or indeterminate.

If we perform calculations similar to the above (see Problem 6) for $\Delta\tau < 0$ and combine the results, we get, for $\tau_E - \tau_L = 2q\tau_{res} = 2\tau_p$.

$$e = \begin{cases} \operatorname{sgn}(\Delta\tau)\dfrac{4\tau_p|\Delta\tau| - 3\Delta\tau^2}{2\tau_p^2 - \Delta\tau^2}\dfrac{\tau_p}{2} & |\Delta\tau| \le \tau_p \\ \operatorname{sgn}(\Delta\tau)\dfrac{\tau_p}{2} & \tau_p < |\Delta\tau| \le 2\tau_p\,. \\ \text{indeterminate} & |\Delta\tau| > 2\tau_p \end{cases} \tag{4.46}$$

Figure 4.10 contains plots of discriminator curves for summing and sampling gate discriminators and a 1 μs, unmodulated pulse. The sampling gate discriminator used a $q$ of ½ and the summing gate discriminator used a $q$ of 1. The discriminator curve for the sampling gate discriminator becomes indeterminate for $|\Delta\tau| > 1.5$ μs while the discriminator curve for the summing gate discriminator doesn't become indeterminate until $|\Delta\tau| > 2$ μs. Also, the nonzero slope region of the summing gate discriminator curve extends over a larger range of $\Delta\tau$ than the non-zero slope region of the sampling gate discriminator. One drawback of the summing gate discriminator is that the slope of the discriminator curve is not constant. This could impact the transient response, and noise performance, of the range tracker in that the tracker will have a variable open-loop gain.

Figure 4.11 contains plots of summing gate discriminator curves for $q = ½$, 1, and 1.5. Increasing $q$ from 1 to 1.5 increases the extent of $\Delta\tau$ before the discriminator output becomes indeterminate. The case of $q = ½$ is interesting. In that case, the discriminator curve looks like an expanded form of the sampling gate discriminator curve for $q = ½$. To obtain a slope of 1.0 for the $q = ½$ case, $S_\tau$ was changed to $3\tau_p/4$.

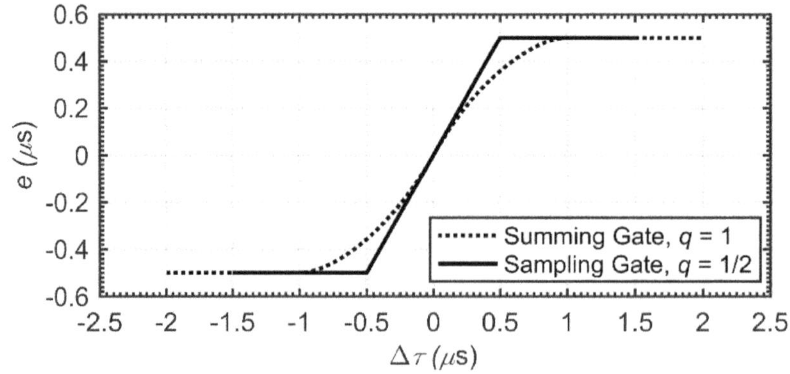


**Figure 4.10** Summing and sampling gate discriminator curve examples—1 μs, unmodulated pulse.

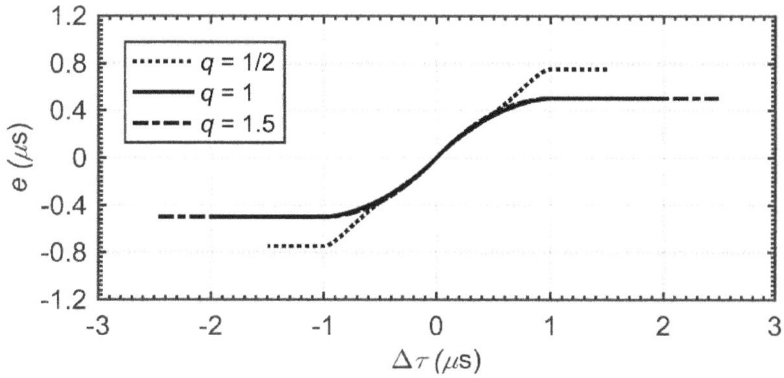


**Figure 4.11** Summing gate discriminator curves for an unmodulated, 1 μs, unmodulated pulse and three values of $q$.

### 4.3.2 LFM Pulse

As another example, Figure 4.12 contains discriminator curves for a 15 μs, 1 MHz LFM pulse for summing and sampling gate discriminators. $q$ values of 1 and 2 were used for the summing gate discriminators, and $q = ½$ was used for the sampling gate discriminator. The discriminator curves for the sampling gate discriminator and the summing gate discriminator, with $q = 1$, are similar. However, the shape of the discriminator curve for the summing gate discriminator is more rounded than the discriminator curve for the sampling gate discriminator. When $q$ is increased to 2 for the summing gate discriminator, the discriminator curve extends further before decreasing into what would be the indeterminate region.

### 4.3.3 Barker Coded Pulse

As a final example, Figure 4.13 contains summing and sampling gate discriminator curves for a 13 chip, Barker coded pulse with a chip width of 1 μs. $q$ values of 1 and 2 were used for the summing gate discriminators, and $q = ½$ was used for the sampling gate discriminator. As for the LFM pulse, the discriminator curves for the sampling gate discriminator (with $q = ½$) and the summing gate discriminator (with $q = 1$) are similar, except the shape of the discriminator curve for the summing gate discriminator is more rounded than the discriminator curve for the sampling gate discriminator. When $q$ is increased to 2 for the summing gate discriminator, the discriminator curve extends further before decreasing toward zero.

### 4.3.4 Digital Matched Filter Implications

In an analog receiver, the matched filter output is a continuous time signal, and the integrators of the summing gate discriminator are integrators. In a digital receiver, the matched filter output is a digital signal and summers must replace the integrators. The use of summers raises the question of how often the matched filter output should be sampled.

Figure 4.14 contains plots of discriminator curves for a summing gate discriminator with different values of the spacing between matched filter samples. For a spacing of $\tau_{res}$, which is the best spacing for a sampling gate discriminator, the response significantly deviates from the analog case. At a spacing of $\tau_{res}/2$, the discriminator curve looks like a slightly distorted version of the sampling gate discriminator curve. At a spacing of $\tau_{res}/4$, the summing gate discriminator curve approaches the S-shape of the analog discriminator curve. Reducing the spacing to $\tau_{res}/4$ could have an impact on the matched filter implementation in that the matched filter hardware (or software) will need to have the ability to process four or more times the samples than it would in a sampling gate discriminator. As a note, the value of $S_\tau$ needed to produce a discriminator slope of unity is different for the various values of sample spacing. In the curves of Figure 4.14, the values of $S_\tau$ were determined experimentally.

It is expected the above summing gate discussions also apply to LFM or phase coded pulses. Verification, or refutation, of this conjecture is left as an exercise (see Problems 9 and 10).

As a note, the selection of sample spacing does not affect the length of the flat regions of the discriminator curves. Those are still determined by $q$.

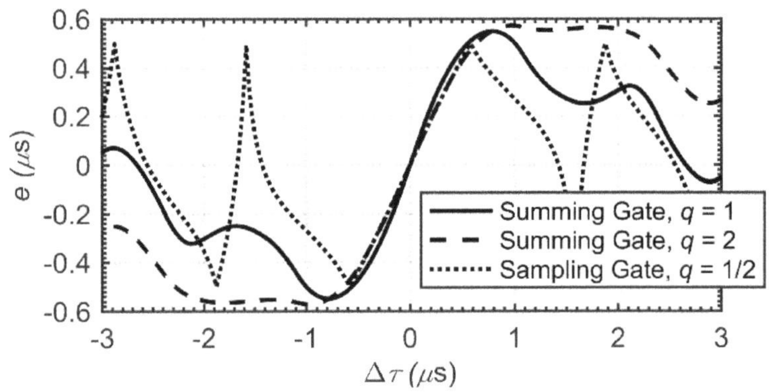


**Figure 4.12** Discriminator curves for summing and sampling gate discriminators and a 15 μs, 1 MHz LFM pulse.

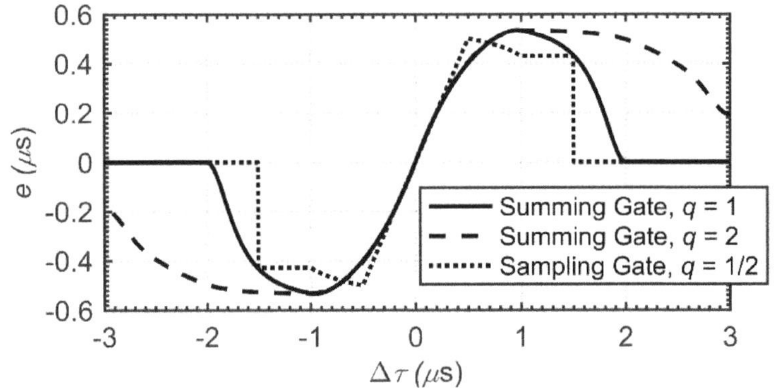


**Figure 4.13** Discriminator curves for summing and sampling gate discriminators and a 13 chip Barker coded pulse with 1 μs chip widths.

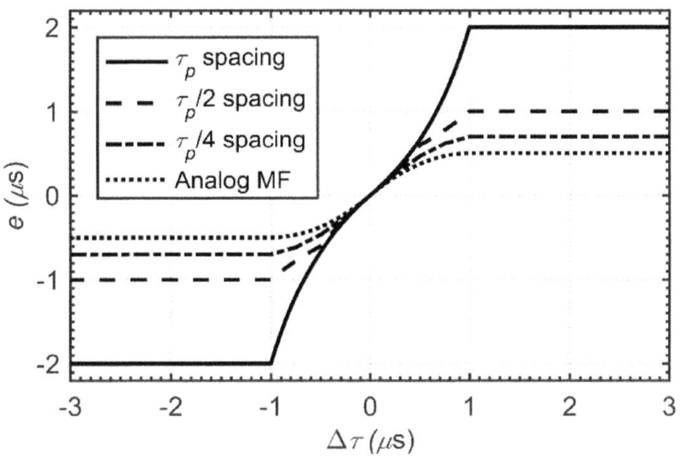


**Figure 4.14** Discriminator curves for a summing gate discriminator with different matched filter output spacing – unmodulated pulse, $\tau_{res} = \tau_p = 1$ μs, $q = 2$.

Summing gate discriminators are most appropriate in radars whose compressed pulse width is significantly smaller than the pulse repetition interval (PRI). In those cases, the range extent of the summing, or integration, window will not encroach on the transmit pulses. In radars where the compressed pulse width is a significant percentage of the PRI (more than about 10%) the use of a summing gate discriminator is questionable because the extent of the summing window could encroach on the transmit pulses. In those instances, sampling range gates would be more appropriate.

In this book, summing of range samples occurs after the matched filter. In some radars that use unmodulated pulses, the summing gate discriminator also performs the matched filter operation [17].

## 4.4 DIRECT RANGE MEASUREMENT

With the direct measurement method, the output of the envelope detector is sampled at regularly spaced time intervals that are typically separated by, at most, the range resolution of the waveform. This is illustrated in Figure 4.15, which contains a plot of the envelope detector output for an unmodulated pulse with a width of $\tau_p$. The noisy looking baseline is the noise at the envelope detector output, and the distorted triangle is the signal plus noise at the detector output. The circles represent the aforementioned samples (the range samples), which are spaced $\tau_p$ apart in this example.

The range measurement algorithm selects the range sample in the range measurement window (the set of points in Figure 4.15) with the largest amplitude. It then uses that sample and the largest adjacent sample to estimate the time of the peak of the matched filter output, $\tau_{Rest}$. $\tau_{Rest}$ is, ideally, the target range delay, $\tau_R$. As a point of clarification, the range measurement algorithm only has knowledge of the sample amplitudes, not the signal and noise amplitude between samples. Thus, the range measurement algorithm can only use the samples to compute $\tau_{Rest}$.

A way to compute $\tau_{Rest}$ is as a weighted average of the times of the two samples used in the range measurement. In particular,

$$\tau_{Rest} = \frac{V_E \tau_E + V_L \tau_L}{V_E + V_L} \tag{4.47}$$

where $\tau_E$ and $\tau_L$ are the times of the samples and $V_E$ and $V_L$ are the magnitudes of the envelope detector output at $\tau_E$ and $\tau_L$, respectively. $\tau_E$, $\tau_L$, $V_E$, and $V_L$ have the same meaning as in the discussion of sampling gate discriminators. That is, $\tau_E$ and $\tau_L$ are the early and late sample times, and $V_E$ and $V_L$ are the early and late signal amplitudes. Equation (4.47) is an exact equation for an unmodulated rectangular pulse and no noise. That is, it perfectly predicts $\tau_R$. It also perfectly predicts $\tau_R$ for a phase coded pulse, and no noise. If the matched filter contains noise, (4.47) provides an approximation of $\tau_R$ whose accuracy depends on SNR. Equation (4.47) provides only an approximation for LFM pulses because the main lobe of the matched filter output for an LFM pulse is not a triangle, which is the basis of (4.47).

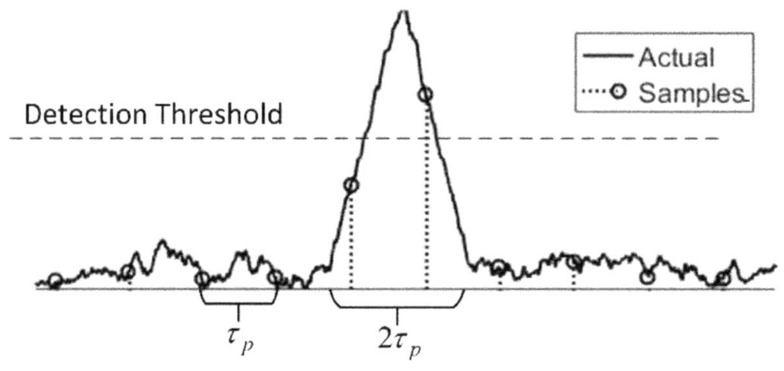


**Figure 4.15** Envelope detector output for an unmodulated pulse.

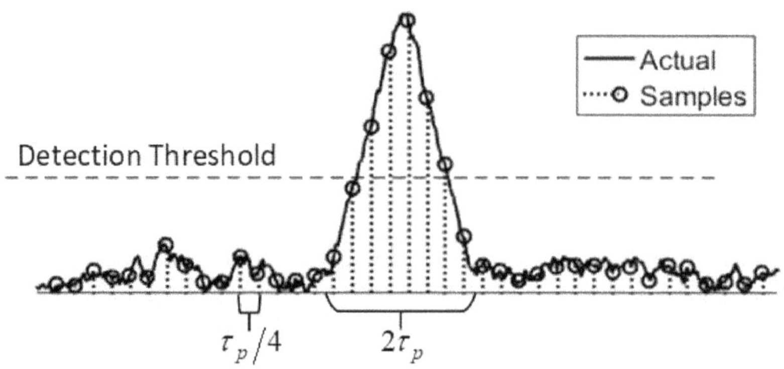


**Figure 4.16** Envelope detector output for an unmodulated pulse with 4 times oversampling.

As with the sampling gate discriminator, the direct measurement technique suffers a SNR loss. This loss is termed a *range straddling loss* [6, p. 253; 15] and is nominally 3 dB for samples spaced $\tau_{res}$ apart.

Sometimes it could be helpful to space the range samples considerably closer than $\tau_{res}$. This is illustrated in Figure 4.16, where the range samples are spaced at $\tau_{res}/4$. As can be seen, it is possible to get up to 8 samples that are on the main lobe of the matched filter response.

The advantage of using such close spacing is that two of the samples will be near the peak of the matched filter output. Thus, the SNR of these samples will be closer to the peak SNR than for more widely spaced samples. This will reduce the range straddling loss and may improve measurement accuracy.

To determine $\tau_{Rest}$, the two largest amplitude samples could be used in (4.47). Alternately, $\tau_{Rest}$ could be computed as the weighted average of all range samples for which the envelope detector output is above a threshold. That is, $\tau_{Rest}$ could be computed from

$$\tau_{Rest} = \frac{\sum_{k=1}^{N} \tau_k V_k}{\sum_{k=1}^{N} V_k} \tag{4.48}$$

where $V_k$ are the outputs of the envelope detector that are above the threshold.

## 4.5 RANGE TRACKER MODELING

A good way to gain insight into the operation of closed loop range trackers is through simulation. Because of that, this will be the next topic we consider. In Chapter 3, we developed several track filter models, and in the preceding sections of this chapter, we developed models for sampling and summing gate discriminators, and the direct measurement technique. The remaining models we need to discuss are target and noise models.

In an actual radar, target return signals and environment noise combine at the antenna face. Thermal noise is then added by the receiver and propagated, along with the signal, through the receiver and matched filter to the envelope detector of the range discriminator (see Figures 4.1 and 4.7). Because of this, strictly speaking, we should combine simulated target and noise signals at the front of the radar and propagate them through the receiver and matched filter. However, if we make the reasonable assumption that all of the receiver components up to, and including, the matched filter, are linear, we can invoke superposition and need only model the signal and noise at the output of the matched filter. Specifically, we can separately model the target and noise signals at the matched filter output, and then sum them. This concept is illustrated in Figure 4.17. The top part of the figure shows the target and noise signals being added at the receiver input, and then propagated through the receiver and matched filter. The bottom portion shows the target and noise signals being separately propagated through the receiver and matched filter, and then added.

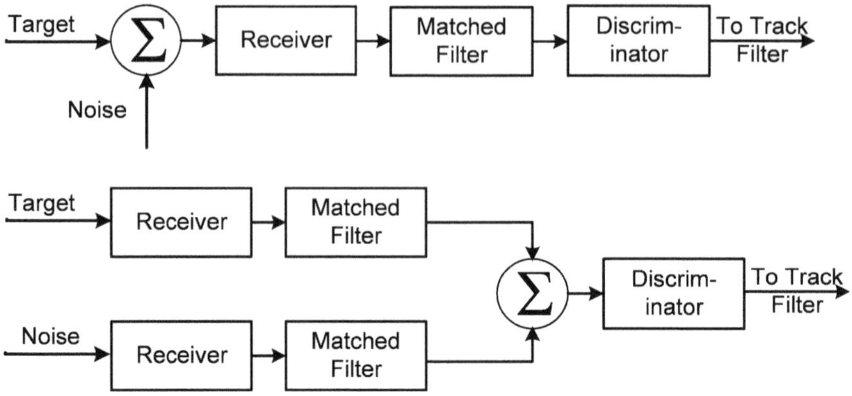


**Figure 4.17** An illustration of superposition.

We can mathematically justify superposition as follows. Let $v_{inT}(t)$ and $v_{inN}(t)$ represent the target and noise voltages at the receiver input, and let $h(t)$ represent the impulse response of the receiver and matched filter. We can write the combined signal at the matched filter output as

$$v_{MF}(t) = h(t) * \left[v_{inT}(t) + v_{inN}(t)\right] = \int_{-\infty}^{\infty} h(t-\tau)\left[v_{inT}(\tau) + v_{inN}(\tau)\right] d\tau \,. \qquad (4.49)$$

If we separate the integral into two integrals we get

$$\begin{aligned} v_{MF}(t) &= \int_{-\infty}^{\infty} h(t-\tau) v_{inT}(\tau) dt + \int_{-\infty}^{\infty} h(t-\tau) v_{inN}(\tau) dt \\ &= h(t) * v_{inT}(t) + h(t) * v_{inN}(t) \\ &= v_{MFT}(t) + v_{MFN}(t) \end{aligned} \qquad (4.50)$$

Equations (4.49) and (4.50) provide a mathematical statement of superposition, and the claim that the summation of signals can take place at either the input to the receiver and matched filter, or at the output of the receiver and matched filter.

We now proceed to develop models of signal and noise at the output of the matched filter.

### 4.5.1 Signal Model

Let the normalized transmit pulse be represented by

$$v_t(t) = e^{j2\pi f_{RF} t} e^{j\phi(t)} \text{rect}\left[\frac{t}{\tau_p}\right] \qquad (4.51)$$

where $f_{RF}$ is the carrier frequency, $\phi(t)$ is the phase modulation on the pulse, and $\tau_p$ is the pulse width. For an unmodulated pulse, $\phi(t) = 0$, and for an LFM pulse $\phi(t) = \pi\alpha t^2$. For a phase coded pulse, such as the previously considered Barker coded pulse, $\phi(t)$ is the phase modulation [6, 7, 15].

The return signal can be written as

$$v_r(t) = \sqrt{P_s} e^{j2\pi f_{RF}(t-\tau_{Rn})} e^{j\phi(t-\tau_{Rn})} \text{rect}\left[\frac{t-\tau_{Rn}}{\tau_p}\right]. \tag{4.52}$$

where $\tau_{Rn}(t) = 2R_n/c$. $R_n$ is the target range at the beginning of the $n^{th}$ track update period, and $c$ is the speed of light. $P_s$ is a term we use to scale the return signal level at the matched filter output to provide the proper signal-to-noise ratio (SNR). It will be discussed shortly.

We assume the various down conversion stages (heterodyning stages) of the receiver remove the carrier term to leave the baseband signal, at the matched filter input, of

$$v_r(t) = \sqrt{P_s} e^{j\psi_{Rn}} e^{j2\pi f_d t} e^{j\phi(t-\tau_{Rn})} \text{rect}\left[\frac{t-\tau_{Rn}}{\tau_p}\right] \tag{4.53}$$

where $f_d = -2R_n'/\lambda$ and $R_n'$ is the target range rate at the beginning of the $n^{th}$ track update period. $f_d$ is the target Doppler frequency [15].

The phase term

$$\psi_{Rn} = -2\pi f_{RF}\tau_{Rn} = -4\pi R_n/\lambda \tag{4.54}$$

is important when modeling multiple targets because it captures interference effects between the target return signals (when they overlap in range and angle). The change in $R_n$, and thus $\psi_{Rn}$, will be different for different targets because their motion properties are generally not the same. Because of this, the signals will combine differently from track update to track update. On some updates, the signals could have the same phase, and thus add. On other updates, they could have opposite phases, and thus subtract.

We assume the matched filter takes the form shown in Figure 4.18. The mixer tunes the matched filter to the frequency $f_m$. Ideally, $f_m$ is selected to cancel the target Doppler frequency, $f_d$. In practice, there will be a residual frequency of

$$f = f_d - f_m, \tag{4.55}$$

which is termed the Doppler mismatch.

We assume $h(t)$ is matched to the transmit pulse. As such its impulse response is

$$h(t) = e^{-j\phi(t)} \text{rect}\left[\frac{t}{\tau_p}\right]. \tag{4.56}$$

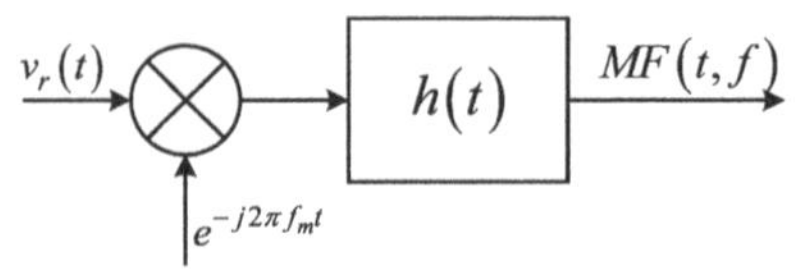


**Figure 4.18** Matched filter block diagram.

From matched filter theory, we can write the output of the matched filter as [15]

$$MF(\tau,f)=e^{-j2\pi f_m\tau}v_r(\tau)*h(\tau). \tag{4.57}$$

where we changed the time variable to the range delay variable, $\tau$, normally associated with the matched filter output. As a note, $MF(\tau, f)$ is related to the ambiguity function $\chi(\tau, f)$ by [4, 6, 15, 18]

$$\chi(\tau,f)=\left|MF(\tau,f)\right|. \tag{4.58}$$

For the special case of an unmodulated pulse, evaluation of (4.57) yields

$$\begin{aligned}MF(\tau,f)=&\sqrt{P_s}e^{j\psi_{Rn}}e^{j\pi f(\tau_{Rn}+\tau)}\left(\tau_p-\left|\tau-\tau_{Rn}\right|\right)\\&\times\operatorname{sinc}\left(f\left[\tau_p-\left|\tau-\tau_{Rn}\right|\right]\right)\operatorname{rect}\left[\frac{\tau-\tau_{Rn}}{2\tau_p}\right].\end{aligned} \tag{4.59}$$

For an LFM pulse, the matched filter output is

$$\begin{aligned}MF(\tau,f)=&\sqrt{P_s}e^{j\psi_{Rn}}e^{j\pi f(\tau_{Rn}+\tau)}\left(\tau_p-\left|\tau-\tau_{Rn}\right|\right)\\&\times\operatorname{sinc}\left(\left[f-\alpha(\tau-\tau_{Rn})\right]\left[\tau_p-\left|\tau-\tau_{Rn}\right|\right]\right)\operatorname{rect}\left[\frac{\tau-\tau_{Rn}}{2\tau_p}\right].\end{aligned} \tag{4.60}$$

Note that the magnitude of (4.59), for $f = 0$, degenerates to the matched Doppler range cut of the ambiguity function, or

$$\chi(\tau,0)=\left|MF(\tau,0)\right|=\sqrt{P_s}\left(\tau_p-\left|\tau-\tau_{Rn}\right|\right)\operatorname{rect}\left[\frac{\tau-\tau_{Rn}}{2\tau_p}\right] \tag{4.61}$$

This is the equation of the triangle of Figure 4.2 if we let $\sqrt{P_s} = 1/\tau_p$. Similarly, (4.60) degenerates to (4.32) for $f = 0$ and $\sqrt{P_s} = 1/\tau_p$.

In general, the equation for the matched filter response for a phase coded pulse is not a simple function. However, if we take advantage of the fact that the main lobe of the matched filter response has a triangular shape with a width of approximately twice the chip width, and the matched range, Doppler cut of the ambiguity function is a sinc function, we can write an approximate expression for the matched filter output as [7, 15, 19]

$$MF(\tau,f)=\sqrt{P_S}e^{j\psi_{Rn}}\tau_p e^{j\pi f(\tau_{Rn}+\tau)}$$
$$\times\left\{\left[\frac{1-SL}{\tau_c}\left(\tau_c-|\tau-\tau_{Rn}|\right)\text{sinc}\left(f\left[\tau_p-|\tau-\tau_{Rn}|\right]\right)\right]\text{rect}\left[\frac{\tau-\tau_{Rn}}{2\tau_c}\right]\right. \quad (4.62)$$
$$\left.+(SL)\text{sinc}\left(f\left[\tau_c-|\tau-\tau_{Rn}|\right]\right)\text{rect}\left[\frac{\tau-\tau_{Rn}}{2\tau_p}\right]\right\}$$

In (4.62), the first term accounts for the main lobe of the matched filter response, and the second term provides an approximation of the sidelobe region. $\tau_c$ is the chip width, and $SL$ is an estimate of the average sidelobe level.

### 4.5.2 Noise Model

We now turn our attention to the modeling noise; specifically, the noise at the output of the matched filter. An important consideration when modeling noise at the matched filter output is that the noise samples are correlated. Our noise model needs to account for this.

We start by assuming the noise into the matched filter is complex, wide-sense stationary (WSS), zero-mean, white, Gaussian noise with a power spectral density of $N_o$. The assumption of white noise derives from matched filter theory wherein it is assumed the noise into a matched filter is white [20, 21]. Radar designers try to satisfy this condition by assuring the spectrum of the noise into the matched filter is wider than the frequency spectrum of the matched filter, and that the spectrum is reasonably flat. The Gaussian assumption is standard. Without it, much of the mathematics used in radar theory becomes quite difficult. Also, the central limit theorem tells us the noise will be approximately Gaussian unless there is a nonlinearity late in the receiver chain [21, 22]. The zero-mean restriction is satisfied because the amplifiers in radars are band-pass amplifiers. The WSS assumption means the power spectral density and mean are constant; a reasonable assumption. Finally, the complex assumption is consistent with the fact the signal and noise must be characterized in terms of their amplitude and phase.

We represent the noise into the matched filter as

$$n(t)=\sqrt{\frac{N_o}{2}}\left[n_I(t)+jn_Q(t)\right]. \quad (4.63)$$

The standard assumption is that $n_I(t)$ and $n_Q(t)$ are independent. From random processes theory, the autocorrelation function of $n(t)$ is

$$R(\tau)=E\left\{n(t+\tau)n^*(t)\right\}=N_o\delta(\tau) \quad (4.64)$$

where $\delta(t)$ is the Dirac delta function [23].

The matched filter impulse response is

$$h(t) = e^{-j\phi(t)} \text{rect}\left[\frac{t}{\tau_p}\right]^{14}, \tag{4.65}$$

and the noise out of the matched filter is

$$n_o(t) = h(t) * n(t) = \int_{-\infty}^{\infty} h(\gamma) n(t-\gamma) d\gamma . \tag{4.66}$$

The autocorrelation of the noise at the matched filter output is

$$\begin{aligned} R_o(\tau) &= E\left\{n_o(t+\tau) n_o^*(t)\right\} \\ &= E\left\{\int_{-\infty}^{\infty} h(\gamma) n(t+\tau-\gamma) d\gamma \int_{-\infty}^{\infty} h^*(\beta) n^*(t-\beta) d\beta\right\}. \\ &= \int_{-\infty}^{\infty}\int_{-\infty}^{\infty} h(\gamma) h^*(\beta) E\left\{n(t+\tau-\gamma) n^*(t-\beta)\right\} d\beta d\gamma \end{aligned} \tag{4.67}$$

From (4.64), we recognize the expected value in the integrand as

$$E\left\{n(t+\tau-\gamma) n^*(t-\beta)\right\} = N_o \delta(\tau-\gamma+\beta). \tag{4.68}$$

Substituting this into (4.67) gives

$$R_o(\tau) = \int_{-\infty}^{\infty}\int_{-\infty}^{\infty} h(\gamma) h^*(\beta) N_o \delta(\tau-\gamma+\beta) d\beta d\gamma . \tag{4.69}$$

Evaluating the integral with respect to $\gamma$ gives

$$R_o(\tau) = N_o \int_{-\infty}^{\infty} h(\tau+\beta) h^*(\beta) d\beta . \tag{4.70}$$

Substituting (4.65) into (4.70) results in (see Problem 14)

---

[14] We ignored the mixer in front of $h(t)$ because it does not affect the matched filter output when the input is noise.

$$R_o(\tau) = N_o(\tau_p - |\tau|)\text{rect}\left[\frac{\tau}{2\tau_p}\right] \tag{4.71}$$

for the unmodulated pulse,

$$R_o(\tau) = N_o(\tau_p - |\tau|)\text{sinc}\left(\alpha\tau(\tau_p - |\tau|)\right)\text{rect}\left[\frac{\tau}{2\tau_p}\right] \tag{4.72}$$

for an LFM pulse, and

$$R_o(\tau) \approx N_o\tau_p\left\{\left[\frac{1-SL}{\tau_c}(\tau_c - |\tau|)\right]\text{rect}\left[\frac{\tau}{2\tau_c}\right] + (SL)\text{rect}\left[\frac{\tau}{2\tau_p}\right]\right\} \tag{4.73}$$

for the phase coded pulse approximation. Comparing (4.71), (4.72), and (4.73) to (4.59), (4.60), and (4.62) leads to the observation:

$$R_o(\tau) = MF(\tau, 0), \tag{4.74}$$

with $\tau_{Rn} = 0$, $\psi_{Rn} = 0$, and $P_s = 1$.

We can make some observations from (4.71), (4.72), and (4.73). For the unmodulated pulse, $R_o(\tau) = 0$ for noise samples separated by $\tau_p$ or greater. Since $R_o(\tau)$ is zero, and since $n_o(t)$ is zero mean (because $n(t)$ is zero mean), the autocovariance of $n_o(t)$ is zero for noise samples separated by greater than $\tau_p$. Since the autocovariance is zero for such noise samples, the noise samples are uncorrelated[15]. Since the noise samples are also Gaussian, they are independent.[16]

For the LFM pulse, $R_o(\tau)$ will be small for noise sample separations greater than $\tau_{res,}$ and the noise samples will be "almost independent". Thus, it may be possible to assume they are independent and not introduce significant error. However, a safe approach would be to assume they are not independent and choose them accordingly. A similar argument applies to phased coded pulses.

---

[15] The fact that the autocovariance equals the autocorrelation for zero-mean random process derives from the equation,

$$\begin{aligned} C_o(\tau) &= E\left\{\left(n(t+\tau) - \eta(t+\tau)\right)\left(n(t) - \eta(t)\right)^*\right\} = E\left\{n(t+\tau)n^*(t)\right\} - \eta(t+\tau)\eta^*(t) \\ &= R_o(\tau) - \eta(t+\tau)\eta^*(t) \end{aligned}.$$

[16] As a reminder, independent random variables are also uncorrelated. However, in general, uncorrelated random variables are not independent. An exception to this rule is that Gaussian random variables that are uncorrelated, are also independent.

#### 4.5.2.1 Generating Correlated Noise Samples

The discussion of the previous section leads to the problem of how to choose appropriately correlated noise samples. One way of doing this would be to use a detailed matched filter simulation. However, this is computationally inefficient. Since many modeling applications require only a few noise samples, a computationally efficient method of generating correlated noise samples can be derived from random variable theory [21, 22, 24].

Let $\eta$ represent an $N$ element vector of the noise samples at the matched filter output. From the above discussions, these samples are zero-mean, correlated, complex, Gaussian, random variables. The autocorrelation (and autocovariance) matrix for the vector is

$$R = E\left\{\eta\eta^*\right\} \tag{4.75}$$

where the * notation represents the conjugate-transpose operation. If the elements of $\eta$ are $n_1$, $n_2$, …,$n_N$, $R$ is of the form

$$R = \begin{bmatrix} E\left\{n_1 n_1^*\right\} & E\left\{n_1 n_2^*\right\} & \cdots & E\left\{n_1 n_N^*\right\} \\ E\left\{n_2 n_1^*\right\} & E\left\{n_2 n_2^*\right\} & \cdots & E\left\{n_2 n_N^*\right\} \\ \vdots & \vdots & \ddots & \vdots \\ E\left\{n_N n_1^*\right\} & E\left\{n_N n_2^*\right\} & \cdots & E\left\{n_N n_N^*\right\} \end{bmatrix}. \tag{4.76}$$

We can relate the elements of $R$ to $R_o(\tau)$ as

$$\begin{aligned} E\left\{n_k n_k^*\right\} &= R_o(0) = \sigma_k^2 \\ E\left\{n_k n_{k+m}^*\right\} &= R_o(m\delta\tau) \\ E\left\{n_{k+m} n_k^*\right\} &= R_o(-m\delta\tau) = R_o(m\delta\tau) \end{aligned} \tag{4.77}$$

where $\Delta\tau$ is the spacing between noise samples. As an example, $\delta\tau = \tau_L - \tau_E$ for the sampling gate discriminator.

The last equality derives from the fact, $R_o(\tau)$ is a real, even function, see (4.71), (4.72), and (4.73). Also, the form of $R_o(\tau)$ implies the $n_k$ are samples from a WSS random process.

$R$ is a real, positive definite, Toeplitz matrix [25], and we can perform a Cholesky decomposition on it to yield [21, 26]

$$R = LL^* \tag{4.78}$$

where $L$ is a real, lower triangular matrix.

Let $X$ be an $N$ element vector of zero-mean, unit variance, independent, complex, Gaussian random variables and form the vector

$$Y = LX \ . \tag{4.79}$$

The correlation of matrix of $Y$ is

$$R_Y = E\left\{YY^*\right\} = E\left\{LX\left(LX\right)^*\right\} = E\left\{LXX^*L^*\right\} = LE\left\{XX^*\right\}L^* . \tag{4.80}$$

Since the elements of $X$ are zero-mean, unit variance and independent,

$$E\left\{XX^*\right\} = I , \tag{4.81}$$

an $N\times N$ identity matrix, and

$$R_Y = LL^* = R . \tag{4.82}$$

Thus, the elements of $Y$ have the same statistics as the elements of $\eta$, and $Y$ can be used to represent samples of the noise out of the matched filter. If we normalize $R_o(\tau)$ by dividing by $R_o(0)$, $R$ will have ones on the diagonal, and the elements of $\eta$ will be unit variance.

With the above, we can outline an algorithm for computing properly correlated noise samples as

1. Compute $R_o(k\delta\tau)$ for $k = 0, 1, \ldots, N-1$ where $\delta\tau$ is the desired spacing between range samples, and $N$ is the number of range samples needed.
2. Form the correlation matrix as defined by (4.76) and (4.77), and divide all elements by $R_o(0)$.
3. Compute the Cholesky decomposition of $R$ to produce the lower triangular matrix $L$.
4. Generate an $N$–element vector, $X$, of zero-mean, unit variance, independent, complex, Gaussian random numbers.
5. Compute a vector of appropriately correlated random numbers using $\eta = LX$. As a note, the elements of $\eta$ will have a variance of unity.

There are standard computer programs that can be used to form a Toeplitz matrix and its Cholesky decomposition. The Matlab functions are `toeplitz` and `chol`.

When only two noise samples are needed, such as with the sampling gate discriminator, they can be generated using the simplified algorithm:

1. Generate a pair of zero-mean, independent, unit variance, Gaussian random numbers $n_I$ and $n_Q$ and form the complex random number $n_1 = (n_I + jn_Q)/\sqrt{2}$.
2. Generate another complex, zero-mean, unit variance, Gaussian random number, $x$, using the same method.
3. The second noise sample, $n_2$, is then computed using

$$n_2 = rn_1 + x\sqrt{1-r^2}$$

where $r = R_o(\delta\tau)/R_o(0)$ and $\delta\tau = \tau_L - \tau_E$.

As a reminder, if $\delta\tau \geq \tau_p$, $r$ will be zero and the noise samples, $n_1$ and $n_2$, will be uncorrelated, and independent since they are Gaussian.

### 4.5.3 Scaling the Signal and Noise

Now that we have signal and noise models, we discuss how to scale each to provide the proper SNR at the matched filter output. In particular, we develop equations for $P_s$ and $N_o$ to give the proper SNR. From (4.59), (4.60), and (4.62), the peak signal "power" (this is in quotes because it is not really a power but a squared signal) at the output of the matched filter occurs at $\tau = \tau_R$ and $f = 0$ and is equal to,

$$\left|MF\left(\tau_R, 0\right)\right|^2 = P_s \tau_p^2 . \tag{4.83}$$

The average noise "power" at the output of the matched filter is, from (4.71), (4.72), and (4.73),

$$R_o\left(0\right) = N_o \tau_p . \tag{4.84}$$

By definition, the SNR is the ratio of peak signal power divided by the average noise power at the matched filter output [15, 27], or

$$SNR = \frac{P_s \tau_p^2}{N_o \tau_p} = \frac{P_t G_T G_R \lambda^2 \sigma \tau_p}{\left(4\pi\right)^3 R^4 k T_s L} \tag{4.85}$$

where the last term is the radar range equation [15, 27, 28].

Equation (4.85) results in one equation and two unknowns. However, since absolute power levels are not of interest here, we can arbitrarily choose $N_o = 1/\tau_p$ so that $R_o(0) = N_o \tau_p = 1$. This allows us to compute $P_s$ as

$$P_s = \frac{SNR}{\tau_p^2} . \tag{4.86}$$

Thus, to scale the signal and noise we use

$$\sigma^2 = R_o\left(0\right) = 1 \tag{4.87}$$

when generating noise samples and (4.86) when generating the signal samples. As a note, the losses, $L$, in (4.85) should not include range straddling losses. These are accounted for in the matched filter model.

### 4.5.4 Signal and Noise Generation Algorithm

An algorithm for generating signal and noise samples is:

- Decide on a waveform (unmodulated, LFM or phase coded).
- Decide on a spacing between samples of signal and noise. Call this $\delta\tau$.
- Generate the desired number of noise samples using the algorithm discussed at the end Section 4.5.2. The noise samples will have a variance of $\sigma^2 = 1$.
- Compute $P_s$ using (4.86) and generate the signal(s) samples at values of $\tau$ from $\tau_E$ to $\tau_L$ with a spacing of $\delta\tau$.
- Add the signal(s) and noise samples. This represents the sum of signal(s) and noise at the output of the matched filter (the (s) on the end of signal was included because there could be multiple signals from multiple targets). A new set of signal(s) plus noise samples would be generated at each track update period.

## 4.6 SIGNAL PROCESSOR CONSIDERATIONS

The discussions above assumed the signal processor consisted of only a matched filter. If the radar uses a coherent integrator after the matched filter (e.g., an MTI or Doppler processor) it can be modeled by its frequency response. To account for it, we would determine the gain of the signal processor at the target Doppler frequency and multiply the *signal* amplitudes by this gain. This would be done separately for each target, and before noise is added. To properly scale the SNR, the gain of the signal processor must not be normalized to unity, as is the common custom when generating frequency responses. Instead, the frequency response must capture the SNR gain through the coherent integrator. Also, the frequency response must capture any frequency tuning before the matched filter. Thus, if the matched filter is tuned to the target Doppler frequency, the coherent signal processor will be operating on $f_d - f_m$, not just $f_d$. The gain of the signal processor must be in volts/volt, not watts/watt.

This concept of applying the signal processor gain to the signals and then summing them and the noise is based on the assumption that superposition holds. This, in turn, is based on the assumption the coherent integrator is linear, which is the standard assumption for coherent integrators.

If the radar contains a non-coherent integrator, it could reside between the envelope detectors and the discriminator, or between the discriminator and track filter. This would depend on how the radar was designed. The only rigorous way to account for the impact of a non-coherent integrator is to actually implement it in the simulation.

## 4.7 EXAMPLES

The examples discussed in this section implement a tracker similar to the block diagram shown in Figure 4.1. However, the specific implementation of the track filter is different than in Figure 4.1. The filter in that figure used $e$ from the current time to compute the predicted range delay, $\tau_{trk}$, for the next update. This approach makes the assumption the time of the next update is known. In applications where track is the highest priority function, and the updates can be scheduled in advance, such an approach is acceptable. It is not suitable for implementations where the update period is not constant, or a track update can be delayed by a higher priority function, or task. In those applications, a better approach is to use (3.19) and (3.22) to compute the predicted range delay at the current time—the time of the next measurement. At the measurement time, the radar transmits the pulse, receives the return and processes it though the discriminator to produce the error signal, $e$ . $e$ is then used in (3.20) to compute the smoothed state vector, $X_s$, which saved until the next track update. In algorithmic form

- Compute the predicted state and range delay using

$$X_p(n) = FX_s(n-1) \text{ and} \tag{4.88}$$

$$\tau_{trk}(n) = H^T X_p(n) . \tag{4.89}$$

- Obtain the measurement and process it through the discriminator to obtain $e$.
- Compute smoothed state using

$$X_s(n) = X_p(n) + K(n)e(n) . \tag{4.90}$$

### 4.7.1 Example 1: Sampling Gate Discriminator and α-β Filter

In the first example, the tracker contains a sampling gate discriminator and a α-β filter, where $\alpha$ and $\beta$ are computed using the Benedict-Bordner technique discussed in Chapter 3 [29]. The parameters used with the first example are contained in Table 4.1.

$SNR_{ref}$, $R_{ref}$, and $\sigma_{ref}$ are used to avoid having to specify the individual parameters of the radar range equation. In terms of the radar range equation [15]

$$SNR_{ref} = \frac{P_T G_T G_R \lambda^2 \sigma_{ref} \tau_p}{(4\pi)^3 R_{ref}^4 k T_s L} = \frac{K_{SNR} \sigma_{ref}}{R_{ref}^4} \tag{4.91}$$

which leads to

$$K_{SNR} = \frac{SNR_{ref} R_{ref}^4}{\sigma_{ref}} . \tag{4.92}$$

With this, the SNR for any range and RCS can be computed from

$$SNR = \frac{K_{SNR}}{R^4} \sigma . \tag{4.93}$$

**Table 4.1**
Simulation Parameters for Example 1

| *Parameter* | *Value* | *Comment* |
|---|---|---|
| $SNR_{ref}$ | 10 dB | The reference SNR |
| $R_{ref}$ | 50 km | The reference range |
| $\sigma_{ref}$ | 1 m$^2$ | The reference RCS |
| $\sigma_{tgt}$ | 1 m$^2$ | The target RCS |
| $R_{init}$ | $R_{ref} + \delta R$ | The initial range for the track filter |
| $\delta R$ | ±100 m | Random number between ±100m |
| $R'$ | −300 m/s | Actual target range rate |
| $f_{RF}$ | 5 GHz | |
| $f_c$ | 1 Hz | Closed loop bandwidth |
| $T$ | 1/10 s | 10 Hz track rate |
| $q$ | ½ | |
| $\tau_{res}$ | 1 μs | 1 μs unmodulated pulse |

$R_{ref}$ is also the initial target range. It, plus a range offset, $\delta R$, is used as the initial position state of the range tracker. The initial velocity state is set to zero, as might be the situation if the range tracker is initialized using a single search or acquisition measurement. If range tracker initialization is based on several search or acquisition measurements, it would be possible, and desirable, to compute an initial value for the velocity state.

The target range rate, $R_n'$, is used to determine the Doppler frequency from

$$f_d = -\frac{2R_n'}{\lambda} \tag{4.94}$$

where $\lambda = c/f_{RF}$ is the wavelength associated with the radar RF (radio frequency), $f_{RF}$, and $c$ is the speed of light. $f_d$ is used to determine the mismatch frequency, $f$, (see (4.59), (4.60), and (4.62)) from $f = f_d - f_m$. We assume $f_m = f_d$ in this simulation. That is, the matched filter is tuned, or matched, to the target Doppler frequency.

The target range rate is also used to compute the true target range from $R_n = R_{ref} + R_n't$, where $t$ is time, and $t = 0$ is at the beginning of the simulation. With this definition of $R_n$, we are assuming (somewhat unrealistically) the target is flying along a radial path toward the radar at a constant range rate (the radar is at the origin of the coordinate system used in the simulation). $R_n$ is used to compute the range delay via $\tau_{Rn} = 2R_n/c$.

The track rate is 10 Hz, and the track filter one-sided, half power bandwidth is $f_c = 1$ Hz. The track filter is a α-β filter of the Benedict and Bordner type (see Section 3.3) [29]. Thus, we use (3.50) to determine α and (3.43) to determine $\beta$. The results of those calculations are

$$\alpha = 0.323 \text{ and } \beta = 0.0622. \tag{4.95}$$

The radar uses a 1 μs, unmodulated pulse.

The specification of $q = ½$ means the early and late samples of the discriminator, $\tau_E$ and $\tau_L$, are separated by $\tau_E - \tau_L = 2q\,\tau_{res} = \tau_p$. This means the noise samples will be uncorrelated (and thus independent—see Section 4.5.2).

The initialization section of the simulation is used to specify the parameters of Table 4.1 and compute the other static simulation parameters such as $\alpha$, $\beta$, $f_d$, and so forth. The initialization section is also used to compute the initial value of $\tau_{trk} = 2(R_{ref} + \delta R)/c$ and the initial smoothed state of the α-β tracker, $X_s$. $X_s$ is set to

$$X_s = \begin{bmatrix} \tau_{trk} \\ 0 \end{bmatrix}. \tag{4.96}$$

The initialization section is also used to define the arrays that hold the parameters to be plotted.

The sequence of operations in the iterative section of the simulation are:

1. Compute the predicted track filter state, $X_p$, at the current time using (4.88).
2. Compute the predicted range delay, $\tau_{trk}$, at the current time using (4.89).
3. Compute the true target range, $R_n$, at the current time.
4. Use $R_n$ to compute $\tau_{Rn} = 2R_n/c$ and $\psi_{Rn} = -4\pi R_n/\lambda$.
5. Compute $\tau_E$ and $\tau_L$.
6. Compute the $SNR$ using (4.93), $P_s$ using (4.86), and the matched filter outputs, $v_E$ and $v_L$, using (4.59).
7. Draw four, independent, zero mean, unit variance, Gaussian random numbers and use them to form the noise samples $n_E = (n_1 + jn_2)/\sqrt{2}$ and $n_E = (n_3 + jn_4)/\sqrt{2}$.
8. Add the noise samples to $v_E$ and $v_L$ and compute their magnitudes to form $V_E$ and $V_L$.
9. Use $V_E$ and $V_L$, along with $S_\tau = (1-q)\tau_p$, in (4.14) to find $e$.
10. Use $e$ to compute the smoothed state, $X_s$, using (4.90).
11. Save the outputs.
12. Update $t$.

Figure 4.19 contains a plot of the simulation output. The top plot contains curves of $\tau_{Rn}$ and $\tau_{trk}$, converted to range, and labeled as $R_n$ and $R_{trk}$. The middle plot is $\Delta\tau$, and the bottom plot is $e$, both in units of meters. We were able to convert $e$ to range error because we chose $S_\tau$ so that the slope of the discriminator curve was unity. In the top plot, $R_n$ and $R_{trk}$ appear to coincide. This is partly because the tracker is working well and partly because the scale is in km.

The plots of $\Delta\tau$ and $e$ show the tracker has reduced the measurement error because the variations in the plot of $\Delta\tau$ are smaller than the variations in the plot of $e$. Both variations decrease over time because the SNR increases as the target approaches the radar

The initial error in the plots of $\Delta\tau$ and $e$ is due to filter transients. In the simulation run that generated Figure 4.19, the predicted range, $\tau_{trk}$, was greater than the actual range, $\tau_{Rn}$, by 0.33 μs, or 50 m. Also, the initial predicted range rate was set to zero. Because of this, and the 1 Hz bandwidth of the track filter, it took about 1.5 seconds for the tracker to recover from the initial error.

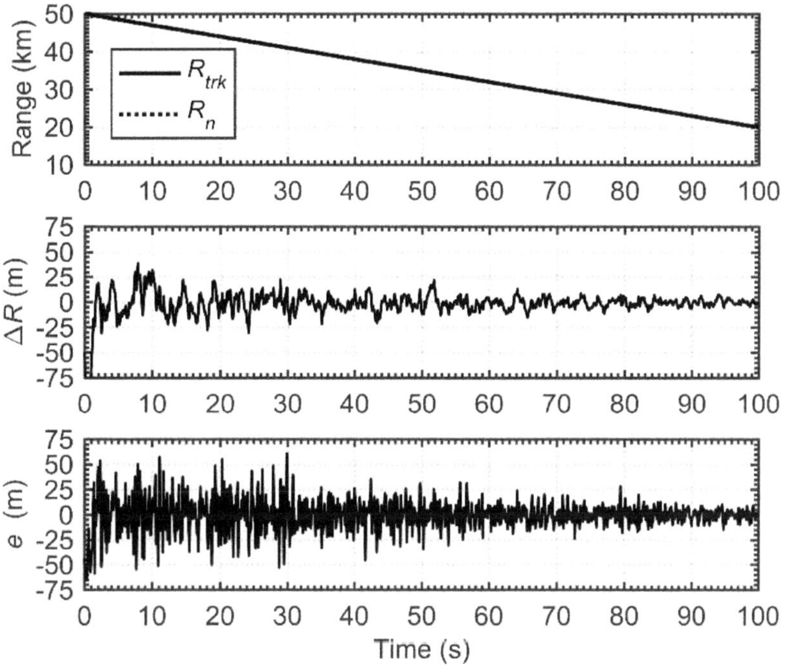


**Figure 4.19** Example 1 simulation results, sampling gate, α-β filter, unmodulated pulse.

The transient behavior is illustrated in Figure 4.20, which is a plot of $\Delta\tau$ from 0 to 5 seconds with an expanded vertical scale. In this example, the error, $\Delta\tau$, starts at –50 m and becomes more negative. At the first update, at $t = 0.1$ s, $\Delta\tau$ decreases to less than –75 m ($-\tau_p/2$), and enters the flat portion of the discriminator curve (see Figure 4.3). This reduces the loop gain to zero and slows the response. However, $\Delta\tau$ is still negative, and this drives the $\tau_{trk}$, albeit slowly, toward $\tau_{Rn}$. Eventually, $\Delta\tau$ enters the linear region of the discriminator curve, the loop speeds up, and the tracker is able to track the target.

In some of the simulation runs, with different noise sequences, the tracker never did establish track on the target. This was due to the aforementioned initial conditions, and the initial SNR of 10 dB. When the initial SNR was increased to 13 dB, the tracker tracked the target much more reliably. At an initial SNR of 15 dB, the track reliability was close to 100%. However, the transient was present in all cases

In some instances, the tracker took a long time to recover from the initial error. An example of this is illustrated in Figure 4.21. Because of the combination of the initial predicted range difference, the initial predicted range rate of zero, and the noise, $\Delta\tau$ was driven into the indeterminate region of the discriminator curve. This means the tracker was coasting because the only input to the track filter was noise. In this particular case, the sequence of noise samples eventually began driving $\tau_{trk}$ toward $\tau_{Rn}$, and $\Delta\tau$ eventually entered the flat, then linear, region of the discriminator curve. Since the SNR was high enough by that time, the tracker was able to acquire and lock-on to the target.

The results illustrated in Figures 4.20 and 4.21 indicate setting the initial range delay rate estimate to zero was probably not a good idea. Figure 4.22 contains a plot of Example 1 simulation results when perfect initialization was used in the tracker. That is, when the initial $\tau_{trk}$ was set to the initial value of $\tau_{Rn}$ and when $X_s$ was set to

$$X_s = \begin{bmatrix} \tau_{trk} \\ \tau'_{trk} \end{bmatrix} \tag{4.97}$$

where $\tau'_{trk} = -2R'_n/\lambda$. The script that generated this figure used the same noise sequence as the one used to generate Figure 4.21. As expected, the use of perfect initialization completely eliminated the initial transient. This variation on Example 1 indicates the importance of setting the initial rate term to a value that is somewhat close to the expected value for the target set of interest.

As a final variation on Example 1, the unmodulated pulse was replaced by a 15 μs, 1 MHz bandwidth LFM pulse. The results of that simulation run are shown in Figure 4.23. The script that generated this figure used the same noise sequence and initialization as the one used to generate Figure 4.19. Not surprisingly, the response of the tracker with the LFM pulse is about the same as with the unmodulated pulse. This is because both pulses have essentially the same range resolution of 1 μs. The slight differences in the plots is due to the slight differences in the matched filter outputs for the two types of pulses.

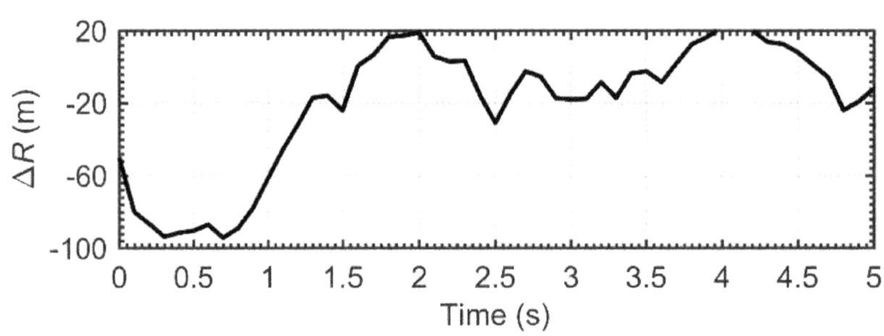


**Figure 4.20** Expanded view of the transient behavior for Example 1.

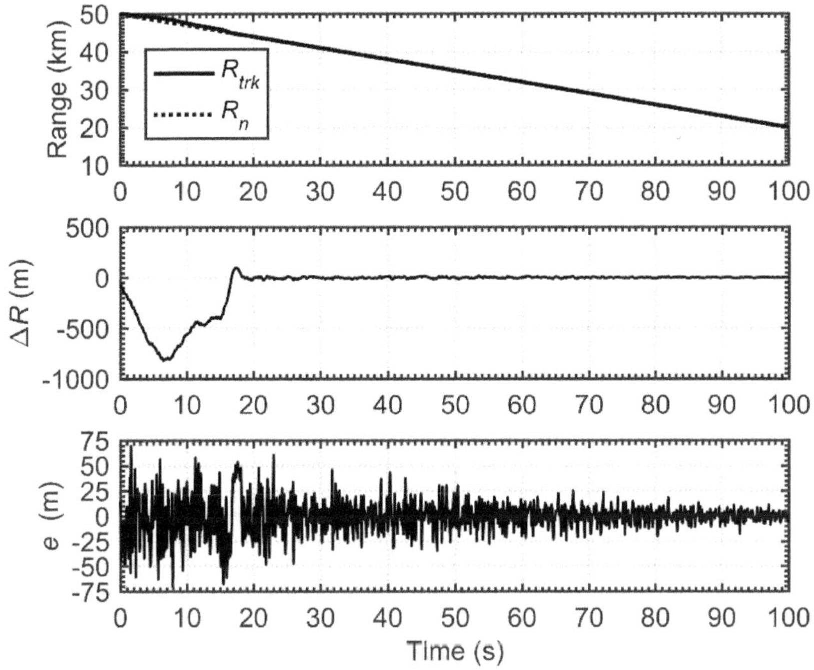


**Figure 4.21** Example 1 simulation results showing delayed lock.

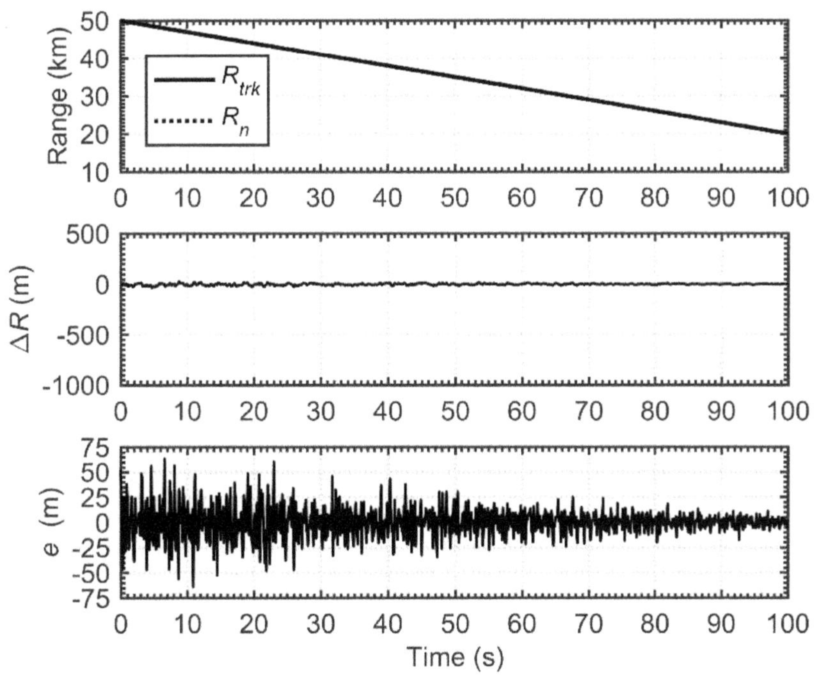


**Figure 4.22** Example 1 with perfect tracker initialization.

### 4.7.2 Example 2: Summing Gate Discriminator and α-β Filter

For the second example, the sampling gate discriminator is replaced with a summing gate discriminator. The unmodulated pulse is used, and $q$ is set to 2 to produce a wide gate. The matched filter output samples are spaced $\delta\tau = \tau_p/4$ apart to provide a discriminator curve that has the approximate S shape associated with integrating gate discriminators (see Section 4.3). With $q = 2$, $\tau_E = \tau_{trk} - q\tau_p = \tau_{trk} - 2\tau_p$ and $\tau_L = \tau_{trk} + q\tau_p = \tau_{trk} + 2\tau_p$ so that $\tau_L - \tau_E = 2q\tau_p = 4\tau_p$. This means 17 matched filter outputs are needed. They start at $\tau_E$ and end at $\tau_L$, and are spaced $\delta\tau = \tau_p/4$ apart. The discriminator scaling is set to $S_\tau = 1.4\tau_p/2$ to give a discriminator slope of unity near $\Delta\tau = 0$.

Seventeen correlated noise samples are needed since the matched filter output is sampled at less than the decorrelation time of the matched filter (see Section 4.5.2). To accommodate the correlated noise, the initialization part of the simulation is expanded to compute the 17×17 lower triangular matrix, $L$, as described in Section 4.5.2. $L$ is used in the iterative part of the simulation to generate correlated noise samples.

The sequence of operations in the iterative part of the simulation are:

1. Compute the predicted track filter state, $X_p$, at the current time using (4.88).
2. Compute the predicted range delay, $\tau_{trk}$, at the current time using (4.89).
3. Compute the true target range, $R_n$, at the current time.
4. Use $R_n$ to compute $\tau_{Rn} = 2R_n/c$ and $\psi_{Rn} = -4\pi R_n/\lambda$.
5. Compute $\tau_E = \tau_{trk} - 2\tau_p$ and $\tau_L = \tau_{trk} + 2\tau_p$.
6. Compute $SNR$ using (4.93), and $P_s$ using (4.86).
7. Use (4.59) to compute the matched filter output, $v$, at 17 values of $\tau$ between $\tau_E$ and $\tau_L$, spaced $\delta\tau = \tau_p/4$ apart.
8. Generate a vector, $X$, of 17, zero mean, unit variance, complex, independent, Gaussian random numbers.
9. Form a vector of 17, correlated random numbers as $\eta = LX$.
10. Add the 17 correlated random numbers (noise samples) to $v$, and compute their magnitudes to form $V$.
11. Sum the first 9 values of $V$ to form $I_E$, and sum the last 9 values to form $I_L$. Note, the value of $V$ at $\tau_{trk}$ is in both sums.
12. Use $I_E$ and $I_L$, along with $S_\tau = \tau_p/2$, in (4.40) to find $e$.
13. Use $e$ to compute the smoothed state, $X_s$, using (4.90).
14. Save the outputs.
15. Update $t$.

Figure 4.24 contains a plot of the simulation output for an initial SNR of 10 dB. The lower two curves are similar to those of Figure 4.19, except the error signal out of the discriminator, $e$, is smaller in Figure 4.24 than in Figure 4.19. This is due to the fact that the summing gate discriminator does not suffer the 6 dB range straddling loss that the sampling gate discriminator does. To check this assertion, we re-ran the simulation that produced Figure 4.19 with an initial SNR of 16 dB, rather than 10 dB. The results of that simulation run are indicated in Figure 4.25. As expected, the errors are now about the same size as in Figure 4.24.

The transients of Figures 4.24 and 4.25 are not as large as in Figure 4.19, because the α-β tracker of this example was initialized to the target range rate of –300 m/s. There is a small transient because the initial range estimate of the tracker was 50 m greater than the actual range, as in Example 1.

### 4.7.3 Example 3: Direct Range Measurement and α-β-γ Filter

For this example, the discriminator is replaced by the direct range measurement algorithm (see Section 4.4) and the α-β filter is replaced by a α-β-γ filter. The range window for the direct range measurement algorithm extends from $\tau_E = \tau_{trk}-3\,\tau_p$ to $\tau_L = \tau_{trk}+3\,\tau_p$, and the samples are spaced one pulse width, or $\tau_p$, apart. An unmodulated, 1-μs pulse is used. Because of this spacing, the noise samples are independent. The algorithm searches the seven samples in the range window and selects the largest. It then uses the largest, and the larger of the two adjacent samples, to estimate the range delay using (4.47). If the largest sample is on either end of the range window, the algorithm uses that sample and the one next to it.

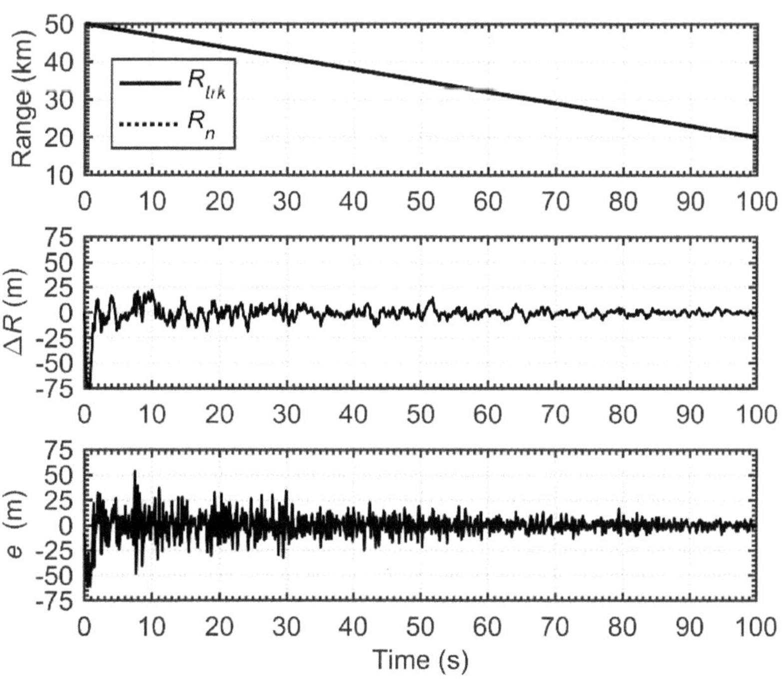


**Figure 4.23** Example 1 simulation results, sampling gate, α-β filter, LFM pulse.

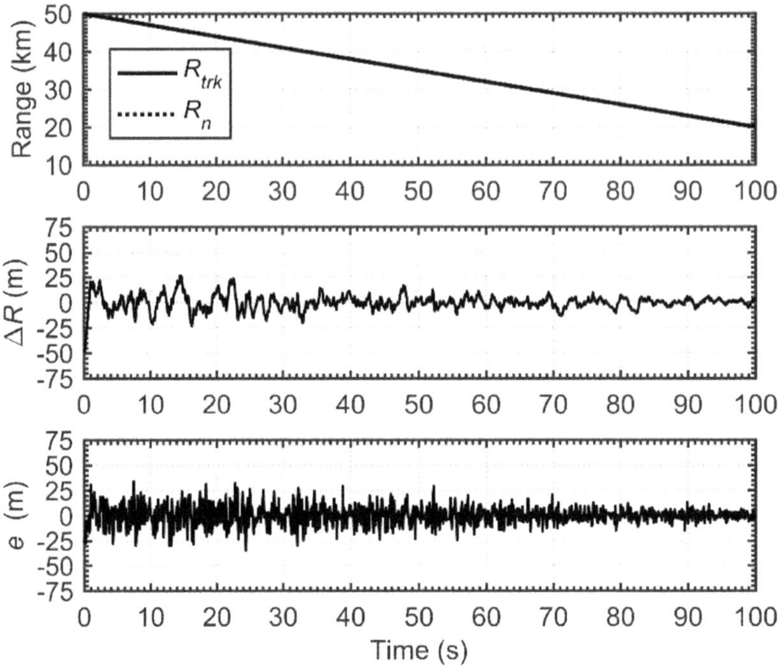


**Figure 4.24** Example 2 simulation results—summing gate, α–β filter, unmodulated pulse.

The α-β-γ filter is the critically damped version discussed in Section 3.8.3.1. As such, the filter parameters are calculated using

$$z_0 = e^{-1.6115 f_c T} \tag{4.98}$$

and

$$\alpha = 1 - z_0^3,\ \beta = \frac{3}{2}(z_0 - 1)^2 (z_0 + 1) \text{ and } \gamma = \frac{1}{2}(1 - z_0)^3. \tag{4.99}$$

The parameters of Table 4.1 are also used in this example and the structure of the simulation is the same as in Examples 1 and 2. Finally, the α-β-γ filter is perfectly initialized. That is, the initial predicted range delay and delay rate are set to the actual values. The initial acceleration was set to zero, which is the true range acceleration since the range rate is constant.

Since the tracker is using direct range measurement, and doesn't have a range discriminator, the input to the track filter is $e = \tau_{Rmeas} - \tau_{trk}$ instead of that given by (4.14) $\Delta\tau$ is still defined as $\Delta\tau = \tau_R - \tau_{trk}$.

This simulation was very seldom able to establish track when the initial SNR was less than 13 dB. Experiments indicated, at low SNRs, the noise samples were occasionally larger than the signal-plus-noise samples. Because of this, the range measurement algorithm occasionally computed a range that was not close to the true range delay. This caused $e = \tau_{Rmeas} - \tau_{trk}$ to be unexpectedly large. As a result, $X_s(n) = X_p(n) + K(n)e(n)$ was unexpectedly perturbed far away from $X_p(n)$, the predicted state. This, in turn, caused a perturbation in the predicted range delay on the next update. This caused the next range measurement to have significant error, which made the subsequent predicted range have significant error, and so forth. Eventually, the true range delay moved out of the range window, and the input to the track filter was based only noise.

Sometimes, the tracker was able to maintain a coarse track until the SNR became large enough for the range measurement algorithm to begin providing accurate range measurements. An example of this is contained in Figure 4.26. In this particular case, the initial SNR was set to 11 dB. During the first 30 seconds, the tracker is barely able to maintain track. After about 40 seconds, the SNR has increased to the point where the tracker can reliably make measurements, once it starts. After that, $\Delta\tau$ and $e$ settle down to reasonable values.

Figure 4.27 contains results for the same simulation set-up used to generate Figure 4.26, except that the initial SNR was set to 15 dB. In this case, the radar only had a few instances where it made bad measurements. As a result, the track was more reasonable.

The results of these experiments indicate it might be better to use a sampling or integrating gate discriminator in a closed loop tracker, instead of direct range measurement. In a tracker that uses a discriminator, the error signal into the track filter, $e$, is limited to $S_\tau$, the saturation value of the discriminator. This, in turn, will limit how severely the tracker is driven off of the true range. In a tracker that uses direct range measurement, $e$ can be quite large. This large measurement error can cause the tracker to be quickly driven off the true range, to the point where it may not be able to recover.

Another alternative would be to compute the measured range only when the largest signal, and possibly signal in the adjacent cell, crossed a detection threshold. If there was no threshold crossing, the error signal, $e$, would be set to zero, and the filter would coast. That would reduce the number of times the tracker was being influenced by excessively large variations in measured range.

Another alternative might be to use a Kalman filter rather than a $\alpha$-$\beta$ or $\alpha$-$\beta$-$\gamma$ filter. In a Kalman filter, the predicted measurement variance can be used to set the width of the window over which the measurement algorithm searches. This may help prevent the situation where a bad measurement drives the track filter away from the true target range delay. Also, the Kalman filter uses SNR to compute the measurement variance. If the SNR is low, the Kalman filter will reduce the emphasis it places on the range measurement.

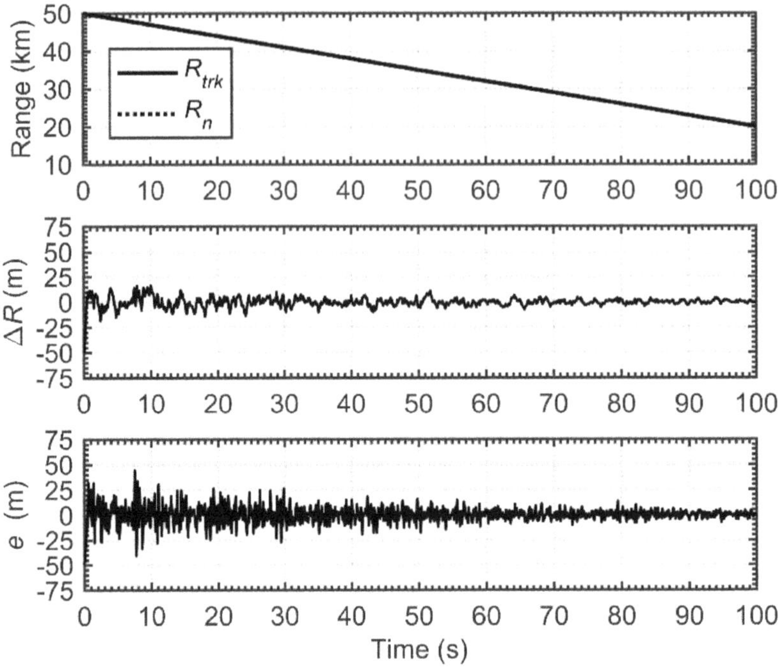


**Figure 4.25** Sampling gate, α-β filter, unmodulated pulse, initial SNR of 16 dB.

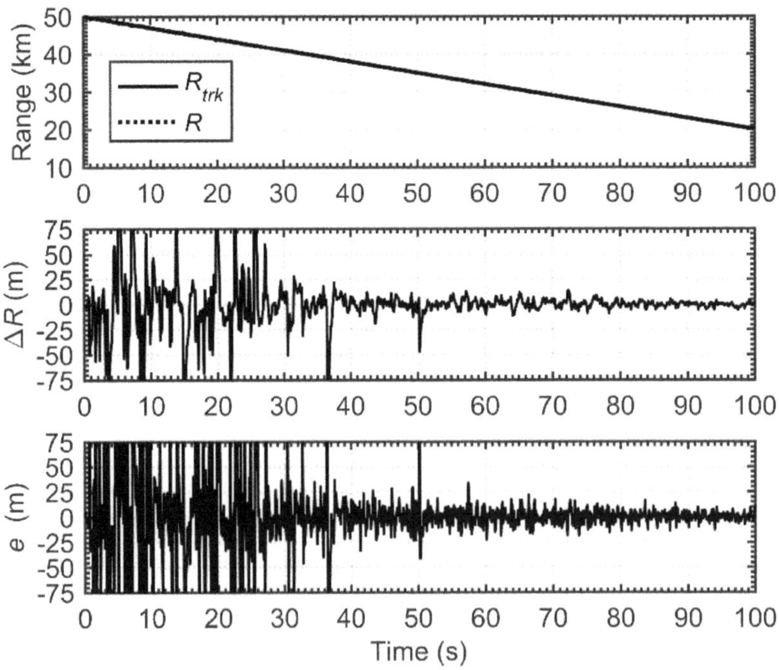


**Figure 4.26** Experiment 3 simulation results—direct range measurement, α-β-γ filter, 11 dB initial SNR.

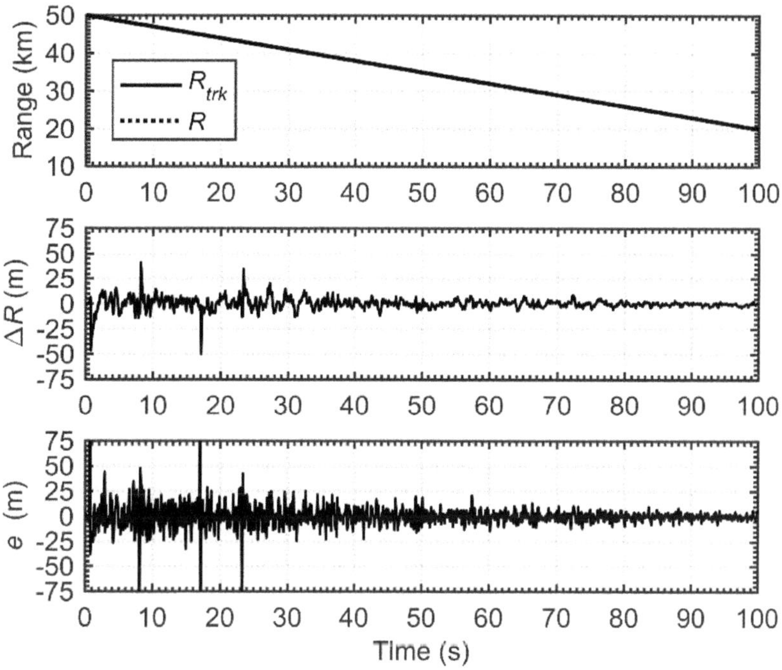


**Figure 4.27** Experiment 3 simulation results—direct range measurement, α-β-γ filter, 15 dB initial SNR.

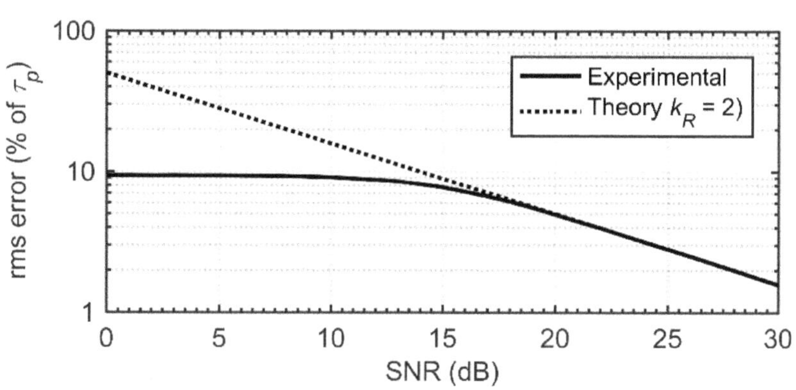


**Figure 4.28** RMS range measurement error for a $\tau_p$ μs, unmodulated pulse

## 4.8 FUNCTIONAL LEVEL ERROR MODEL

In some preliminary or functional level tracking studies, development of a discriminator and tracker model is not desirable. In such studies, the analyst simply wants an estimate of the expected range measurement error as a function of SNR and range resolution. To that end, various authors provide the equation [6, p. 412; 11; 14; 27, p. 318]

$$\sigma_{Rmeas} = \frac{\tau_{res}}{k_R\sqrt{SNR}} \tag{4.100}$$

for the root-mean-square (RMS) range measurement error. This equation is based on the assumptions the SNR is "large", and the absolute values in (4.14) and (4.47) can be ignored. To use the equation, one would draw a zero-mean, Gaussian random number with a variance of $(\sigma_{Rmeas})^2$ and add it to the true range.

Figure 4.28 contains a plot of (4.100) with $k_R = 2$, and a plot of $\sigma_{Rmeas}$ for sampling gate discriminator and the direct measurement technique. The experimental curve was generated for an unmodulated pulse with a width of $\tau_p$. The sampling gate discriminator used $q = ½$ and the direct measurement technique used samples spaced $\tau_p$ apart. As can be seen, (4.100) and the experiment results match well for SNR values above about 13 dB. For SNR values below 13 dB, the two curve deviate because ignoring the absolute values becomes questionable at low values of SNR

The plot of Figure 4.28 also applies to LFM and phase coded pulses with $\tau_p$ replaced by $\tau_{res}$ (see Problem 21). If $q$ is changed, the experimental and theoretical curves deviate at different values of SNR. For $q < ½$ the curves deviate at lower values of SNR than in Figure 4.28. For $q > ½$ the curves deviate for higher values of SNR (see Problems 22 and 23).

## 4.9 EXERCISES

1. Derive the conditions of (4.22) and (4.24), and (4.26) and (4.28).
2. Derive (4.29).
3. Use (4.35) to plot discriminator curves for $q = ¼$, ½, and ¾ and a 1 μs unmodulated pulse. How does your plot compare to Figure 4.4? The matched filter output for an unmodulated pulse is given in (4.59).
4. Repeat Problem 3 for an LFM pulse and the parameters of Figure 4.6.
5. Derive (4.41) and (4.42).
6. Derive (4.46).
7. Generate Figures 4.11 and 4.12.
8. Generate Figure 4.13.
9. Generate plots like Figure 4.15 for a 15 μs, 1 MHz bandwidth LFM pulse.
10. Generate plots like Figure 4.15 for a 13 bit, Barker coded pulse with a chip width of 1 μs.
11. Prove (4.47) perfectly predicts $\tau_R$ for an unmodulated pulse and no noise.
12. Derive (4.59).
13. Derive (4.60).
14. Show $R_o(\tau) = MF(\tau,0)$ with $\tau_{Rn} = 0$, $\psi_{Rn} = 0$ and $P_s = 1$. Derive (4.71),(4.72), and (4.73).
15. Show $n_o(t)$ is zero mean.
16. Verify (4.76) and (4.77).

17. Show (4.81) is correct if $X$ is a vector of independent, zero-mean, unit variance random variables.
18. Reproduce Example 1.
19. Reproduce Example 2.
20. Reproduce Example 3.
21. Produce a plot like Figure 4.28 for an LFM pulse. It may be necessary to change $k_R$.
22. Produce plots like Figure 4.28 for an unmodulated pulse and a sampling gate discriminator with $q = ¼$ and $q = ¾$. It may be necessary to change $k_R$.
23. Repeat Problem 22 for an unmodulated pulse and a summing gate discriminator with $q = 2$ and $\Delta\tau = \tau_p/4$. It may be necessary to change $k_R$.

## References

[1] Y. Bar-Shalom, X.-R. Li and T. Kirubarajan, *Estimation with Applications to Tracking and Navigation*, New York: John Wiley & Sons, Inc., 2001.

[2] M. Barkat, *Signal Detection and Estimation*, 2nd ed., Norwood, MA: Artech House, 2005.

[3] S. M. Kay, Fundamentals of Statistical Signal Processing, Vol. I: Estimation Theory, Upper Saddle River, NJ: Prentice Hall, 1993.

[4] P. M. Woodward, Probability and Information Theory with Applications to Radar, 2nd ed., New York: Pergamon Press, 1953.

[5] H. L. V. *Trees, Detection, Estimation, and Modulation Theory,* Part III: Radar-Sonar Signal Processing and Gaussian Signals in Noise, New York: John Wiley & Sons, Inc., 2001.

[6] D. K. Barton, *Radar Systems Analysis and Modeling*, Boston, MA: Artech House, 2005.

[7] N. Levanon and E. Mozeson, *Radar Signals*, Hoboken, New Jersey: Wiley-Interscience, 2004.

[8] G. Biernson, *Optimal Radar Tracking Systems*, New York: John Wiley & Sons, Inc., 1990.

[9] H. Meikle, *Modern Radar Systems*, 2nd ed., Norwood: Artech House, 2008.

[10] J. Minkoff, Signal Processing - Fundamentals and Applications for Communications and Sensing Systems, Norwood: MA, 2002.

[11] J. L. Eaves and E. K. Reedy, *Principles of Modern Radar*, New York: Van Nostrand Reinhold, 1987.

[12] S. Kingsley and S. Quegan, *Understanding Radar Systems*, Raleigh, NC: Scitech Publishing, 1999.

[13] R. S. Hughes, *Analog Automatic Control Loops in Radar and EW*, Norwood, MA: Artech House, 1988.

[14] П. И. Дудник (P. I. Dudnik), Г. С. Кондратенков (G. S. Kondratenkov), Б. Г. Татарский (B. G. Tatarskii), А. Р. Ильчук (A. R. Ilchik) and А. А. Герасимов (A. A. Gerasimov), Авиационные радиолокационные комплексы и системы (Aviation Radar Complexes and Systems), П. И. Дудкина (P. I. Dudnika), Ed.,

Москва: ВВИА им. проф. Н.Е. Жуковского (Moscow: VVIA named for professor N. E. Zhukovsky), 2006.

[15] M. C. Budge Jr. and S. R. German, *Basic Radar Analysis*, Norwood, MA: Artech House, 2015.

[16] M. I. Skolnik, Ed., *Radar Handbook*, 3rd ed., New York: McGraw-Hill, 2008.

[17] А. Н. Быстров (A. N. Bystrov), "Влияние дискретизации сигналов на характеристики временных дискриминаторов," ("The effect of signal discretization on the characteristics of time discriminators,") in Материалы Международной научнопрактической конференции «Информационные технологии, электронные приборы и системы» (ITEDS-2010) (Materials of the International Scientific and Practical Conference "Information Technologies, Electronic Devices and Systems" (ITEDS-2010)), Беларусь (Belarus), 2010.

[18] F. Le Chevalier, *Principles of Radar and Sonar Signal Processing*, Norwood, MA: Artech House, 2002.

[19] C. E. Cook and M. Bernfeld, *Radar Signals: An Introduction to Theory and Practice*, Norwood, MA: Artech House, 1993.

[20] D. O. North, "An Analysis of the factors which determine signal/noise discrimination in pulsed-carrier systems," *Proceedings of the IEEE,* vol. 51, no. 7, pp. 1016-1027, July 1963.

[21] H. Urkowitz, *Signal Theory and Random Processes*, Dedham, MA: Artech House, 1983.

[22] [1] A. Papoulis, *Probability, Random Vairables, and Stochastic Processes*, 3rd ed., New York: McGraw-Hill, Inc., 1991.

[23] J. A. Cadzow and H. R. Martens, *Discrete-Time and Computer Control Systems*, F. F. Kuo, Ed., Englewood Cliffs, New Jersey: Prentice hall, 1970.

[24] R. N. McDonald and A. D. Whalen, *Detection of Signals in Noise*, Second ed., San Diego, California: Academic Press, 1995.

[25] S. Haykin, *Adaptive Filter Theory*, 3rd ed., New Jersey: Prentice Hall, 1996.

[26] M. S. Grewal and A. P. Andrews, *Kalman Filtering Theory and Practice Using MATLAB*, 2nd ed., NY: Wiley-Interscience, 2001.

[27] M. I. Skolnik, *Introduction to Radar Systems*, 3rd Edition ed., New York: McGraw-Hill, 2001.

[28] D. K. Barton, *Radar Equations for Modern Radar*, Boston: Artech House, 2013.

[29] T. Benedict and G. Bordner, "Synthesis of an optimal set of radar track-while-scan smoothing equations," *IRE Transactions on Automatic Control,* vol. 7, no. 4, pp. 27-32, 1962.

## APPENDIX 4A: DERIVATION OF $V_E$ WHEN $q < ½$ AND $q\tau_P < |\Delta\tau| < (1-q)\tau_P$

The condition $q\tau_p < |\Delta\tau| < (1-q)\tau_p$ with $q < ½$ can occur when $\tau_E$ and $\tau_L$ are on the same side of the triangle of Figure 4.3. If $\tau_E$ and $\tau_L$ are on the left side of the triangle, $V_E$ and $V_L$ become (see (4.4))

$$V_E = \frac{V_R}{\tau_p}\left(\tau_p - \tau_R + \tau_E\right) = \frac{V_R}{\tau_p}\left(\tau_p - \tau_R + \tau_{trl} - q\tau_p\right) = \frac{V_R}{\tau_p}\left[(1-q)\tau_p - \Delta\tau\right] \quad (4A.1)$$

and

$$V_L = \frac{V_R}{\tau_p}\left(\tau_p - \tau_R + \tau_L\right) = \frac{V_R}{\tau_p}\left(\tau_p - \tau_R + \tau_{trl} + q\tau_p\right) = \frac{V_R}{\tau_p}\left[(1+q)\tau_p - \Delta\tau\right]. \quad (4A.2)$$

With this

$$V_L - V_E = \frac{2V_R}{\tau_p} q\tau_p, \quad (4A.3)$$

$$V_L + V_E = \frac{2V_R}{\tau_p}\left(\tau_p - \Delta\tau\right). \quad (4A.4)$$

and

$$e = \frac{V_L - V_E}{V_L + V_E}\tau_D = \frac{q\tau_p}{\tau_p - \Delta\tau}(1-q)\tau_p. \quad (4A.5)$$

When $\tau_E$ and $\tau_L$ are on the right side of the triangle, $V_E$ and $V_L$ become (see (4.4))

$$\begin{aligned} V_E &= -\frac{V_R}{\tau_p}\left(\tau_E - \tau_R - \tau_p\right) = -\frac{V_R}{\tau_p}\left(\tau_{trl} - q\tau_p - \tau_R - \tau_p\right) \\ &= \frac{V_R}{\tau_p}\left[(1+q)\tau_p + \Delta\tau\right] \end{aligned} \quad (4A.6)$$

and

$$\begin{aligned} V_L &= -\frac{V_R}{\tau_p}\left(\tau_L - \tau_R - \tau_p\right) = -\frac{V_R}{\tau_p}\left(\tau_{trl} + q\tau_p - \tau_R - \tau_p\right) \\ &= \frac{V_R}{\tau_p}\left[(1-q)\tau_p + \Delta\tau\right] \end{aligned} \quad . \quad (4A.7)$$

With this

$$V_L - V_E = -\frac{2V_R}{\tau_p} q\tau_p, \tag{4A.8}$$

$$V_L + V_E = \frac{2V_R}{\tau_p}\left(\tau_p + \Delta\tau\right). \tag{4A.9}$$

and

$$e = \frac{V_L - V_E}{V_L + V_E}\tau_D = \frac{-q\tau_p}{\tau_p + \Delta\tau}(1-q)\tau_p. \tag{4A.10}$$

When $\tau_E$ and $\tau_L$ are on the left side of the triangle, $\tau_R > \tau_{trk}$ and $\Delta\tau > 0$. When $\tau_E$ and $\tau_L$ are on the right side of the triangle, $\tau_R < \tau_{trk}$ and $\Delta\tau < 0$. We can use this to combine (4A.5) and (4A.10) and get

$$e = \frac{\operatorname{sgn}(\Delta\tau) q\tau_p}{\tau_p - |\Delta\tau|}(1-q)\tau_p \tag{4A.11}$$

which is the desired result.

# Chapter 5

# Closed Loop Angle Tracking

## 5.1 INTRODUCTION

A closed loop angle tracker is comprised of several functional blocks as illustrated in Figure 5.1. The "input" to the tracker is the angular position of the target. This is denoted in the figure as ($u_T$, $v_T$) or ($Az_T$, $El_T$) where $u$ and $v$ are *sine space* angles [1], and $Az_T$ and $El_T$ are azimuth and elevation angles to the target. Sine space representation is the most commonly used angle notation in radars that employ a phased array antenna and is derived from the convention used to mathematically represent the directive gain properties of phased array antennas [2, 3]. $u$ and $v$ are also termed *direction cosines* [3].

Azimuth and elevation (i.e., $Az$ and $El$) are angles associated with a spherical coordinate system [4]. The location of any point in a 3-Dimension coordinate system can be represented by the triplet ($R$, $Az$, $El$) where $Az$ and $El$ are the aforementioned angles, and $R$ is the slant distance from the origin of the coordinate system to the point in space (i.e., the target).

A right hand Cartesian coordinate system shown in Figure 5.2 is used to define $Az$ and $El$. $Az$ is measured relative to north ($y$) and is positive in the clockwise direction. $El$ is measured from the $x$-$y$ (east-north) plane and is positive counterclockwise. If the phase center of the antenna [5] is located at the origin of the ($x$, $y$, $z$) coordinate system, $R$ is the slant range to the target. The ($Az$, $El$) notation is commonly associated with reflector antennas. However, it can also be used in phased array antennas. A specific example is a search radar that uses a phased array antenna that rotates, or nods up and down. The coordinate system of Figure 5.2 is commonly termed an east-north-up, or ENU, coordinate system.

Sine-space angles are defined in a coordinate system that is centered on the phase center of the antenna as shown in Figure 5.3. Here, $z$ is defined as being normal to, and directed away from, the antenna face. The other two axes ($x$ and $y$) are orthogonal and located in the plane of the antenna face (assuming a planar array). They are also orthogonal to the aforementioned array normal. Using the coordinate system of Figure 5.3, $u$ and $v$ are defined as

$$u = \frac{x}{z}, \quad v = \frac{y}{z}. \tag{5.1}$$

The error sensor of an angle tracker is comprised of three functional blocks to include the antenna, the receiver, and the monopulse processor as shown in Figure 5.1. The antenna produces signals (voltages or currents) that are proportional to the angular location of the target, ($u_T$, $v_T$) or ($Az_T$, $El_T$), relative to the nominal angular location of the antenna beam(s), ($u_{trk}$, $v_{trk}$) or ($Az_{trk}$, $El_{trk}$). These are denoted as ($\Delta u$, $\Delta v$) or ($\Delta Az$, $\Delta El$) in Figure 5.1.

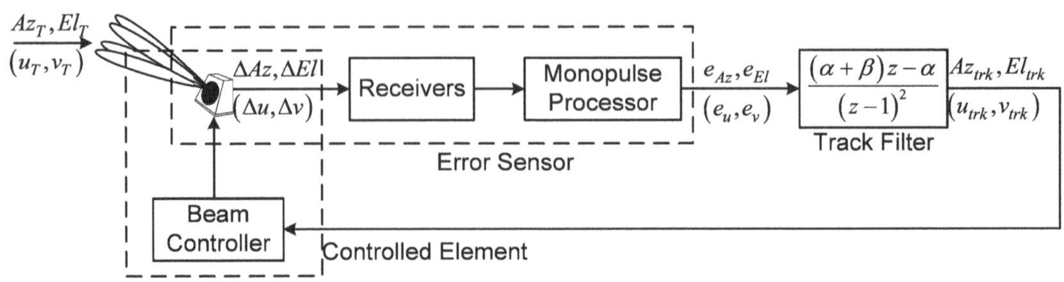


**Figure 5.1** Closed loop angle tracker block diagram.

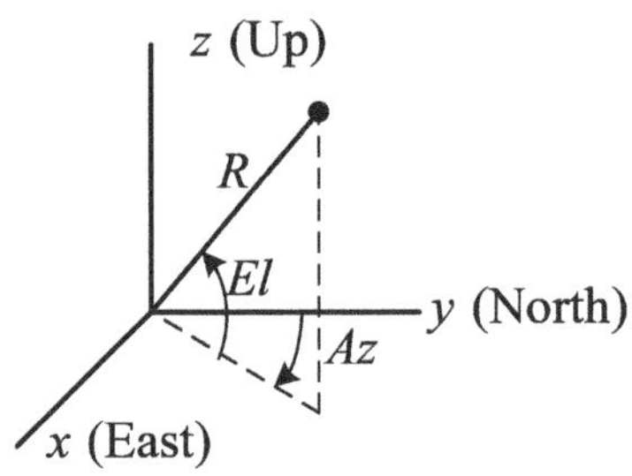


**Figure 5.2** Spherical coordinate system.

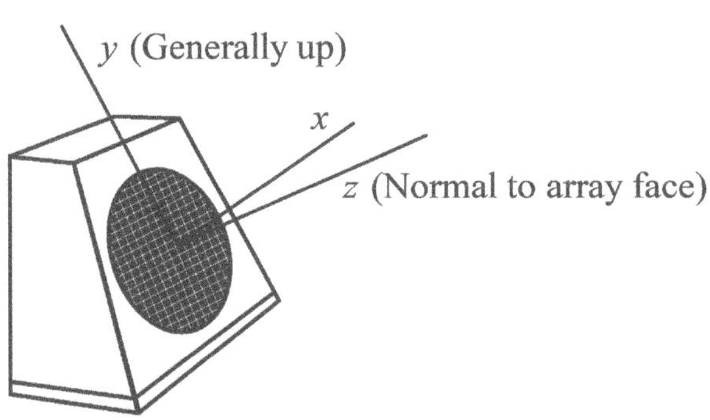


**Figure 5.3** Coordinate system use to define sine space angles.

In this book, we use the phrase "nominal location of the antenna beam(s)" to define closed loop angle tracking where there is one beam on transmit and multiple beams on receive. For example, in some radars, the receive beams point in the same direction as the transmit beam (phase comparison monopulse), and in other cases the receive beams are pointed, or *squinted*, to slightly different angles relative to the transmit beam (amplitude comparison monopulse). The four teardrop shaped lobes radiating from the antenna of Figure 5.1 represent four squinted receive beams, the typical number associated with amplitude comparison monopulse radar.

An integral part of the antenna, in radars that perform closed loop angle tracking, is the *monopulse combiner*, which is also termed a *monopulse comparator*. This is a device that combines signals from the multiple (usually four) receive beams to form the difference signals, ($\Delta u$, $\Delta v$) or ($\Delta Az$, $\Delta El$). It also forms a sum ($\Sigma$) signal by adding the signals from the four beams. This combining occurs at RF prior to the receiver. The sum signal is used, in the monopulse processor, to normalize the difference signals to make the error voltages going to the track filter [($e_{AZ}$, $e_{EL}$) or ($e_u$, $e_v$), see Figure 5.1] independent of signal amplitude associated with the size of the target return.

Radars that use simultaneous multiple beam processing can form error signals like a monopulse feed does or use the information from the multiple beams to form an estimate of the target angle. With the angle measurement approach, the angle errors, ($e_{AZ}$, $e_{EL}$) or ($e_u$, $e_v$), are formed in the track filter, as was the case with direct range measurement discussed in Chapter 4.

The receiver in the closed loop angle tracker of Figure 5.1 converts the RF antenna output signals from the feed to the form needed in the monopulse processor. It heterodynes [6, 7] the radio frequency (RF) signals to a lower operating frequency and provides any necessary amplification. The receivers defined here include the matched filters and signal processors. From an angle tracking perspective, the radar will contain from one to several, identical, receivers operating in parallel. A "true" monopulse radar contains three receivers, two for the difference angle signals (i.e., *Az*, *El* or *u*, *v*) and one for the sum signal. A monopulse radar that uses three receivers is termed a *three channel monopulse radar* or is said to use *three channel monopulse processing*.

There are angle tracking radars that use other than three channels. These include older radars that use conical scanning, and radars that use time multiplexing of the difference signals to reduce the number of required channels to two. These are termed *two channel monopulse* radars. Advanced radars can form many receive beams and use many receiver channels. These radars are said to use *simultaneous multiple beam* techniques. Radars that use conical scan have only one receiver. Strictly speaking, conical scan radars and two channel monopulse radars are not "monopulse" radars because more than one measurement (pulse) is needed to derive angle information. Conical scan and two channel monopulse are discussed further in this chapter.

The *monopulse processor* block (see Figure 5.1) processes the outputs of the receiver(s) to form the error signals, ($e_{AZ}$, $e_{EL}$) or ($e_u$, $e_v$), needed in the track filter. As with range tracking, an important feature of the monopulse processor is that it provides normalization so that ($e_{AZ}$, $e_{EL}$) or ($e_u$, $e_v$), are independent of signal amplitude.

The track filter uses ($e_{AZ}$, $e_{EL}$) or ($e_u$, $e_v$), to generate predicted target angles, ($u_{trk}$, $v_{trk}$) or ($Az_{trk}$, $El_{trk}$). In older single target radars that use analog (continuous time) tracking, it is usually a Type 1 or 2 analog servo. In radars that use digital (discrete time) tracking, the track filter can be of the $\alpha$-$\beta$, $\alpha$-$\beta$-$\gamma$, or Kalman type. If the angle tracker uses $\alpha$-$\beta$ or $\alpha$-$\beta$-$\gamma$ filters, it will contain two filters, one for each angle. If the track filter of the Kalman type is used, it can be implemented as two filters or as a single filter.

The filter shown in Figure 5.1 is an open loop, $\alpha$-$\beta$ filter. It is defined as open loop because it accepts error signals as input. If the processor produces measured target angles instead of errors, the track filters are considered closed loop. These configurations were discussed in Chapter 3.

The *controlled element* in an angle tracker (see Figure 5.1) uses the outputs of the track filters to steer the antenna beam toward the predicted target angles, that is, toward ($u_{trk}$, $v_{trk}$) or ($Az_{trk}$, $El_{trk}$). In a radar that uses a reflector antenna, the controlled element is an electromechanical servo that moves the entire antenna. In a phased array radar, the controlled element consists of a beam steering computer and phase shifters. The beam steering computer uses ($u_{trk}$, $v_{trk}$) to compute the phase shifter settings needed to form a beam in the direction of ($u_{trk}$, $v_{trk}$).

### 5.1.1 Chapter Outline

The remainder of this chapter discusses how the radar antenna forms the RF signals needed to generate the angle information required by the angle tracker. It includes discussions of types of monopulse sensing, monopulse combiners, and the RF hardware needed to form the signals for the receiver and monopulse processor. In the receiver section, we consider two and three channel receivers. In the monopulse processor section, we consider an *exact processor*, a *modified exact processor,* and a *log processor*. The chapter wraps up with several angle tracker examples, plus an example of a combined range and angle tracker.

## 5.2 TYPES OF MONOPULSE SENSING

There are two common types of monopulse sensing: *amplitude comparison* and *phase comparison*. With amplitude comparison monopulse, target angle information is encoded in the amplitude of the signal produced by the antenna. With phase comparison monopulse, target angle information is encoded in the phase of the signal. Amplitude comparison monopulse is the type of sensing used in radars that use reflector antennas or space-fed, planar, phased array antennas. Phase comparison monopulse is the type of sensing used in radars that use planar phased array antennas with corporate or constrained feeds. A radar that uses an active array could use either amplitude or phase comparison monopulse sensing, depending on how the output of the transmit/receive (T/R) modules are combined.

A depiction of a simple, constrained, or corporate, feed planar phased array is shown in Figure 5.4. In this example, the array is divided into four, equal size, subarrays labeled 1, 2, 3, and 4. The phase shifters in the four subarrays are set to steer the receive beams to the same angle; specifically ($u_{trk}$, $v_{trk}$). Since the four subarray beams are steered to the same angle, the magnitude of the signal out of each subarray will be the same. However, because of path length differences from the target to the phase center of each subarray, the relative phases of the signals will be different. Thus, target angle information is encoded into the phases of the four signals.

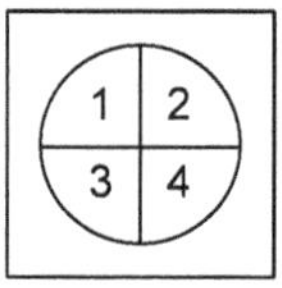


**Figure 5.4** Depiction of a constrained feed phased array antenna using four subarrays.

Figure 5.5 contains a sketch of a simple space fed phased array and an associated four-horn feed. The dashed line from the feed to the array denotes the fact the center line of the feed is aligned with the center of the array. The phase shifters in the array are adjusted to steer the beam (of the entire array) to ($u_{trk}$, $v_{trk}$), when the four-horn feed is replaced by a single-horn feed located on the center line. Said another way, to compute the phase shifts needed to steer the beam to ($u_{trk}$, $v_{trk}$), it is assumed the four-horn feed is replaced by a single-horn feed, whose center is on the center line of Figure 5.5.

Since the four horns of the actual (four-horn) feed are not on the center line, the beams associated with each feed horn will be slightly steered, or squinted, away from ($u_{trk}$, $v_{trk}$). Because of this, the amplitude of the signals from each horn will be different. Thus, target angle information is encoded in the amplitudes of the four signals.[17]

Beam squint in a space fed phased array is related to how the phase shifters of the array are adjusted to correct for the spherical wave front created by the feed [5, 7, 8]. This idea of a spherical wave front is illustrated in Figure 5.6. In that figure, the electric field everywhere on a circular arc (spherical segment in a planar array) centered on the phase center of the feed has a phase of

$$\phi_r = 2\pi r/\lambda \tag{5.2}$$

To form a focused beam normal to the array face (as shown in Figure 5.6), the phase at all radiating elements of the array must be the same. In other words, the spherical phase front from the feed must be corrected to be planar at the rear surface of the array. To accomplish the transition from a spherical wavefront to a planar wavefront, the phase shifters of the array account for the additional phase, $\Delta\phi_{rk}$, caused by the additional propagation length, $\Delta r_k$, shown in Figure 5.6. The process of accounting for the additional phase shifts is termed *spherical correction*. It essentially corrects the spherical wavefront of the feed so that it is a plane wavefront when it reaches the outer array face.

When the feed is moved down, the spherical wavefront (circular wavefront in Figure 5.6) moves with it, as illustrated in Figure 5.7. Since the spherical correction is determined for the feed at the original position, when it is applied to the case where the feed is in the new position, the phase across the array face will not be the same. As will be shown in Section 5.4, it will vary almost linearly across the array face, as long as the feed displacement is small. The linear variation in phase causes the beam to *squint* by an amount

$$\sin\theta_s = \frac{\delta y}{r}. \tag{5.3}$$

A similar phenomenon happens in parabolic reflector antennas. A simple parabolic reflector antenna is formed by rotating a parabola about its axis. A cross section view of such an antenna is illustrated in Figure 5.8. For a parabolic reflector, a feed at the focal point of the parabola causes the beam to point along a line parallel to the axis of the parabola. If the feed is

---

[17] On transmit, the signals in the four parts of the feed are adjusted so that the four-horn feed acts as a single-horn, feed. Thus, the transmit beam is also steered to ($u_{trk}$, $v_{trk}$).

moved below the axis, the beam will squint up. This is illustrated in Figure 5.8. If the feed is moved above the axis, the beam will squint down.

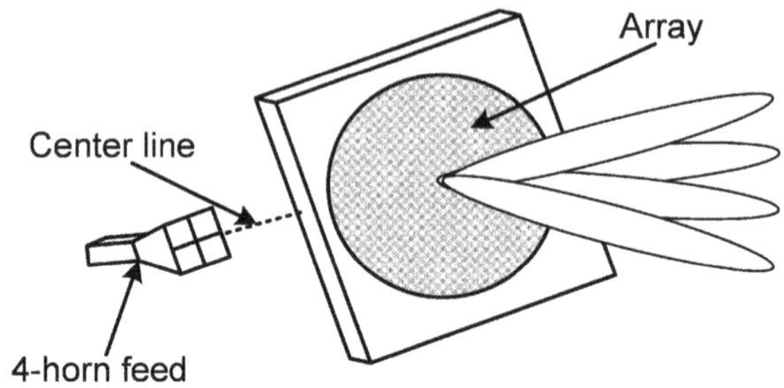


**Figure 5.5** Space-fed phased array.

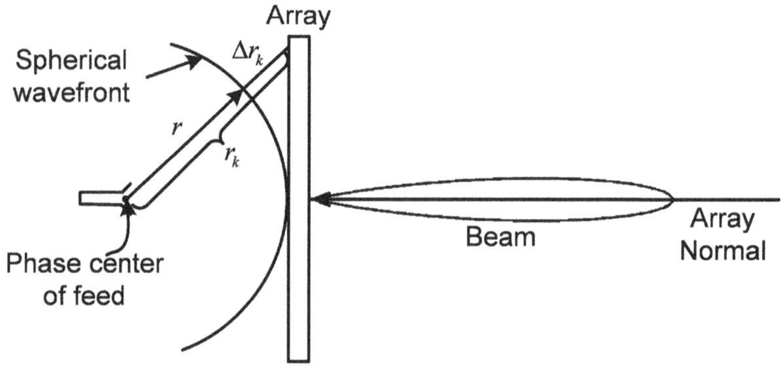


**Figure 5.6** Illustration of spherical phase from correction on space-fed phased arrays.

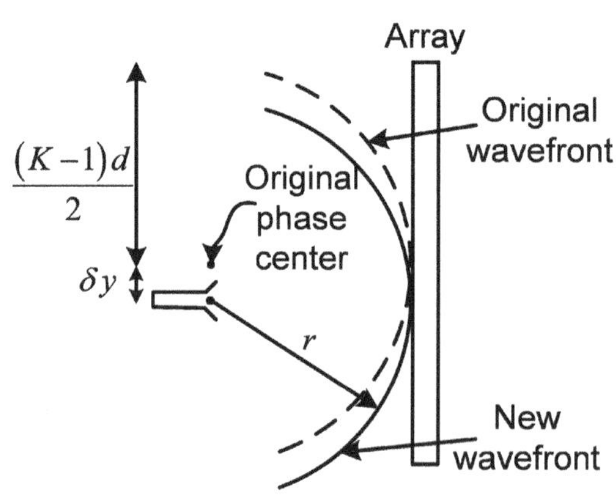


**Figure 5.7** Geometry for shifted feed.

The direction of the beam is determined by the phase of the E-field across a plane termed the *aperture plane*. The aperture plane is a plane that is perpendicular to the parabola axis. It is sometimes described as the plane that passes through the focus of the parabola. It can also be taken to be such that it touches the edge of the parabola, as illustrated in Figure 5.8. The phase at various points in the aperture plane is given by

$$\phi = \frac{2\pi}{\lambda}(d_1 + d_2) \tag{5.4}$$

where $d_1$ is the distance from the phase center of the feed to the reflector, and $d_2$ is the distance from the reflector to the aperture plane, measured along the reflected ray (see Figure 5.8). For the case where the feed is located at the focus of the parabola, the reflected rays are parallel to the axis of the parabola, and $d_1 + d_2$ is constant for all points in the aperture plane. As a result, the phases at all points in the aperture plane are the same, and the beam points along the parabola axis, or 0°.

When the feed is moved away from the focus, $d_1 + d_2$ will not be the same at all points in the aperture plane. As a result, there is a phase gradient across the aperture plane. The phase gradient for the one dimension case (i.e., Figure 5.8), is

$$\phi(y) = K_\phi \left(\frac{\delta y}{f}\right) y, \tag{5.5}$$

where $K_\phi$ is a constant determined by the parameters of the feed and reflector. Equation (5.5) indicates that, to a first order approximation, the phase varies linearly with $y$ and is a function of the feed offset, $\delta y$. The amount of beam squint, $\theta_s$, is given by

$$\sin\theta_s = \frac{\delta y}{f}, \tag{5.6}$$

which is similar to (5.3). In the above equations, $f$ is the focal length of the parabola. This is the distance from the vertex to the focus of the parabola. It is also the distance $r$ of Figures 5.6 and 5.7.

To gain further insight into how different types of antennas encode angle information in the antenna signal, we consider two linear array configurations. The first is representative of a constrained, or corporate, feed arrays, and the second is representative of space-fed phased arrays and parabolic reflector antennas. For the constrained feed example, we develop math that shows the angle information is contained in the relative phases of signals out of the subarrays of the array. For the space-fed array, the math demonstrates that the angle information is contained in the relative amplitudes of the signals out of the four offset feed horns. These developments have the added benefit of providing antenna models for the monopulse processor and simulation examples presented later.

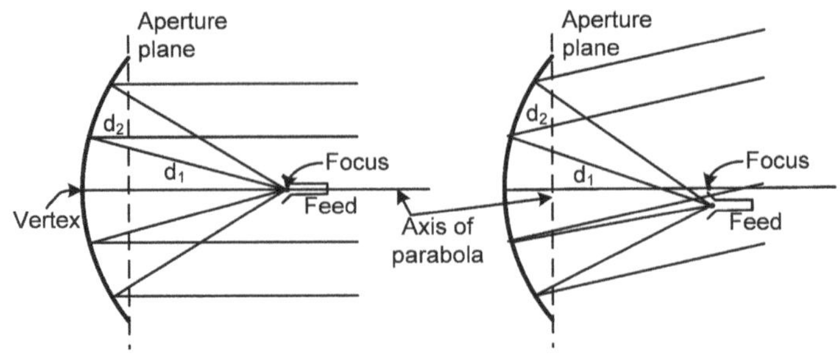


**Figure 5.8** Illustration of beam squint for a parabolic reflector antenna.

## 5.3 PHASE COMPARISON MONOPULSE

For the constrained feed case, we consider the linear array of Figure 5.9. This is a standard depiction of a linear array used in text books that discuss phased arrays [1, 3, 7, 9, 10]. The array is divided into two subarrays that contain $K$ elements each. The column of blocks represents the array. We term the $a_k$'s and $b_k$'s in the block the array *weights*. They are complex numbers that are used to create and steer the antenna beam, and to control sidelobe levels of the antenna radiation patterns. The phases of the $a_k$ and $b_k$ are used to form and steer the beam, and the amplitudes are used to control the radiation pattern sidelobes.

The symbols $\prec$ on each block represent the array elements, which are equally spaced, with a separation of $d$. The lines and summers to the left of the column of blocks represent the combining network of the array. It is usually a complicated waveguide network. However, in small arrays, it could be implemented with microstrip technology [10, 11]. The amplitude weighting, that is, the magnitude portion of $a_k$ and $b_k$, are usually applied in the combining network.

In this analysis, we assume the physical path length (and hence phase) from the blocks to the summers (i.e., the lines in Figure 5.9) are the same. Because of this, the phase shifts imposed on the signals as they propagate from the blocks to the summers will be the same. In practice, the path lengths will be different, which means the $K$ signals of each subarray will experience different phase shifts. The different phase shifts are compensated for in the phase shifters. Our assumption of equal path lengths means we can ignore this compensation.

Assume the target is located at infinity and an angle of $\theta_T$ relative to array normal. Since the array of Figure 5.9 is vertical, $\theta_T$ would represent the target elevation angle, $El_T$, or $v_T$ in sine space. If we would have drawn Figure 5.9 as a horizontal array, $\theta_T$ would represent $Az_T$, or $u_T$ in sine space. Henceforth, we will use $\theta$ to represent both $El$ and $Az$ and $s$ to represent $u$ and $v$. Where necessary, we will revert to the plane-specific notation.

We assume the target is a point. As such, it reradiates a spherical E-field, which for long range targets, approximates a planar E-field, a plane wave, when it reaches the array [7, 12, 13]. The significance of this, in terms of this development, is that the E-field everywhere on the line labeled *plane wave* can be represented by

$$E(t) = E_o e^{j2\pi f_c t} \tag{5.7}$$

where $E_o$ is the E-field magnitude, and $f_c$ is the carrier frequency.

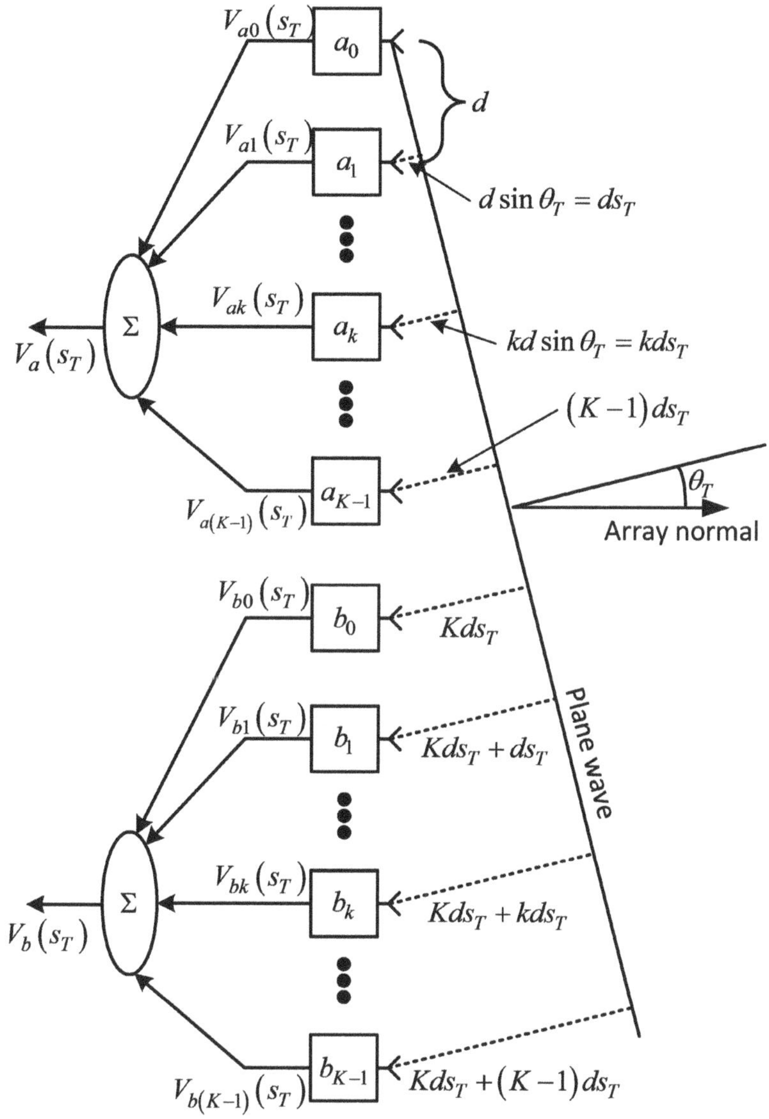


**Figure 5.9** Constrained feed, linear array.

The times required for the E-field to propagate from the plane wave line to the $k^{\text{th}}$ elements of the upper and lower arrays are

$$\tau_{ak} = k\frac{d}{c}\sin\theta_T = k\frac{d}{c}s_T \tag{5.8}$$

and

$$\tau_{bk} = (k+K)\frac{d}{c}\sin\theta_T = k\frac{d}{c}s_T + K\frac{d}{c}s_T. \tag{5.9}$$

Because of this, the E-field at these elements will be

$$E_{ak}(t)=E_o e^{j2\pi f_c(t-\tau_{ak})}=E_o e^{j2\pi f_c t}e^{-j\phi_{aTk}} \tag{5.10}$$

and

$$E_{bk}(t)=E_o e^{j2\pi f_c(t-\tau_{bk})}=E_o e^{j2\pi f_c t}e^{-j\phi_{bTk}} \tag{5.11}$$

where

$$\phi_{aTk}=2\pi f_c\tau_{ak}=2\pi f_c k\frac{d}{c}s_T=2\pi k\frac{d}{\lambda}s_T \tag{5.12}$$

and

$$\phi_{bTk}=2\pi f_c\tau_{bk}=2\pi f_c(k+K)\frac{d}{c}s_T=2\pi K\frac{d}{\lambda}s_T+2\pi k\frac{d}{\lambda}s_T, \tag{5.13}$$

and $\lambda=c/f_c$ is the wavelength associated with $f_c$. $c$ is the speed of light.

The array elements convert the E-field to voltages, which are then multiplied by $a_k$ and $b_k$. Thus, the voltages out of the $k^{th}$ blocks of the upper and lower subarrays are

$$\begin{aligned}V_{ak}(t)&=a_k V_o e^{j2\pi f_c t}e^{-j2\pi k\frac{d}{\lambda}s_T}\\&=V_o e^{j2\pi f_c t}a_k V_{ak}(s_T)\end{aligned} \tag{5.14}$$

and

$$\begin{aligned}V_{bk}(t)&=b_k V_o e^{j2\pi f_c t}e^{-j2\pi K\frac{d}{\lambda}s_T}e^{-j2\pi k\frac{d}{\lambda}s_T}\\&=V_o e^{j2\pi f_c t}b_k V_{bk}(s_T)\end{aligned}\,.[18] \tag{5.15}$$

The outputs of the two summers are

[18] The carrier term, $V_o\exp(j2\pi f_c t)$, is common to all of the terms in (5.14) and (5.15). Because of this, we omitted it from Figure 5.9. This essentially means we are using baseband, complex signal notation.

$$\begin{aligned}V_a(t) &= \sum_{k=0}^{K-1} V_{ak}(t) \\ &= V_o e^{j2\pi f_c t} \sum_{k=0}^{K-1} a_k e^{-j2\pi k \frac{d}{\lambda} s_T} \\ &= V_o e^{j2\pi f_c t} V_a(s_T)\end{aligned} \tag{5.16}$$

and

$$\begin{aligned}V_b(t) = \sum_{k=0}^{K-1} V_{bk}(t) = V_o e^{j2\pi f_c t} \sum_{k=0}^{K-1} b_k e^{-j2\pi K \frac{d}{\lambda} s_T} e^{-j2\pi k \frac{d}{\lambda} s_T} \\ = V_o e^{j2\pi f_c t} V_b(s_T)\end{aligned} . \tag{5.17}$$

The receiver will remove the carrier frequency to produce a baseband, complex signal at the receiver location of interest to us; specifically, at the monopulse processor. This means we can work with $V_a(s_T)$ and $V_b(s_T)$ rather than $V_a(t)$ and $V_b(t)$. With this, we have, from (5.14) through (5.17),

$$V_a(s_T) = \sum_{k=0}^{K-1} a_k e^{-j2\pi k \frac{d}{\lambda} s_T} \tag{5.18}$$

and

$$V_b(s_T) = \sum_{k=0}^{K-1} b_k e^{-j2\pi K \frac{d}{\lambda} s_T} e^{-j2\pi k \frac{d}{\lambda} s_T} = e^{-j2\pi K \frac{d}{\lambda} s_T} \sum_{k=0}^{K-1} b_k e^{-j2\pi k \frac{d}{\lambda} s_T} . \tag{5.19}$$

We now assume

$$a_k = e^{j\phi_{atrk}} = e^{j2\pi k \frac{d}{\lambda} s_{trk}} \tag{5.20}$$

and

$$b_k = e^{j\phi_{btrk}} = e^{j2\pi K \frac{d}{\lambda} s_{trk}} e^{j2\pi k \frac{d}{\lambda} s_{trk}} . \tag{5.21}$$

That is, we assume uniform amplitude weighting and a phase, $\phi_k$, that is a linear function of $k$. We chose this particular form of $\phi_k$ because we know the end result we want. Specifically, we want to be able to cancel the $k$-dependent terms of $\phi_{aTk}$ and $\phi_{bTk}$.

Substituting $a_k$ and $b_k$ into (5.18) and (5.19) yields

$$V_a\left(s_T - s_{trk}\right) = \sum_{k=0}^{K-1} e^{-j2\pi k\frac{d}{\lambda}\left(s_T - s_{trk}\right)} \tag{5.22}$$

and

$$V_b\left(s_T - s_{trk}\right) = e^{-j2\pi K\frac{d}{\lambda}\left(s_T - s_{trk}\right)} \sum_{k=0}^{K-1} e^{-j2\pi k\frac{d}{\lambda}\left(s_T - s_{trk}\right)} . \tag{5.23}$$

We changed the argument of $V_a$ and $V_b$ to emphasize the dependency of the signals on the angle error, $s_T - s_{trk}$.

In (5.22) and (5.23), we note that

$$V_b\left(s_T - s_{trk}\right) = e^{-j2\pi K\frac{d}{\lambda}\left(s_T - s_{trk}\right)} V_a\left(s_T - s_{trk}\right). \tag{5.24}$$

If we form the magnitude of both sides of (5.24), we note that

$$\left|V_b\left(s_T - s_{trk}\right)\right| = \left|V_a\left(s_T - s_{trk}\right)\right|, \tag{5.25}$$

which tells us we cannot obtain angle information from the magnitudes of the two voltages since they are the same. This means the angle information is not encoded in the amplitudes of the subarray outputs. It is encoded in the phases. We need a way to extract the angle information in a form that will be useful in an angle tracker.

From previous tracking studies (see Chapter 4), we would like to combine the subarray outputs, $V_a(s_T - s_{trk})$ and $V_b(s_T - s_{trk})$, such that the combination is proportional $s_T - s_{trk}$. A means of doing this is to use

$$V_\Delta\left(s_T - s_{trk}\right) = V_a\left(s_T - s_{trk}\right) - V_b\left(s_T - s_{trk}\right). \tag{5.26}$$

With (5.24), (5.26) becomes

$$\begin{aligned} V_\Delta\left(s_T - s_{trk}\right) &= V_a\left(s_T - s_{trk}\right) - e^{-j2\pi K\frac{d}{\lambda}\left(s_T - s_{trk}\right)} V_a\left(s_T - s_{trk}\right) \\ &= \left(1 - e^{-j2\pi K\frac{d}{\lambda}\left(s_T - s_{trk}\right)}\right) V_a\left(s_T - s_{trk}\right) \\ &= 2je^{-j\pi K\frac{d}{\lambda}\left(s_T - s_{trk}\right)} \sin\left(\pi K\frac{d}{\lambda}\left(s_T - s_{trk}\right)\right) V_a\left(s_T - s_{trk}\right) \end{aligned} \tag{5.27}$$

We term $V_\Delta(s_T - s_{trk})$ the difference voltage.

The appearance of the sine in (5.27) is encouraging because, for small $s_T - s_{trk}$, that is, track errors near zero, the sine will reduce to $\pi Kd(s_T - s_{trk})/\lambda$. That is, it will be proportional to $s_T - s_{trk}$. However, we need to examine the form of $V_a(s_T - s_{trk})$ to see if it affects this relation.

If we apply the geometric progression relation [14]

$$\sum_{k=0}^{K-1} x^k = \frac{1-x^K}{1-x} \tag{5.28}$$

to (5.22) we get (see Problem 12)

$$V_a\left(s_T - s_{trk}\right) = e^{-j\pi(K-1)\frac{d}{\lambda}(s_T - s_{trk})} \frac{\sin\left(\pi K \frac{d}{\lambda}\left(s_T - s_{trk}\right)\right)}{\sin\left(\pi \frac{d}{\lambda}\left(s_T - s_{trk}\right)\right)}. \tag{5.29}$$

With this (5.27) becomes

$$\begin{aligned} V_\Delta\left(s_T - s_{trk}\right) &= 2je^{-j\pi(2K-1)\frac{d}{\lambda}(s_T - s_{trk})} \\ &\times \sin\left(\pi K \frac{d}{\lambda}\left(s_T - s_{trk}\right)\right) \frac{\sin\left(\pi K \frac{d}{\lambda}\left(s_T - s_{trk}\right)\right)}{\sin\left(\pi \frac{d}{\lambda}\left(s_T - s_{trk}\right)\right)}. \end{aligned} \tag{5.30}$$

For small $s_T - s_{trk}$, the (5.30) reduces to (see Problem 1)

$$V_\Delta\left(s_T - s_{trk}\right) \approx 2je^{-j\pi(2K-1)\frac{d}{\lambda}(s_T - s_{trk})}\left[2\pi K^2 \frac{d}{\lambda}\left(s_T - s_{trk}\right)\right], \tag{5.31}$$

which, except for the $j$ and exponential terms, is what we want.

If we add the outputs of the two summers we get (see Problem 2)

$$\begin{aligned} V_\Sigma\left(s_T - s_{trk}\right) &= V_a\left(s_T - s_{trk}\right) + e^{-j2\pi K\frac{d}{\lambda}(s_T - s_{trk})} V_a\left(s_T - s_{trk}\right) \\ &= \left(1 + e^{-j2\pi K\frac{d}{\lambda}(s_T - s_{trk})}\right) V_a\left(s_T - s_{trk}\right) \\ &= 2e^{-j\pi K\frac{d}{\lambda}(s_T - s_{trk})} \cos\left(\pi K \frac{d}{\lambda}\left(s_T - s_{trk}\right)\right) V_a\left(s_T - s_{trk}\right) \end{aligned} \tag{5.32}$$

or with (5.29)

$$V_\Sigma(s_T - s_{trk}) = 2e^{-j\pi(2K-1)\frac{d}{\lambda}(s_T - s_{trk})} \times \cos\left(\pi K\frac{d}{\lambda}(s_T - s_{trk})\right)\frac{\sin\left(\pi K\frac{d}{\lambda}(s_T - s_{trk})\right)}{\sin\left(\pi\frac{d}{\lambda}(s_T - s_{trk})\right)} = e^{-j\pi(2K-1)\frac{d}{\lambda}(s_T - s_{trk})}\frac{\sin\left(2\pi K\frac{d}{\lambda}(s_T - s_{trk})\right)}{\sin\left(\pi\frac{d}{\lambda}(s_T - s_{trk})\right)} \tag{5.33}$$

which reduces to

$$V_\Sigma(s_T - s_{trk}) \approx 2e^{-j\pi(2K-1)\frac{d}{\lambda}(s_T - s_{trk})}K \tag{5.34}$$

for small $s_T - s_{trk}$. We term $V_\Sigma(s_T - s_{trk})$ the sum voltage.

Figure 5.10 contains plots of the center portions of $V_\Delta(s_T - s_{trk})$ and $V_\Sigma(s_T - s_{trk})$, without the exponential and $j$ terms. To generate the plot, we used $2K = 50$ and $d = \lambda/2$.

It is worth noting that (5.18) and (5.19) are in the form of a discrete Fourier transform of the $a_k$ and $b_k$. Thus, we could have used discrete Fourier transform theory to compute $V_\Delta(s_T - s_{trk})$ and $V_\Sigma(s_T - s_{trk})$.

In summary, we learned we can extract angle information from a linear, corporate fed phased array by subtracting the complex signals from the two subarrays. If we extend this to the planar array depicted in Figure 5.4, we would subtract the outputs of the left two subarrays (subarrays 1 and 3) from the outputs of the right two subarrays (subarrays 2 and 4) to extract azimuth, or $u$, angle error information. To extract elevation, or $v$, angle error information, we would subtract the output of the lower two subarrays (subarrays 3 and 4) from the upper two subarrays (subarrays 1 and 2).

We still need to deal with the $j$ and exponential terms of (5.31). We will consider these two terms when we discuss monopulse combiners and monopulse processors. For now, we turn our attention to space-fed phased arrays and amplitude comparison monopulse.

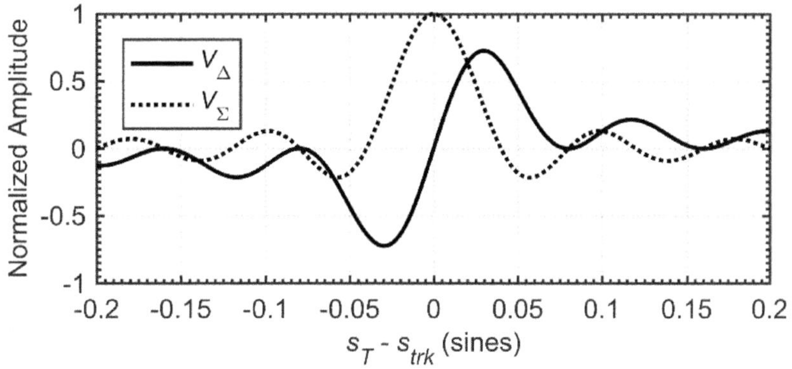


**Figure 5.10** Sum and difference voltage for a constrained feed array.

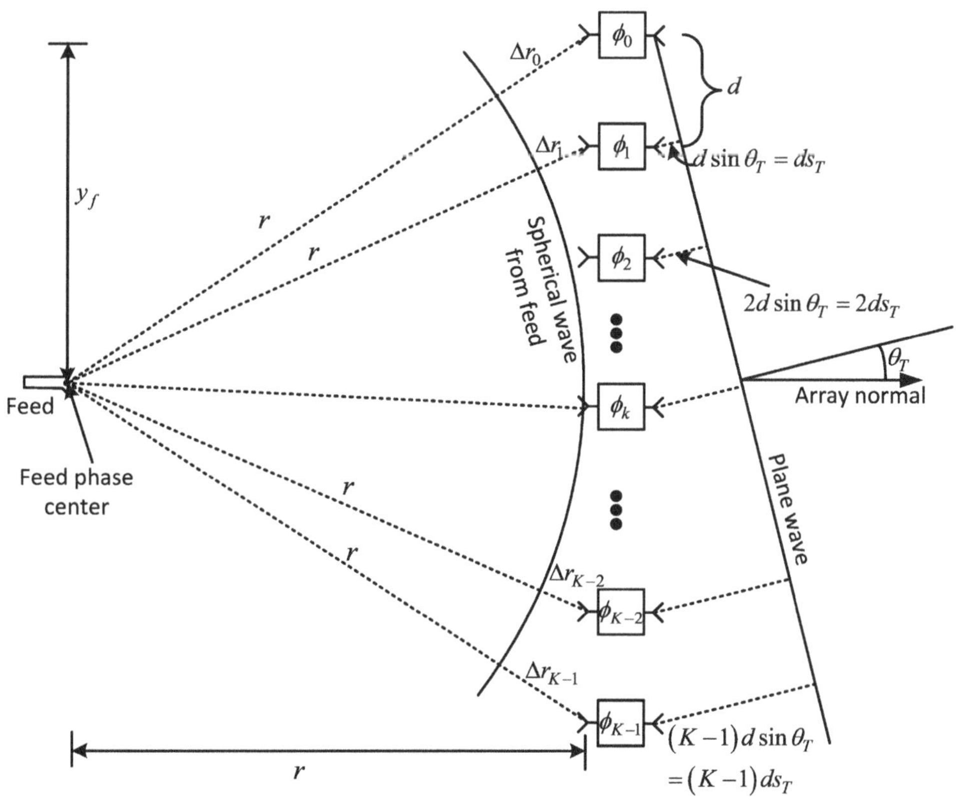


**Figure 5.11** Space-fed, linear array.

## 5.4 AMPLITUDE COMPARISON MONOPULSE

For the space-fed case, we consider the linear array of Figure 5.11. The column of blocks represents the array. The blocks represent the phase shifters, and the $\phi_k$'s represent the values of the phase shifts. The sideways >– and –< on each side of the blocks represent the array elements, which are equally spaced with a separation of $d$. The feed phase center is located a distance of $r$ behind the array, and, vertically, in the center of the array.

The target is located at infinity and an angle of $\theta_T$, or $s_T = \sin\theta_T$ in sine space, relative to array normal. The E-field at every point on the line labeled plane wave is

$$E(t) = E_o e^{j2\pi f_c t}. \tag{5.35}$$

As discussed in Section 5.3, the E-field at the $k^{\text{th}}$ element of the array is

$$E_k(t) = E_o e^{j2\pi f_c(t-\tau_k)} = E_o e^{j2\pi f_c t} e^{-j\phi_{Tk}} \tag{5.36}$$

where

$$\phi_{Tk} = 2\pi f_c \tau_k = 2\pi f_c k \frac{d}{c} s_T = 2\pi k \frac{d}{\lambda} s_T. \tag{5.37}$$

The $k^{\text{th}}$ phase shifter imposes a phase shift of $\phi_k$ so that the E-field radiated by the $k^{\text{th}}$ element on the back of the array (the feed side) is

$$E_{Bk}(t) = E_k(t) e^{j\phi_k} = E_o e^{j2\pi f_c t} e^{j(\phi_k - \phi_{Tk})}. \tag{5.38}$$

As $E_{Bk}(t)$ travels from the $k^{\text{th}}$ element to the feed, it undergoes a phase shift of

$$\phi_{rk} = 2\pi \frac{r + \Delta r_k}{\lambda} = 2\pi \frac{r}{\lambda} + 2\pi \frac{\Delta r_k}{\lambda}. \tag{5.39}$$

Because of the radiation pattern of the feed horn, it experiences an amplitude gain (amplitude taper) of $a_k$. Thus, the E-field at the (phase center) of the feed, from the $k^{\text{th}}$ element, is

$$E_{fk}(t) = E_o e^{j2\pi f_c t} a_k e^{j(\phi_k - \phi_{Tk} - \phi_{rk})}. \tag{5.40}$$

The E-fields from the elements combine at the feed, and the feed converts the combined E-field to a voltage. Thus, the voltage at the feed output is

$$V(t) = \sum_{k=0}^{K-1} V_{fk}(t) = V_o e^{j2\pi f_c t} \sum_{k=0}^{K-1} a_k e^{j(\phi_k - \phi_{Tk} - \phi_{rk})} = V_o e^{j2\pi f_c t} V(s_T). \tag{5.41}$$

The receiver will remove the carrier frequency to produce a baseband, complex signal at the receiver location of interest to us; specifically, at the monopulse processor. This means we can work with $V(s_T)$ rather than $V(t)$.

Thus far, we have

$$V(s_T) = \sum_{k=0}^{K-1} a_k e^{j(\phi_k - \phi_{Tk} - \phi_{rk})} . \tag{5.42}$$

We choose $\phi_k$ as (because we know the final answer we want)

$$\phi_k = 2\pi k \frac{d}{\lambda} s_{trk} + \phi_{rk} \tag{5.43}$$

where $\phi_{rk}$ is given by (5.39). The phase, $\phi_{rk}$ is included in $\phi_k$ to cancel the phase shift due to propagation of the E-field from the back of the array to the feed (i.e., the $-\phi_{rk}$ in (5.42)). Its inclusion is the spherical correction discussed in Section 5.2. The term

$$2\pi k \frac{d}{\lambda} s_{trk} = \phi_{trk} \tag{5.44}$$

is used to steer the beam to $s_{trk} = \sin\theta_{trk}$, the predicted angle from the tracker. With (5.43), (5.42) becomes (see Problem 3)

$$V(s_T) = \sum_{k=0}^{K-1} a_k e^{-j2\pi k \frac{d}{\lambda}(s_T - s_{trk})} , \tag{5.45}$$

which is the desired result.

We now consider what happens when we move the feed vertically by an amount of $\delta y$. From Figure 5.11, we can write

$$(r + \Delta r_k)^2 = r^2 + (y_f - kd)^2 \tag{5.46}$$

for the original feed position. When we move the feed to $y_f + \delta y$, the right side of (5.46) will change to $r^2 + (y_f + \delta y - kd)^2$. To maintain equality, the left side will also need to change. This leads to

$$(r + \Delta r_k + \delta r_k)^2 = r^2 + (y_f + \delta y - kd)^2 . \tag{5.47}$$

Manipulating (5.46) and (5.47) leads to

$$r^2 + 2r\Delta r_k + (\Delta r_k)^2 = r^2 + (y_f - kd)^2 \tag{5.48}$$

and

$$\begin{aligned} &r^2 + 2r(\Delta r_k + \delta r_k) + (\Delta r_k + \delta r_k)^2 \\ &= r^2 + (y_f - kd)^2 + 2(y_f - kd)\delta y + (\delta y)^2 \end{aligned} . \tag{5.49}$$

Subtracting (5.48) from (5.49) results in, after some manipulation (see Problem 4),

$$2r\delta r_k + \delta r_k (2\Delta r_k + \delta r_k) = 2(y_f - kd)\delta y + (\delta y)^2 . \tag{5.50}$$

We assume $\delta r_k(2\Delta r_k + \delta r_k)$ and $(\delta y)^2$ are small relative to the rest of the terms (see Problem 5). With this we get

$$\delta r_k = \frac{y_f - kd}{r}\delta y = y_f \frac{\delta y}{r} - kd\frac{\delta y}{r} . \tag{5.51}$$

Since the vertical position of the original feed is the center of the array,

$$y_f = \frac{(K-1)d}{2} \tag{5.52}$$

and (5.51) becomes

$$\delta r_k = \left(\frac{K-1}{2}\right) d\frac{\delta y}{r} - kd\frac{\delta y}{r} . \tag{5.53}$$

When the feed is moved, the path length from the $k^{th}$ element to the feed is $r + \Delta r_k + \delta r_k$, and the associated phase is

$$\begin{aligned} \phi'_{rk} &= 2\pi \frac{(r + \Delta r_k + \delta r_k)}{\lambda} \\ &= 2\pi \frac{(r + \Delta r_k)}{\lambda} + 2\pi \frac{\delta r_k}{\lambda} . \\ &= \phi_{rk} + \delta\phi_{rk} \end{aligned} \tag{5.54}$$

In other words, vertically moving the feed causes a phase shift differential of,

$$\delta\phi_{rk} = 2\pi \frac{\delta r_k}{\lambda} = 2\pi \left(\frac{K-1}{2}\right) \frac{d}{\lambda}\frac{\delta y}{r} - 2\pi k \frac{d}{\lambda}\frac{\delta y}{r} \tag{5.55}$$

where we made use of (5.53). If we use $\phi'_{rk}$ in place of $\phi_{rk}$ in (5.42), we get

$$V'(s_T)=\sum_{k=0}^{K-1}a_k e^{j(\phi_k-\phi_{Tk}-\phi'_{rk})}=\sum_{k=0}^{K-1}a_k e^{j(\phi_k-\phi_{Tk}-\phi_{rk}-\delta\varphi_{rk})}\ . \tag{5.56}$$

Recall that we chose $\phi_k$ as (see (5.43))

$$\phi_k=2\pi k\frac{d}{\lambda}s_{trk}+\phi_{rk} \tag{5.57}$$

to implement the spherical correction. Substituting this in (5.56) and manipulating gives (see Problem 6)

$$V'(s_T)=\sum_{k=0}^{K-1}a_k e^{-j2\pi k\frac{d}{\lambda}(s_T-s_{trk})}e^{-j\delta\phi_{rk}} \tag{5.58}$$

or, with $\delta\phi_{rk}$ from (5.55)

$$\begin{aligned}V'(s_T)&=\sum_{k=0}^{K-1}a_k e^{-j2\pi k\frac{d}{\lambda}(s_T-s_{trk})}e^{-j\left(2\pi\left(\frac{K-1}{2}\right)\frac{d}{\lambda}\frac{\delta y}{r}-2\pi k\frac{d}{\lambda}\frac{\delta y}{r}\right)}\\&=e^{-j\pi(K-1)\frac{d}{\lambda}\frac{\delta y}{r}}\sum_{k=0}^{K-1}a_k e^{-j2\pi k\frac{d}{\lambda}\left(s_T-s_{trk}-\frac{\delta y}{r}\right)}\end{aligned}\ . \tag{5.59}$$

Since $\delta y/r$ is of the same form as $s_{trk}$ and $s_T$, we define it as

$$s_s=\frac{\delta y}{r} \tag{5.60}$$

and term it the *squint angle*. With this we get

$$V'(s_T)=e^{-j\pi(K-1)\frac{d}{\lambda}s_s}\sum_{k=0}^{K-1}a_k e^{-j2\pi k\frac{d}{\lambda}(s_T-s_{trk}-s_s)}\ . \tag{5.61}$$

We note that, when $\delta y=0$, $s_s=0$ and $|V'(s_T)|$ will be maximum when $s_T=s_{trk}$. When $\delta y\neq 0$, the maximum will occur at $s_T=s_{trk}-s_s$, or at some squint angle, $s_s$.

At this point, we make the unrealistic assumption that $a_k=1$ for all $k$. This assumption is unrealistic because it assumes the feed can provide uniform illumination. This will be the case, only if the feed is a point source radiation. However, the assumption makes the subsequent math simpler.

If we make use of (5.28), we can write (5.61) (with $a_k = 1$) as

$$V'(s_T) = e^{-j\pi(K-1)\frac{d}{\lambda}(s_T - s_{trk})} \frac{\sin\left(\pi K \frac{d}{\lambda}(s_T - s_{trk} - s_s)\right)}{\sin\left(\pi \frac{d}{\lambda}(s_T - s_{trk} - s_s)\right)} = V(s_T - s_{trk}) \tag{5.62}$$

where we changed the notation on $V$ to be consistent with the notation we used for the constrained feed antenna.

If we use two feeds located at $\pm|\delta y|$ we would get

$$V_a(s_T - s_{trk}) = e^{-j\pi(K-1)\frac{d}{\lambda}(s_T - s_{trk})} \frac{\sin\left(\pi K \frac{d}{\lambda}(s_T - s_{trk} - s_s)\right)}{\sin\left(\pi \frac{d}{\lambda}(s_T - s_{trk} - s_s)\right)} \tag{5.63}$$

and

$$V_b(s_T - s_{trk}) = e^{-j\pi(K-1)\frac{d}{\lambda}(s_T - s_{trk})} \frac{\sin\left(\pi K \frac{d}{\lambda}(s_T - s_{trk} + s_s)\right)}{\sin\left(\pi \frac{d}{\lambda}(s_T - s_{trk} + s_s)\right)}. \tag{5.64}$$

From these two equations, we note that the phases of $V_a$ and $V_b$ are the same, but the amplitudes are different. Thus, we say the angle information is encoded in the amplitude of the antenna output for a space fed phased array. This is in contrast to (5.24) where the amplitudes were the same, and the phases were different.

Figure 5.12 contains normalized plots of $|V(s_{trk} - s_T - s_s)|$ and $|V(s_{trk} - s_T + s_s)|$ versus $s_{trk} - s_T$ for a $K = 50$ element, space fed, phased array with an element spacing of $d = \lambda/2$. We arbitrarily assume the distance from the back array to the phase center of the feeds is $r = 1.5(K - 1)d = 1.5(K - 1)\lambda/2$. (This gives an $f/D$ ratio of 1.5, which is close to the value of 1.6 Sherman indicates for the AN/FPQ-6 radar [5, p. 107].) The two feeds were located $\pm\delta y = \pm\lambda/2$ relative to the vertical center of the array (i.e., at $y_f \pm \lambda/2$). This results in a squint angle of

$$|s_s| = \frac{|\delta y|}{r} = \frac{\lambda/2}{1.5(K-1)\lambda/2} = \frac{1}{1.5(K-1)} = 0.0136 \text{ sines} \tag{5.65}$$

or

$$|\theta_s| = \sin^{-1}(|s_s|) = 0.78^\circ . \tag{5.66}$$

The beam width of this antenna is about 0.035 sines, or 2°, so that the squint angle is about 0.4 beam widths, which is close to the optimum value of 0.453 derived by Sherman [5, p. 112].

Figure 5.13 contains normalized plots of the center portion of

$$V_{\Delta}\left(s_T - s_{trk}\right) = V_a\left(s_T - s_{trk}\right) - V_b\left(s_T - s_{trk}\right) \tag{5.67}$$

and

$$V_{\Sigma}\left(s_T - s_{trk}\right) = V_a\left(s_T - s_{trk}\right) + V_b\left(s_T - s_{trk}\right) \tag{5.68}$$

without the exponential terms. The plots look similar to those for the constrained feed array (see Figure 5.10) but are not exactly the same. The sum pattern for the space fed array is slightly wider, and the sidelobes are slightly lower. The peak of the difference pattern is not as large for the space fed array as for the constrained feed array. Also, the difference pattern sidelobes of the space fed array cross the zero line, while those of the constrained feed array do not.

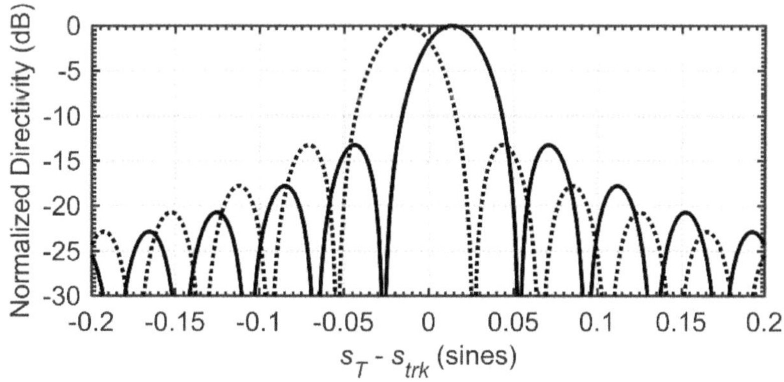


**Figure 5.12** Example of squinted beams for a 50-element, space-fed phased array.

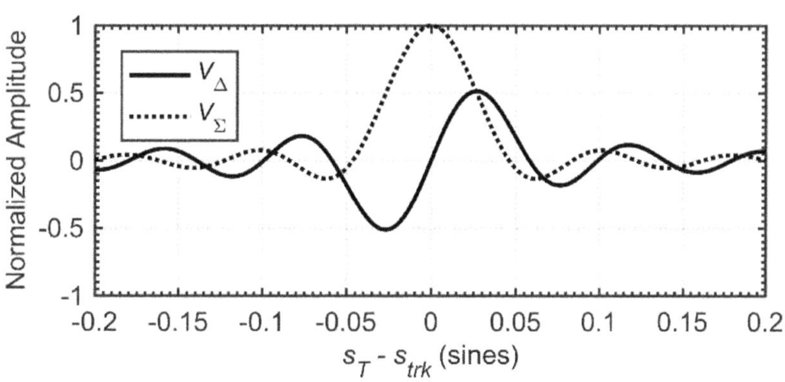


**Figure 5.13** Sum and difference voltage for a space-fed array.

## 5.5 MONOPULSE COMBINERS

A monopulse combiner (or comparator) [5, 15, 16] is a device that converts the antenna outputs to a form needed by the receiver and monopulse processor of Figure 5.1. Specifically, it creates the $\Delta$ and $\Sigma$ signals, or patterns, discussed in Sections 5.3 and 5.4. A block diagram of a classic monopulse combiner is shown in Figure 5.14 [5]. The inputs are the signals from the four subarrays of a constrained feed array (see Figure 5.4), or the four feed horns of a space fed phased array, or a reflector antenna (see Figure 5.5). The output is a sum ($\Sigma$) signal and two difference signals ($\Delta u$, $\Delta v$). The sum signal is the sum of all four subarray or horn outputs, and the difference signals are the differences of appropriate pairs of subarray or feed horn signals. The fourth output is termed a $\Delta\Delta$ signal [5,15] that is seldom used. It is the difference between sums of diagonal subarray or horn outputs.

The four boxes of Figure 5.14 are hybrid junctions [5, 10]. They are 4-port RF devices configured so that one output is the sum of the two inputs, and the other output is the difference between the two inputs. Three common types of hybrid junctions are the magic tee, the rat race (also known as a hybrid ring junction), and the 3-dB coupler [10].

### 5.5.1 Magic Tee

A drawing of a magic tee, which is a waveguide device, is shown in Figure 5.15. Different references show it used in different configurations [5, 7, 17]. In the configuration shown, the inputs are ports 2 and 3, and the outputs are ports 1 and 4. Port 1 contains the $\Sigma$ signal, and Port 4 contains the $\Delta$ signal. To explain how this occurs, we assume the signals injected into ports 2 and 3 are such that the E-fields they create are in the same direction. If we assume the lengths of ports 2 and 3 are the same, the directions of the two E-fields will also be the same when they get to Port 1. Since the E-fields are in the same direction, the signals associated with them add (algebraically). Thus, Port 2 is the sum, or $\Sigma$, port.

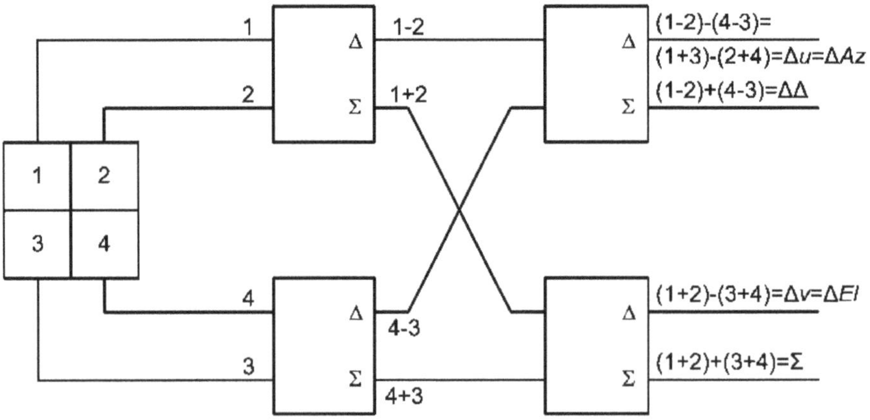


**Figure 5.14** Monopulse combiner block diagram.

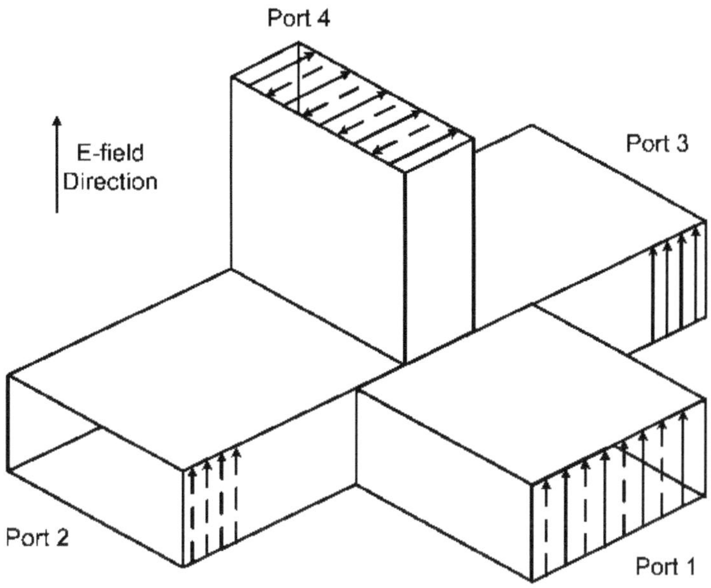


**Figure 5.15** Magic tee.

When the E-field in Port 2 transitions to Port 4, it will be oriented from right to left, assuming it is pointing up at the junction between ports 2 and 4. To explain this we note that, as an E-field "turns a corner" in a waveguide, the tip of the arrow representing its direction will always point to the same side of the waveguide [10, 18]. Thus, if the tip of the E-field arrow in Port 2 is pointing up at the junction with Port 4, it will point to the left when it enters Port 4. By a similar argument, if the E-field in Port 3 is pointing up at the junction with Port 4, it will point to the right when it enters Port 4. Since the E-fields from ports 2 and 3 are in opposite directions in Port 4, the signals associated with them will subtract (algebraically). Thus, Port 4 is the difference, or Δ, port.

In the alternate configuration, the input port are ports 1 and 4, and the output ports are ports 2 and 3. It is left as a homework assignment to show how this works (Problem 7).

### 5.5.2 Rat Race

A schematic drawing of a rat race, which can be implemented with waveguide, microstrip, or stripline techniques [11], is shown in Figure 5.16. The total length of the circular path of the rat race is $1.5\lambda_g$ long, where $\lambda_g = v_g/f_c$ with $v_g$ equal to the velocity of propagation of the signal in the rat race and $f_c$ equal to the carrier frequency. The distances between ports 1 and 2, 2 and 3, and 3 and 4 is $\lambda_g/4$. The distance between ports 1 and 4 is $3\lambda_g/4$[19]. Each distance of $\lambda_g/4$ causes a $\pi/2$ phase shift. We assume the lengths of all four ports are the same so that they will cause the same amount of phase shift. Thus, the port lengths do not need to be considered.

---

[19] All of the differences could be increased by a multiple of $\lambda_g/2$ without affecting the operation of the rat race (see Problem 9).

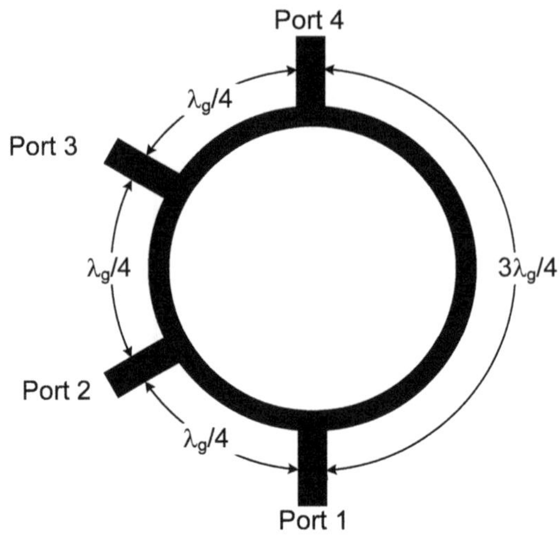


**Figure 5.16** Rat race.

Let the signals input at ports 1 and 3 be

$$v_1(t) = V_1 e^{j(2\pi f_c t + \phi_1)} \tag{5.69}$$

and

$$v_3(t) = V_3 e^{j(2\pi f_c t + \phi_3)}. \tag{5.70}$$

The signals at ports 2 and 4, due to $v_1(t)$ will be

$$v_{21}(t) = V_1 e^{j(2\pi f_c t + \phi_1 + \pi/2)} \tag{5.71}$$

and

$$v_{41}(t) = V_1 e^{j(2\pi f_c t + \phi_1 + 3\pi/2)}. \tag{5.72}$$

The signals at ports 2 and 4, due to $v_3(t)$ will be

$$v_{23}(t) = V_3 e^{j(2\pi f_c t + \phi_3 + \pi/2)} \tag{5.73}$$

and

$$v_{43}(t) = V_3 e^{j(2\pi f_c t + \phi_3 + \pi/2)}. \tag{5.74}$$

The signal at Port 2, the Σ port, due to $v_1(t)$ and $v_3(t)$ will be

$$\begin{aligned} v_\Sigma(t) &= v_{21}(t) + v_{23}(t) = e^{j\pi/2}\left(V_1 e^{j(2\pi f_c t+\phi_1)} + V_3 e^{j(2\pi f_c t+\phi_3)}\right) \\ &= e^{j\pi/2}\left(v_1(t) + v_3(t)\right) \end{aligned}, \tag{5.75}$$

or the sum of the two input signals, with an attendant $\pi/2$ phase shift.

The signal at Port 4, the Δ port, due to $v_1(t)$ and $v_3(t)$ will be

$$\begin{aligned} v_\Delta(t) &= v_{41}(t) + v_{43}(t) = e^{j\pi/2}\left(V_3 e^{j(2\pi f_c t+\phi_3)} - V_1 e^{j(2\pi f_c t+\phi_1)}\right) \\ &= e^{j\pi/2}\left(v_3(t) - v_1(t)\right) \end{aligned}, \tag{5.76}$$

or the difference between the two signals, again with the attendant $\pi/2$ phase shift. The details of the math associated with (5.71) through (5.76) is left as an exercise (Problem 8).

As with the magic tee, the roles of the ports of the rat race can be interchanged. That is, the inputs could be ports 2 and 4, and the outputs could be ports 1 and 3 (see Problem 10).

### 5.5.3 3-dB Coupler

A schematic diagram of a 3-dB coupler is shown in Figure 5.17. It consists of two waveguides with slots between them. The slots allow transfer of a desired amount of power, ½ in the case of the 3-dB coupler, between the waveguides. If a signal with a power of $P$ is injected into one side of either the upper or lower waveguides (Port 2 in Figure 5.17), the power at the other end of each waveguide will be $P/2$ (ports 3 and 4 in Figure 5.17). The signals at the same side of the other waveguide (Port 1 in Figure 5.17) will be zero (ideally). The spacing between the two slots is $\lambda_g/4$ so that the signal undergoes a phase shift of $\pi/2$ as it propagates between the slots.

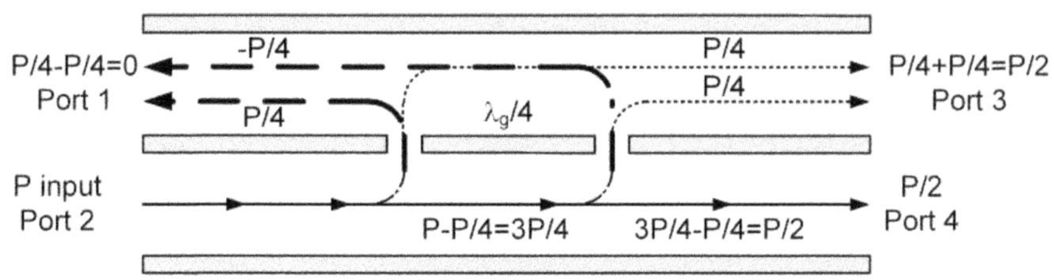


**Figure 5.17** 3-dB coupler.

The operation of a 3-dB coupler as a hybrid junction is considerably more complicated to explain than the magic tee and the rat race. Such an explanation can be found in [5]. If the inputs to ports 1 and 2 are

$$v_1(t) = V_1 e^{j(2\pi f_c t + \phi_1)} \tag{5.77}$$

and

$$v_2'(t) = e^{j\pi/2} v_2(t) = e^{j\pi/2} V_2 e^{j(2\pi f_c t + \phi_2)}, \tag{5.78}$$

the outputs from ports 3 and 4, the Σ and Δ ports, will be

$$v_3(t) = v_\Sigma(t) = v_1(t) + v_2(t) = V_1 e^{j(2\pi f_c t + \phi_1)} + V_2 e^{j(2\pi f_c t + \phi_2)} \tag{5.79}$$

and

$$\begin{aligned} v_4(t) &= v_\Delta(t) = e^{j\pi/2}\left[v_1(t) - v_2(t)\right] \\ &= e^{j\pi/2}\left[V_1 e^{j(2\pi f_c t + \phi_1)} - V_2 e^{j(2\pi f_c t + \phi_2)}\right]. \end{aligned} \tag{5.80}$$

An important thing to note about the above relations is that the 3-dB coupler must include a $\pi/2$ phase shift on one of its inputs, and the difference output is shifted by $\pi/2$.

In passive phased array and reflector antennas, the feed, and accompanying monopulse combiner, can be considerably different than the classic four horn feed of Figure 5.5. An example of this is a feed that consists of a pair of horns that will support both the $TE_{10}$ and $TE_{20}$ modes of propagation [12]. In this instance, the $TE_{20}$ mode provides the difference (Δ) signal for one plane while the difference between the outputs of the two horns provide the difference signal in the other plane. The sum of the two horn outputs provide the sum (Σ) signal [8, 15].

In some instances, a feed with more than four horns is used in an attempt to optimize the sidelobe levels of both the sum and difference antenna patterns. A somewhat extreme example of this is the experimental 12-horn feed developed by Lincoln Laboratories [5, 19, 20]. According to [5], the 12-horn feed provided good control of azimuth and elevation sidelobes. However, because of its complexity, it has not been adopted for general use. A schematic drawing of the 12-horn feed, and its monopulse combiner, is contained in Figure 5.18. Other multi-horn feeds are a 4-horn multimode feed [8], and a 5-horn feed mentioned in [5].

Other examples of more complex feeds include 6-horn feed used in the Russian FLAP LID radar [21] and a five-layer, multimode feed used in the Patriot radar [5, p. 122].

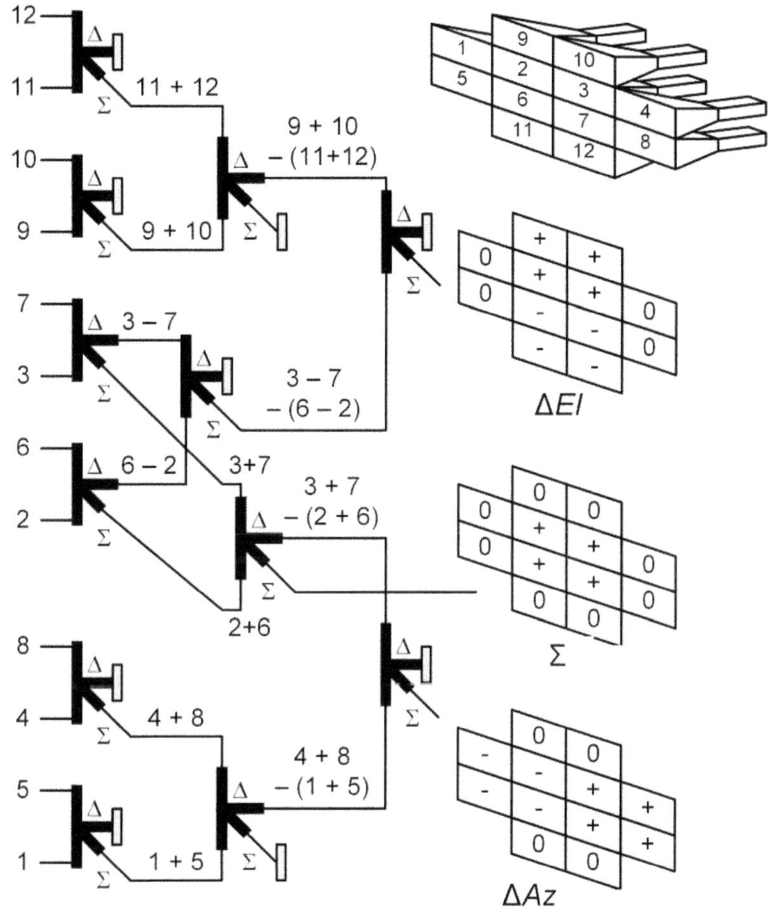


Figure 5.18 12-horn feed and monopulse combiner.

This discussion raises an issue not addressed in Sections 5.3 and 5.4. Specifically, the use of an amplitude taper to control antenna sidelobe levels. In space-fed arrays and reflector antennas, the amplitude taper is provided by the radiation intensity pattern of the feed. A simple 4-horn feed, or a 2-horn multimode feed, can be designed to optimize sidelobe levels of either the sum or difference patterns, but not both. This stems from the fact that the radiation intensity patterns of the horns are the same, assuming the 4 or 2 horns of the feed are the same. Since feed output combinations are fixed by the requirement to form the sum and two difference signals, there is no flexibility to provide independent control of the sum and difference pattern sidelobes. Antennas that use more than 4 horns, or 2 horns in a multimode feed (such as those indicated in the previous paragraph), can provide some control of the sum and difference sidelobes because the extra feeds provide increased degrees of freedom.

A similar problem occurs in constrained feed phased arrays. It is possible to configure the waveguide network of a constrained feed phased array to achieve a single type of taper. This is done by adjusting the waveguide junctions to vary how much each element contributes to the total signal in the waveguide (think X-dB couplers). To simultaneously optimize the sum and difference antenna patterns, three sets of waveguide networks would be needed. This could make the array quite complicated, as indicated in Figure 5.19, which contains a photo of the AN/APG-65 radar array. The back of the array is shown in the photo. The horizontal and vertical bars are waveguides from several subarrays of the antenna. Points where the waveguides appear to cross are the aforementioned various combiners. This particular array uses many more than 4 subarrays whose outputs are combined in different ways to obtain control of the sidelobe levels of the sum and difference patterns.

A similar argument applies to active phased arrays. In this case, the receiver portions of the T/R modules would require multiple attenuators and outputs to allow implementation of different tapers for the sum and difference patterns. An example of a radar that does this is the THAAD radar [22].

Figures 5.20 contains plots of sum and difference patterns for the 50-element, linear, constrained feed array considered earlier. In this case, a 30-dB, $\bar{n} = 6$, Taylor weighting is applied across the array. As expected, the weighting reduced the sidelobes of the sum pattern to −30 dB. However, the sidelobes of the difference pattern are still somewhat large. As indicated above, to reduce the difference pattern sidelobes, an additional weighting (e.g., Bayliss [3, 23]) would need to be included in the waveguide network that forms the difference pattern.

Figure 5.21 contains plots of sum and difference patterns for the 50-element, linear, space-fed array considered earlier. In this case, we assumed the feed provided a cosine amplitude weighting across the array. The parameters of the weighting were chosen so that the power delivered to the edge elements was 20 dB below the power delivered to the center elements. This is termed cosine weighting with a 10-dB edge taper. The weighting reduced the first sidelobe levels of the sum pattern to almost 25 dB. The sidelobe levels of the difference pattern is not as good but is still reasonable. The difference pattern sidelobe levels for this array are much lower than for the constrained feed array.

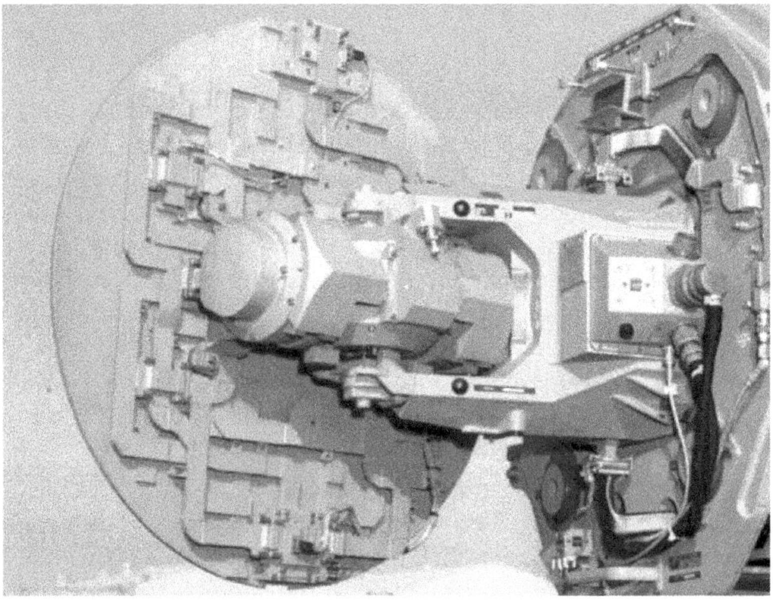

**Figure 5.19** AN/APG-65 array (Source: Artech House).

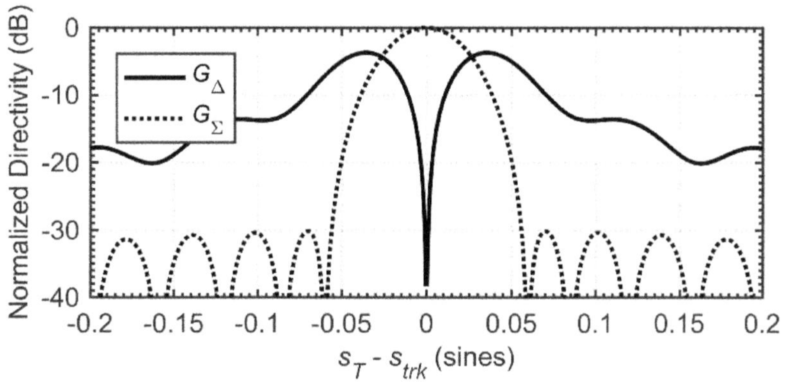


**Figure 5.20** Sum and difference antenna patterns of a 50-element, constrained feed, linear array with Taylor weighting.

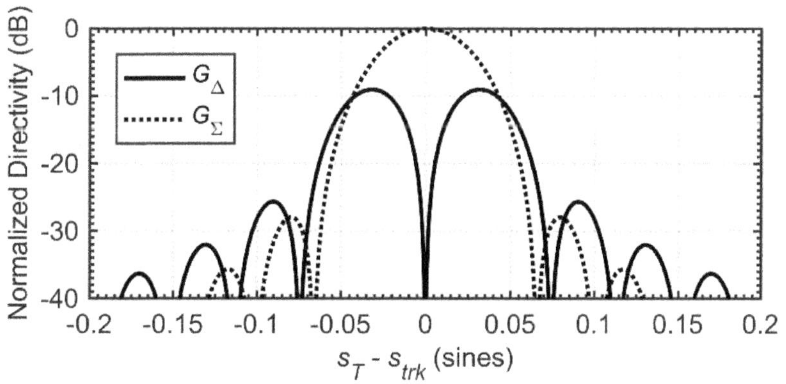


**Figure 5.21** Sum and difference patterns of a 50-element, space-fed, linear array with a cosine amplitude taper, with a 10-db edge taper.

As a side note, while the addition of an amplitude taper reduces the antenna sidelobes, it also broadens the main beam and causes a loss in directivity [7].

## 5.6 MONOPULSE RECEIVERS

We now turn our attention to monopulse receivers. The main function of the monopulse receiver is to convert the RF signals from the monopulse combiner to a form suitable for the monopulse processor. In particular, the signals are heterodyned from RF to baseband or, in some applications, to a lower intermediate frequency (IF). The receiver also includes the matched filter and signal processor. In radars that use digital monopulse processors, the receiver also converts the signals from analog to digital. We begin the discussions with three-channel monopulse receivers. We then discuss two channel receivers and conclude with a discussion of single channel receivers and conical scan.

### 5.6.1 Three-Channel Monopulse Receiver

Figure 5.22 contains a functional block diagram of a three-channel monopulse receiver. This particular configuration is sometimes termed a *digital receiver.* It uses analog components through the IF section. The outputs of the IF portions of the receivers are then converted to digital signals by the ADCs, and the rest of the processing is performed digitally. If we would have delayed the conversion to digital after the signals were heterodyned to baseband, we would have what we term an *analog receiver*.

The outputs of the monopulse combiner are routed to three, separate RF amplifiers, whose primary function are to set the receiver noise level. The outputs of the RF amplifiers are then heterodyned down to an IF by mixing the RF amplifier outputs with a local oscillator (LO) signal. Although only one IF stage is shown in Figure 5.22, the radars commonly have two or more IF stages. This is normally dependent on the operating bandwidth of the radar and other factors [7].

After IF amplification, the three signals are digitized by the ADCs, then heterodyned to baseband using a numerically controlled oscillator (NCO). The baseband signals are shown as complex to preserve both the amplitude and phase information that was present in the received signal. The preservation of amplitude and phase is needed for the signal processor since it usually distinguishes targets from interference on the basis of frequency. The phase information is also needed in the monopulse processor since it forms the error signal as the real (in amplitude comparison) or imaginary (in phase comparison) part of the difference-to-sum ratio.

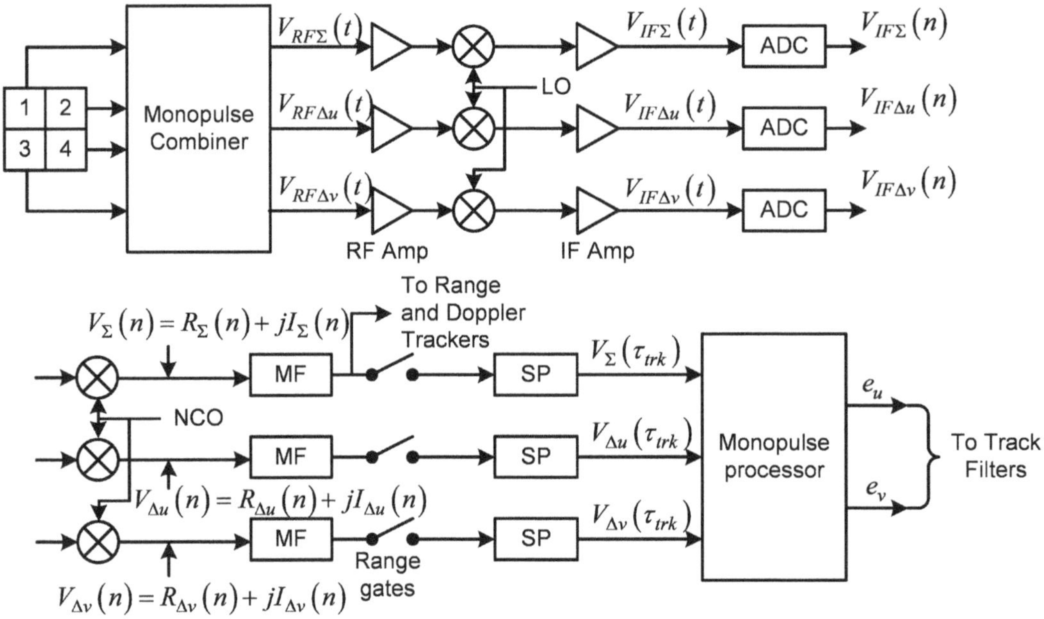


**Figure 5.22** Three-channel monopulse processor.

The three switches following the matched filters (MF) represent the range gate, which is controlled by the range tracker. It samples the matched filter output at $\tau_{trk}$, the tracked range from the range tracker (see Chapter 4). These samples are taken once per PRI and sent to the signal processor (SP). The signal processor is used in radars that coherently process multiple pulses for purposes of mitigating clutter or providing SNR improvement [7, 24-29]. The time duration of pulses used in the signal processor is termed a coherent processing interval (CPI). It is also sometimes referred to as a *dwell*. If the radar does not process multiple pulses, the CPI, or dwell, is equal to the PRI.

If the signal processor is a high pass filter, such as an MTI, it could be implemented before the range gates, and even before the matched filter. In those instances, the sum channel used by the angle tracker, the range tracker, and even the search function could share the same signal processor. This can be done because a high pass filter is, by definition, a wide bandwidth filter and will preserve almost all of the pulse modulation, or the high frequency components of the matched filter output.

If the signal processor contains a coherent integrator, the coherent integrator must be placed after the range gate because it will not preserve pulse modulation.

At the end of the CPI, or dwell, the signal processor output is sent to the monopulse processor and, at the next track update time, the error signals out of the monopulse processor are sent to the angle track filters

If the radar is all analog, as might be the case for older, single target track radars, the ADC would not be present, and the matched filter, range gate, signal processor and monopulse processor would be implemented with analog hardware.

The output of the sum channel matched filter also goes to the range tracker and, if the radar uses one, the Doppler tracker.

The monopulse processor will, in one way or another, form the ratio $V_\Delta(\tau_{trk})/V_\Sigma(\tau_{trk})$. The specifics of how the ratio is formed will be discussed in the section on monopulse processors

The receiver, etc. is designed to be linear up to the monopuolse processor. Specifically, the relative amplitudes and phases of the sum and difference signals out of the monopulse combiner must be preserved through the entire receiver. If the relative amplitudes and phases are not preserved, the monopulse processor would introduce biases in the error signal, or would have a discriminator curve slope that differs from the design intent. Either one of these effects will degrade the performance of the angle trackers.

This requirement to preserve relative amplitudes and phases means calibration of monopulse radars is very important. Because of this, some monopulse radars set aside time in their scheduler to automatically perform receiver calibration by inserting a test signal, called a *pilot pulse* [5, 7, 30], in the receiver front end, and use the monopulse processor output to calibrate the sum and two difference channels. Calibration is a requirement to account for variations in phase and amplitude, usually in analog components that fluctuate, with temperature etc. Once the signal is digital, the phase and amplitude is fixed. As a result of more digital designs, calibration requirements have lessened in newer radar receivers.

### 5.6.2 Two-Channel Monopulse Receiver

To reduce the number of receiver components, some radars use only two channels by multiplexing the difference signals out of the monopulse combiner. We will discuss two methods of doing this: continuous multiplexing and discrete multiplexing [5, 9, 24].

#### 5.6.2.1 Continuous Multiplexing

A functional block diagram of a continuous multiplexing configuration is shown in Figure 5.23. The monopulse combiner is the same as discussed in the three-channel example of the previous section. The $V_{RF\Delta u}$ and $V_{RF\Delta v}$ signals out of the combiner are routed to a mechanical resolver that quadrature modulates them onto a low frequency carrier. A drawing of the mechanical resolver from the Chubb, et. al. patent describing continuous multiplexing is shown in Figure 5.24. It is a waveguide device where the $V_{RF\Delta u}$ and $V_{RF\Delta v}$ (azimuth and elevation) signals are input to a circular waveguide via rectangular waveguides located 90° apart; thus the quadrature relation. A probe (the hooked line coming from the motor in Figure 5.24) is rotated by a motor and thus samples varying combinations of $V_{RF\Delta u}$ and $V_{RF\Delta v}$. The action of the resolver is represented in Figure 5.23 by the two multipliers and summer between the monopulse combiner and the hybrid junction. Mathematically, the output of the resolver is

$$V_{RF\Delta}(t) = V_{RF\Delta u}(t)\cos\omega_r t + V_{RF\Delta v}(t)\sin\omega_r t \tag{5.81}$$

where $\omega_r = 2\pi f_r$ is the motor scan rate, the multiplexing rate. $f_r$ is typically in the range of 10s to 100s Hz [31, 32].

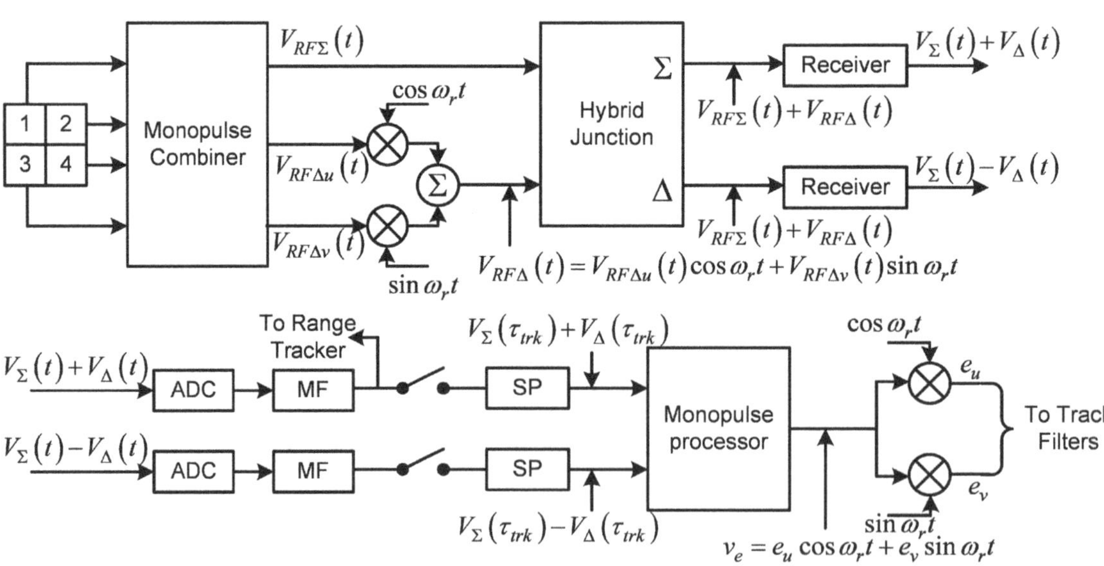


**Figure 5.23** Continuous multiplexing configuration.

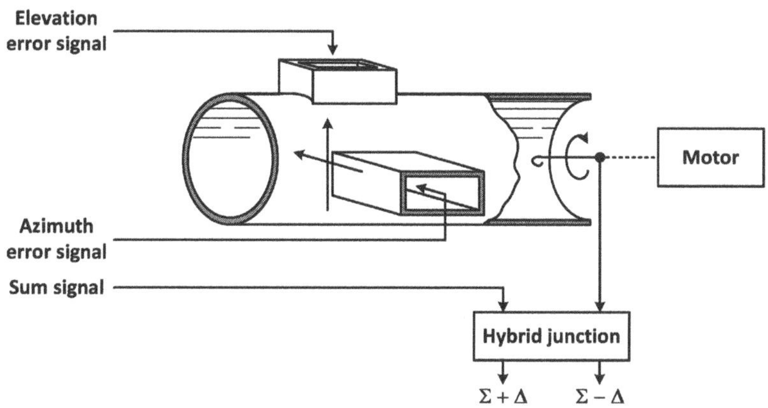


**Figure 5.24** Continuous multiplexing resolver (After [33]).

The hybrid combiner, which can be any of the types discussed in Section 5.5, combines the sum signal from the monopulse combiner, $V_{RF\Sigma}(t)$, with the multiplexed difference signals to produce their sum and difference; that is, to produce $V_{RF\Sigma}(t) + V_{RF\Delta}(t)$ and $V_{RF\Sigma}(t) - V_{RF\Delta}(t)$. These are subsequently sent through two receiver/signal processor chains to the monopulse processor.

In normal operation, the output of the monopulse processor is the combined error signal

$$e = e_u \cos\omega_r t + e_v \sin\omega_r t \, . \tag{5.82}$$

The individual error signals are recovered by synchronously detecting $e$. This is represented by the pair of multipliers following the monopulse processor.

The synchronous detectors can be mathematically represented by the operations

$$e_u = LPF\left\{e\cos\omega_r t\right\} \tag{5.83}$$

and

$$e_v = LPF\left\{e\sin\omega_r t\right\} \tag{5.84}$$

where the notation $LPF\{\ \}$ represents the low-pass filter part of the synchronous detector output [7, 9]. In practical applications, the variation in $e_u$ and $e_v$ is much lower than $2f_r$, the double frequency term that is produced by the multiplication. Thus, the double frequency terms are rejected by the low-pass filter.

One of the claims of Chubb, et. al. patent is that continuous multiplexing is insensitive to receiver (and processor) gain and phase imbalances [33]. At some point, the monopulse processor forms the differences

$$\begin{aligned} V_{diff} &= G_1\left(V_\Sigma + V_\Delta\right) - G_2\left(V_\Sigma - V_\Delta\right) \\ &= \left(G_1 - G_2\right)V_\Sigma + \left(G_1 + G_2\right)V_\Delta \end{aligned} \, . \tag{5.85}$$

In the case where the receiver channels are balanced, $G_1 = G_2$ and the $V_\Sigma$ term disappears. If $G_1 \neq G_2$, $V_{diff}$ will contain some of $V_\Sigma$. In this case, $e$ will be of the form

$$e = v_\Sigma + e_u \cos\omega_r t + e_v \sin\omega_r t \tag{5.86}$$

where we used $v_\Sigma = (G_1 - G_2)V_\Sigma$ for convenience. If the variation in $v_\Sigma$ is slow relative to $f_r$, it will be eliminated by the synchronous detector. To see this, we consider the leg of the synchronous detector that produces $e_u$. If we use (5.86) in (5.83) we get

$$\begin{aligned} e_u &= LPF\left\{e\cos\omega_r t\right\} = LPF\left\{\left(v_\Sigma + e_u \cos\omega_r t + e_v \sin\omega_r t\right)\cos\omega_r t\right\} \\ &= LPF\left\{v_\Sigma \cos\omega_r t + e_u \cos^2\omega_r t + e_v \sin\omega_r t \cos\omega_r t\right\} \end{aligned} \tag{5.87}$$

which, with some trig manipulations reduces to

$$e_u = LPF\left\{v_\Sigma \cos\omega_r t + \tfrac{1}{2}e_u + \tfrac{1}{2}e_u \cos 2\omega_r t + \tfrac{1}{2}e_v \sin 2\omega_r t\right\}. \tag{5.88}$$

Assuming the cutoff frequency of the LPF is less than $f_r$, but greater than the variations in $v_\Sigma$, $e_u$, and $e_v$, the LPF output will contain only $e_u$. In other words, the channel imbalance will be mitigated.

An extreme example of channel imbalance is where one receiver channel fails. In that case, $G_1$ or $G_2$ will be zero, but, by the above math, the $V_\Sigma$ signal should still be filtered by the LPF. However, because of the fact that $V_\Sigma$ is much larger than $e_u$ and $e_v$, some of it could "leak" through the LPF and degrade tracker performance.

A potential issue with continuous multiplexing is coordination between the multiplexing operation and signal processing. In particular, the designer must take care to insure one doesn't interfere with the operation of the other.

Continuous multiplexing was invented in the era of single-target track radars and appears to work well in them.[20] However, it is not clear continuous multiplexing is suitable for modern, multiple-target tracking radars. The problem with the application of continuous multiplexing in such radars is that the multiplexing cycle would need to be synchronized with the track update period, and the framed waveform structure used in such radars. A way around this potential problem would be to replace the continuous multiplexing with discrete multiplexing that is implemented with switches. Some authors refer to this technique as sequential lobing [5, 8, 9, 24].

A functional block diagram of a discrete multiplexing configuration is shown in Figure 5.25.[21] It is similar to the continuous multiplexing configuration with the mechanical resolver

---

[20] As examples, the first edition of Skolnik's radar handbook mentions that the AN/SPG-55 and the AN/FPQ-10 radars use continuous multiplexing [41].

[21] The continuous and discrete multiplexing diagrams show analog receivers as opposed to the digital receiver shown in Figure 5.21 because these types of multiplexing are used in older radars that had most likely not progressed to the digital receivers used in modern radars. The matched filter, etc. is shown as digital to bridge the gap between all analog and all digital radars.

and synchronous detectors replaced by a pair of single pole, double throw (SPDT) switches. These are termed *plane switch* in Figure 5.24. The switch between the monopulse combiner and hybrid junctions is some type of RF switch, such as a ferrite switch [34]. The switch after the monopulse processor is an electronic switch that operates on low frequency (essentially DC) signals. In radars that use a digital monopulse processor, it is a software switch that chooses between the two error signals.

When the switches are in the up position, the receiver/SP/monopulse processor operates on $V_{RF\Sigma}(t) \pm V_{RF\Delta}(t)$ signals to produce the $e_u$ error signal. When the switches are in the down positon, the receiver/SP/monopulse processor operates on $V_{RF\Sigma}(t) \pm V_{RF\Delta}(t)$ to produce the $e_v$ error signal. An example timing arrangement for any one track update on any one target would be as follows: The plane switches would stay in the up ($\Delta u$) position for one or more signal processing intervals (e.g., coherent processing intervals—CPIs) and then switch to the down position for an equal number of signal processing intervals. This would produce the $e_u$ and $e_v$ signals needed to update the track filters for that target and track update. It would also provide the data needed to update the range and Doppler track filters. Synchronization between signal processing intervals and the plane switches should be straight forward since they could be controlled by the radar synchronizer.

Unlike continuous multiplexing, the discrete multiplexing configuration of Figure 5.24 does not mitigate channel balance issues, or provide for operation if one of the receiver channels fails. The upper receiver/SP chain always processes $V_\Sigma + V_\Delta$, and the lower chain always processes $V_\Sigma - V_\Delta$. If the gains of the two chains are $G_1$ and $G_2$, the result of the subtraction operation that takes place in the monopulse processor would be

$$\begin{aligned} V_{diff} &= G_1\left(V_\Sigma + V_\Delta\right) - G_2\left(V_\Sigma - V_\Delta\right) \\ &= \left(G_1 - G_2\right)V_\Sigma + \left(G_1 + G_2\right)V_\Delta \end{aligned}. \tag{5.89}$$

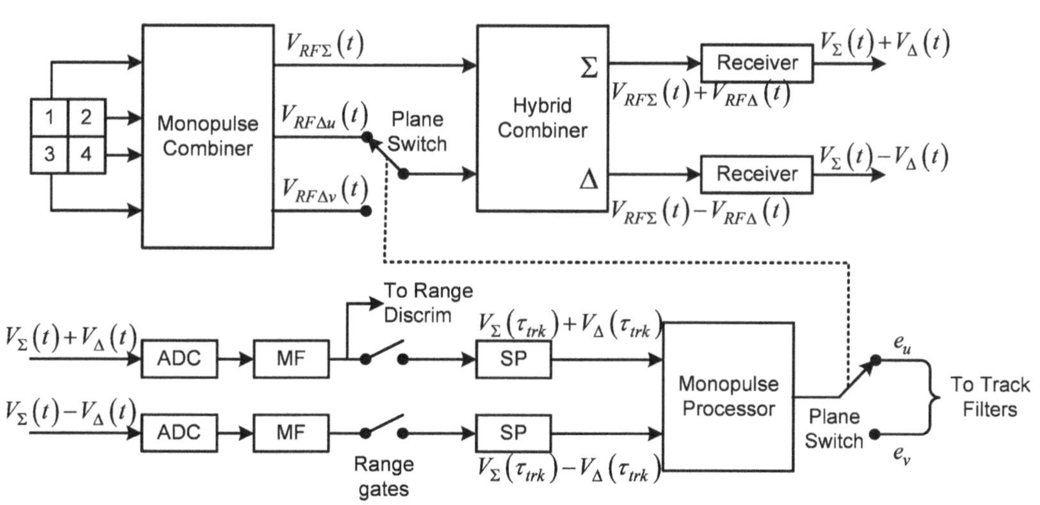


**Figure 5.25** Discrete multiplexing configuration.

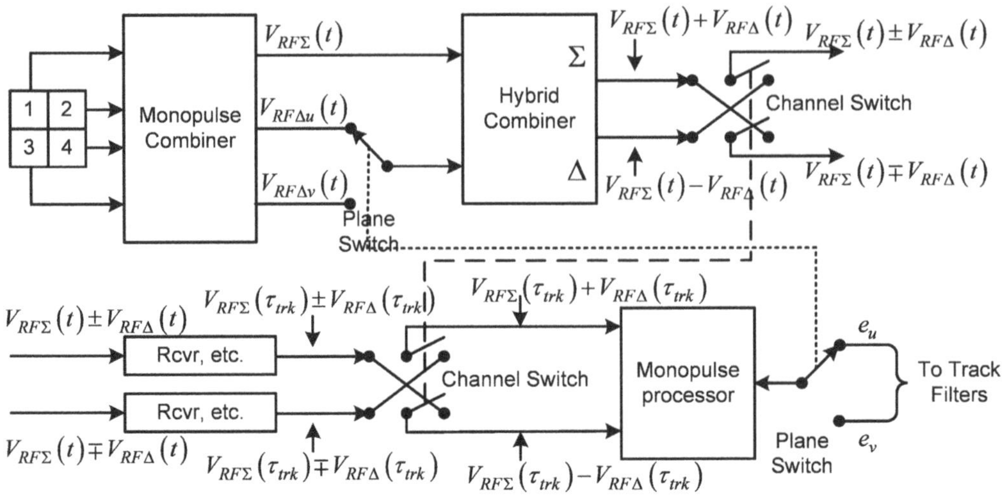


**Figure 5.26** Two-channel monopulse receiver—discrete multiplexing—with channel commutation.

If the channels are balanced, meaning $G_1 = G_2$, $V_{diff}$ would be $(G_1 + G_2)V_\Delta$, which is desired. If $G_1 \neq G_2$, $V_{diff}$ will contain $(G_1 - G_2)V_\Sigma$, which would lead to bias errors in the angle tracker.

Continuous multiplexing mitigates channel balance issues by effectively causing each receiver/SP chain to alternately carry $V_\Sigma + V_\Delta$ and $V_\Sigma - V_\Delta$. The same thing can be accomplished in the discrete multiplexing configuration by including a pair of channel switches as shown in Figure 5.26. For any one plane switch position, the first channel switch would alternately connect the output of the hybrid junction to alternate receiver/SP chains. The switch before the monopulse processor would undo this alternation so that the inputs to the monopulse processor would always be $V_\Sigma + V_\Delta$ and $V_\Sigma - V_\Delta$.

When the channel switches are in the left position, the inputs to the receiver/SP chains will be $V_\Sigma + V_\Delta$ and $V_\Sigma - V_\Delta$. If the receiver gains are $G_1$ and $G_2$, the difference operation of the monopulse processor will yield

$$\begin{aligned} V_{diffL} &= G_1\left(V_\Sigma + V_\Delta\right) - G_2\left(V_\Sigma - V_\Delta\right) \\ &= \left(G_1 - G_2\right)V_\Sigma + \left(G_1 + G_2\right)V_\Delta \end{aligned}. \tag{5.90}$$

When the switches are in the right position, the receiver inputs will be $V_\Sigma - V_\Delta$ and $V_\Sigma + V_\Delta$, and the difference operation will yield

$$\begin{aligned} V_{diffR} &= G_2\left(V_\Sigma + V_\Delta\right) - G_1\left(V_\Sigma - V_\Delta\right) \\ &= \left(G_2 - G_1\right)V_\Sigma + \left(G_1 + G_2\right)V_\Delta \end{aligned}. \tag{5.91}$$

This means the Σ component of $V_{diff}$ will alternate between $\pm(G_1 - G_2)V_\Sigma$ while the Δ component will consistently be $(G_1 + G_2)V_\Delta$. If the monopulse processor includes a summer, the net $V_{diff}$ will be

$$V_{diff} = V_{diffL} + V_{diffR} = 2(G_1 + G_2)V_\Delta . \tag{5.92}$$

Both continuous and discrete multiplexing must process several pulses or dwells (CPIs) to obtain both $u$ and $v$ angle information. Because of this, they are not true monopulse implementations. It appears as if the primary motivation for their use is to reduce the number of receiver channels to two, and reduce channel balancing requirements. In early radars, which used electronic tubes, the receivers were large, consumed a large amount of power and were difficult to keep aligned. Modern radars use discrete, and/or integrated, solid state receivers. As a result, they are much smaller and consume much less power. Also, they are much more stable, which would ease the channel balancing problem.

Strictly speaking, two channel monopulse is not true monopulse because both azimuth and elevation (or $u$ and $v$) angle information are not derived from a single pulse or CPI. To obtain both planes of angle information, two pulses or CPIs are needed, if channel reversing is not used. If channel reversing is used, four pulses or CPIs may be needed. This requirement of multiple pulses or CPIs to obtain angle information is a time tradeoff between three channel monopulse and two channel monopulse. Specifically, two channel monopulse requires more of the radar time line than does three channel monopulse.

In the two channel monopulse receiver, the range track signal can be taken from the output of the matched filter of either receiver/signal processor channel since they both contain valid range information. During search, it may be necessary to disable the difference signal into the hybrid junction so that only the sum signal appears at the output of the combiner. The reason for this is that the difference beams could cause an effective widening of the receive beam, which could degrade angle measurement during search.

### 5.6.3 Simultaneous Multiple Beams/Digital Beam Forming

Newer radars that use solid-state transmit/receive (T/R) modules have the potential of using multiple simultaneous beams to create the error signals needed in the angle track loops. The move toward multiple simultaneous beams is facilitated by the decreasing size and cost of receiver hardware to the point where it is becoming feasible for radars to have a large number of receivers, as opposed to the two or three used in two and three channel monopulse radars. Radars that use T/R modules are termed active phased array radars or radars that use active, electronically scanned antennas (AESAs).

Radars that use AESAs can use either phase or amplitude comparison monopulse. With the former, the array is divided into four quadrants as with constrained feed phased arrays. In such an arrangement, the angle information is contained in the relative phase of the signals from the four quadrants.

With amplitude comparison monopulse, the outputs of the T/R modules would, conceptually, go to four sets of phase shifters that form four beams that are squinted relative to the nominal pointing direction of the transmit beam (which is set by the phase shifters in the T/R modules). Since the beams are squinted, angle information is contained in the relative amplitudes of the signals from the four beams.

If the array is divided into a large number of relatively small subarrays, and a receiver and signal processor is attached to each subarray output, many more than four beams can be formed in the digital computer, or special purpose digital hardware. In this case, we would say the radar is using digital beam forming. The digital beam former could be used to form four, parallel beams for use in standard phase comparison monopulse, or four squinted beams for use in standard amplitude comparison monopulse. They could also be used to form a fairly large number of adjacent beams that could be used to measure $u$ and $v$ via some interpolation scheme.

If we extend the idea of multiple beams, we can postulate the possibility of being able to simultaneously track multiple targets and conduct search, all on the same pulse or burst of pulses. Of course, the transmit beam would need to be wide enough to cover the angle space covered by the multiple receive beams.

## 5.7 CONICAL SCAN

The two and three channel techniques discussed in the previous sections use a single transmit beam and multiple receive beams. The transmit beam points toward the track angles $(u_{trk}, v_{trk})$. Early angle track radars used a single transmit/receive beam that was offset from the track angle. The offset beam was then stepped between the angles $(u_{trk} \pm \Delta u,\ v_{trk} \pm \Delta v)$ or continuously rotated between them. Radars that used the stepped approach were called *switched scan* radars and radars that used continuous rotation were termed conical scan radars [5]. We will discuss the latter.

Conical and stepped scan was used in early track radars because of hardware limitations. Specifically, receivers were large, vacuum tube devices that were difficult to maintain. Because of this, early radars used only one receiver. It is also likely the use of conical and stepped scan was due to the fact that radar engineers did not fully understand monopulse feeds and the associated combiners.

The premise of conical scan is illustrated in Figure 5.27. In this figure, the single beam is squinted relative to the track axis, and the feed or entire antenna assembly is rotated about the track axis so that the beam traces out a cone.[22] This is the origin of the name *conical scan.*

A view of this looking down the track axis toward the antenna is shown in Figure 5.28. In this figure, the dashed circle is the locus of the center of the beam as it rotates about the track axis, which is denoted by the cross. The solid circle about the beam center (represented by the triangle) denotes the 3-dB contour of the beam. It will be noted that the 3-dB contour extends past the track axis. This means that even when the target is on the track axis, that is, $\Delta\theta = 0$, the radar will still see a reasonably strong signal.

In Figure 5.28, $(\alpha_b, \varepsilon_b)$ is the center of the beam relative to the track axis. The angular distance of beam center from the track axis is denoted by $\theta_b$, the squint angle. The target location relative to the track axis is $(\Delta\alpha, \Delta\varepsilon)$, and the solid angle of the target relative to the track axis is $\Delta\theta$. These are the error angles the tracker tries to drive to zero.

Since the beam rotates about the track axis, we can write equations for $\alpha_b$ and $\varepsilon_b$ as

---

[22] The drawing of Figure 5.27 is exaggerated. In actual conical scan radars, the beam is squinted approximately one-half beamwidth. That is $\theta_b \approx \theta_B/2$. Also, in this figure, we are using the angles $\alpha$ (azimuth) and $\varepsilon$ (elevation) instead of $u$ and $v$ since this is the convention one normally encounters when discussing conical scan.

$$\alpha_b = \theta_b \cos \omega_r t \tag{5.93}$$

and

$$\varepsilon_b = \theta_b \sin \omega_r t \tag{5.94}$$

where $\omega_r$ is the scan rate in radians per second, and $f_r = \omega_r/2\pi$ is the scan frequency in Hz. $f_r$ is the range of 10's of Hz. It needs to be larger than the closed loop track bandwidth so that the tracker will help smooth out the amplitude variations due to the scanning.

The location of the target relative to the center of the beam is

$$\begin{aligned}\theta_T(t) &= \sqrt{(\alpha_b - \alpha_T)^2 + (\varepsilon_b - \varepsilon_T)^2} \\ &= \sqrt{(\theta_b \cos \omega_r t - \alpha_T)^2 + (\theta_b \sin \omega_r t - \varepsilon_T)^2}\end{aligned}. \tag{5.95}$$

It will be noticed that $\theta_T(t)$ is a non-linear sinusoidal-like function of $t$.

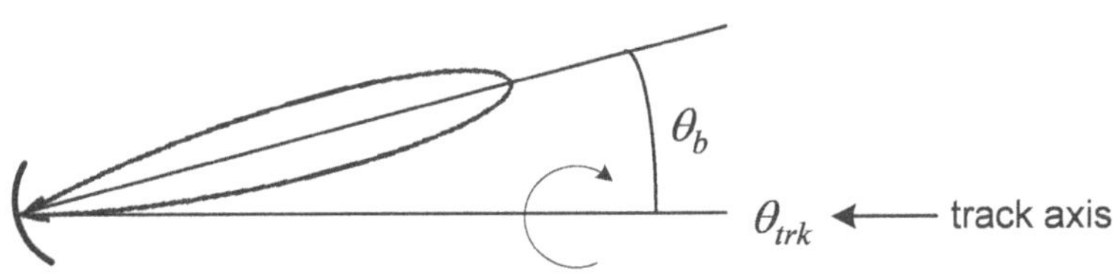


**Figure 5.27** Conical scan configuration.

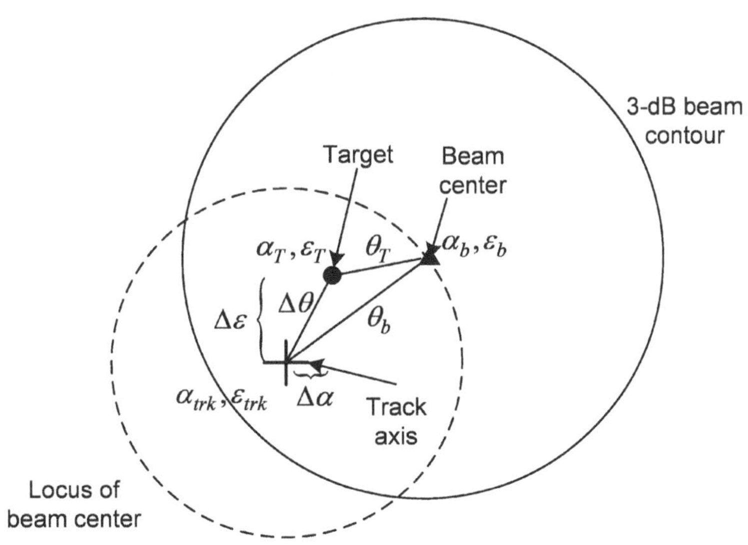


**Figure 5.28** Conical scan geometry.

Figure 5.29 contains a functional block diagram of a conical scan receiver, monopulse processor, and track filter. For purposes of this analysis, we assume the antenna is a uniformly illuminated, parabolic reflector with a circular aperture. As such, we can write its normalized directive gain pattern as

$$G(\theta_T) = \left( 2\frac{J_1(3.233\theta_T/\theta_B)}{3.233\theta_T/\theta_B} \right)^2 \tag{5.96}$$

where $J_1(x)$ is the first order Bessel function of the first kind [35].

The receiver is single channel and consists of a RF section and an IF section. The gain of the IF amplifier is controlled by an automatic gain control (AGC) circuit whose input is the output of the matched filter. The matched filter (MF) includes an envelope detector and range gate that samples the matched filter output at the tracked range, $\tau_{trk}$. The range gate holds the sample for a PRI (i.e., until the next pulse). Although not shown, the radar contains a range tracker. We assume the CPI (coherent processing interval) is the PRI, and there is one pulse per PRI. That is, we assume no coherent processing, other than pulse compression by the matched filter.

The AGC regulates the IF amplifier gain to cancel target related amplitude variations, but preserve the conical scan modulation. It will cancel variations that are much lower frequency than that of the scan modulation. In this way, the AGC also performs the monopulse normalization.

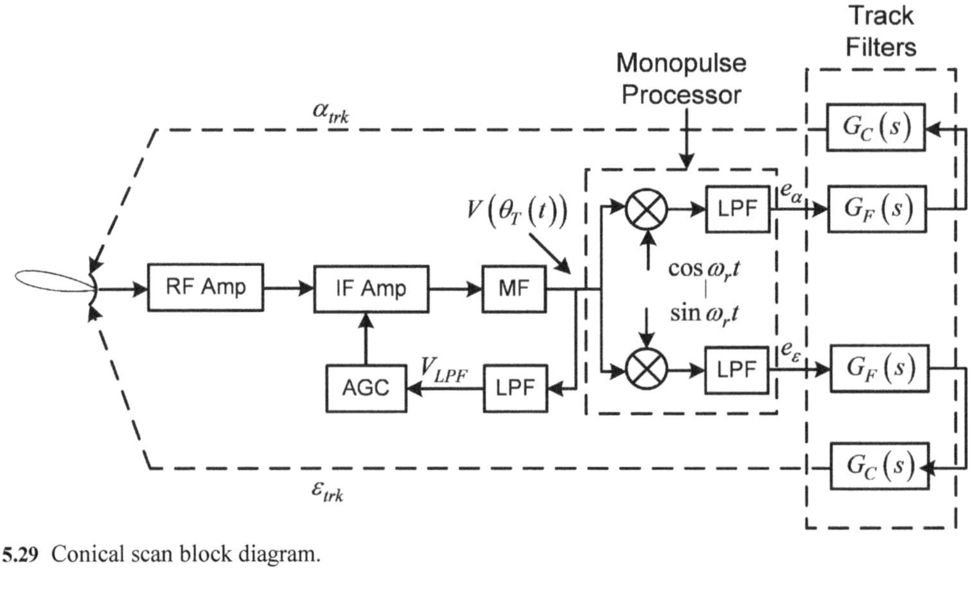


**Figure 5.29** Conical scan block diagram.

With this, we can write the normalized output of the matched filter as

$$V(t)=\left(2\frac{J_1\left(3.233\theta_T(t)/\theta_B\right)}{3.233\theta_T(t)/\theta_B}\right)^2 \tag{5.97}$$

where $\theta_T(t)$ is given by (5.95). Figure 5.30 contains a plot of $V(t)$ versus $\theta_T(t)$ for the case where $\Delta\varepsilon = 0$, $\Delta\alpha = \theta_B/2$, $\theta_b = 0.4\theta_B$, and $f_r = 10$ Hz.

The monopulse processor consists of a synchronous detector similar to the one discussed in Section 5.6.2.1. (The synchronous detector is labeled monopulse processor in Figure 5.29.) The reference signals into the two legs of the synchronous detector are $\cos\omega_r t$ and $\sin\omega_r t$. With this, the outputs of the two legs are

$$e_\alpha = LPF\left\{V(t)\cos\omega_r t\right\} \tag{5.98}$$

and

$$e_\varepsilon = LPF\left\{V(t)\sin\omega_r t\right\}. \tag{5.99}$$

The bandwidths of the LPFs are such that $e_\alpha$ is the average of $V(t)\cos\omega_r t$, and $e_\varepsilon$ is the average of $V(t)\sin\omega_r t$.

Figure 5.31 contains plots of the inputs and outputs of the LPFs for the previous example $\Delta\varepsilon = 0$, $\Delta\alpha = \theta_B/2$, $\theta_b = 0.4\theta_B$, and $f_r = 10$ Hz). Figure 5.32 contains similar plots for the case where $\Delta\varepsilon = 0$, $\Delta\alpha = -\theta_B/2$, $\theta_b = 0.4\theta_B$, and $f_r = 10$ Hz. (sign of $\Delta\alpha$ reversed). As hoped, $e_\varepsilon = 0$ (denoted average on the figures) in both cases because $\Delta\varepsilon = 0$. When $\Delta\alpha > 0$, so is $e_\alpha$, and when $\Delta\alpha < 0$, so is $e_\alpha$, as hoped. Figure 5.33 contains a plot of $e_\alpha$ versus $\Delta\alpha/\theta_B$ for this conical scan example. It is the discriminator curve of the conical scan tracker.

The error signals, $e_\alpha$ and $e_\varepsilon$ out of the monopulse processor are sent to the track filters and controlled elements, which moves the antenna to zero the error signals. In conical scan radars, the track filter is usually analog, and the controlled element will be part of the track filter. It is likely both the azimuth and elevation track loops are Type 1 or Type 2 servomechanisms.

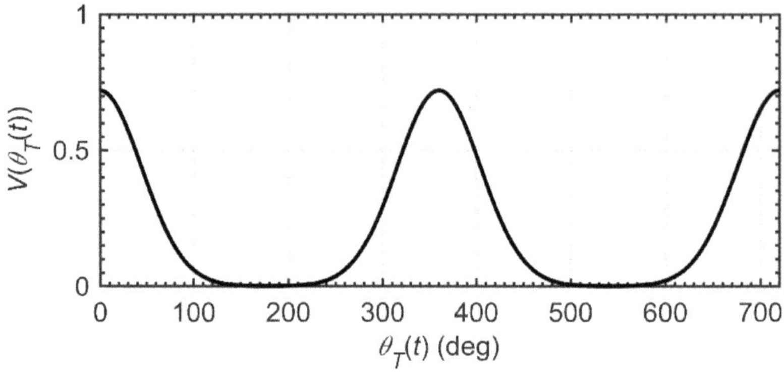


**Figure 5.30** Plot of $V(t)$ from (5.97).

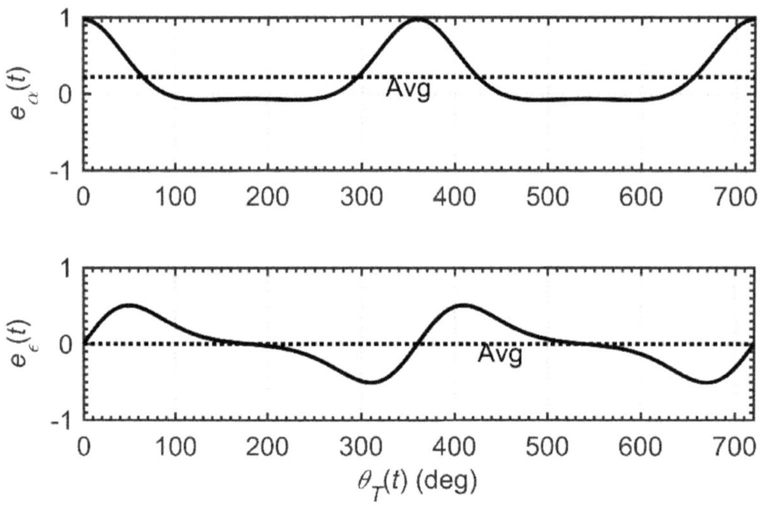


**Figure 5.31** Conical scan plots for $\Delta\alpha > 0$.

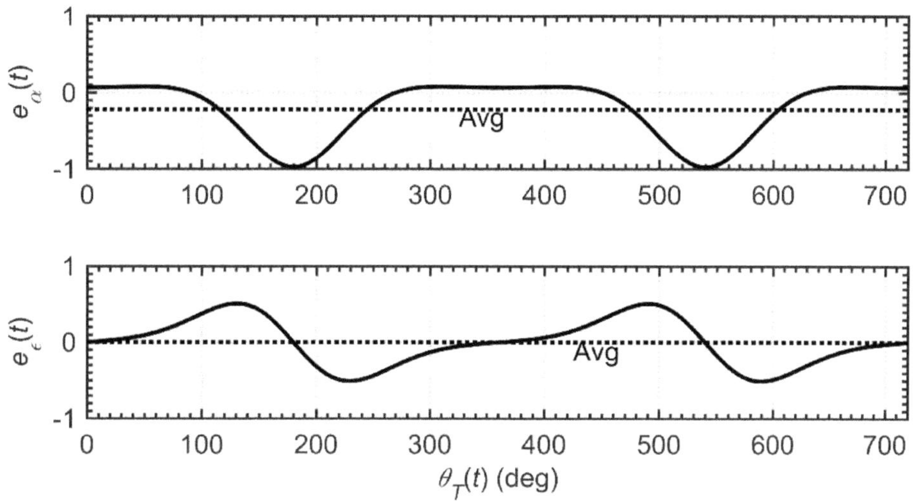


**Figure 5.32** Conical scan plots for $\Delta\alpha < 0$.

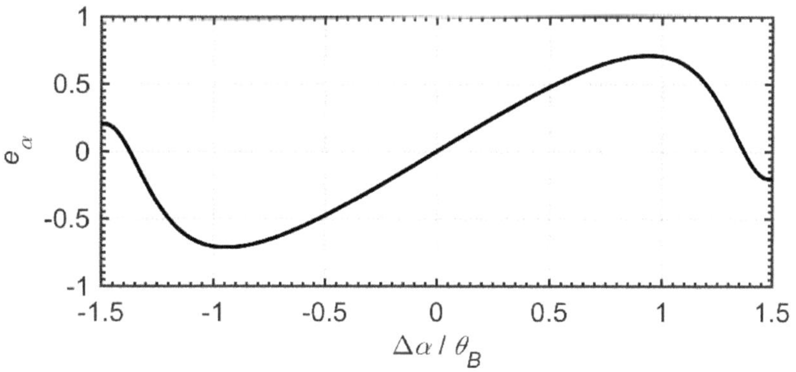


**Figure 5.33** Discriminator curve for conical scan example.

## 5.8 MONOPULSE PROCESSORS

The monopulse processors convert the outputs of the signal processor to a form needed by the angle track filters. Since the antenna forms signals proportional to the difference between the target angles ($u_T$, $v_T$) and the predicted target angles ($u_{trk}$, $v_{trk}$), the primary function of the monopulse processor is to normalize the signals to make the error signals supplied to the angle track filters independent of signal amplitude. It does this by, effectively, dividing the difference ($\Delta$) signals by the sum ($\Sigma$) signal.

One type of monopulse processor, which is termed an *exact processor* [5] does this by dividing $V_\Delta(\tau_{trk})$ by $V_\Sigma(\tau_{trk})$. It then scales the ratio to produce the error signals, $e_u$ and $e_v$, that are sent to the tracker. The scaling changes the units of $e_u$ and $e_v$ to be consistent with the units of the states of the angle track filters.

Exact processors, or variations thereon, are somewhat of a natural fit to radars that use three channel monopulse. This is because three channel monopulse receivers directly provide the sum and difference signals. Exact processors are also suited for radars that use digital processing because of the need to perform division. Exact processors could be used in radars that use two channel monopulse receivers. However, the $\Sigma \pm \Delta$ signals produced by the receivers would need to be preprocessed to separate the $\Sigma$ and $\Delta$ signals needed by the exact processor.

A type of monopulse processor that is a natural fit for two channel monopulse receivers is the *log processor*, because it operates on $\Sigma + \Delta$ and $\Sigma - \Delta$ signals. The log processor uses the properties of logarithms to provide the division of $V_\Delta(\tau_{trk})$ by $V_\Sigma(\tau_{trk})$. The log processor could also be used with radars that use three channel monopulse receivers. However, the $\Sigma$ and $\Delta$ signals would need to be preprocessed to produce the $\Sigma + \Delta$ and $\Sigma - \Delta$ signals. The exact and log processors are discussed in the following sections.

Another method of normalizing $V_\Delta(\tau_{trk})$ by $V_\Sigma(\tau_{trk})$ is through the use of AGC. An AGC circuit adjusts the gain of an amplifier so it is the inverse of the signal amplitude at the input of the amplifier. In three channel monopulse receivers, the AGC is based on signals in the sum channel IF amplifiers, and applied to the IF amplifiers of all three channels. Thus, the AGC divides the inputs to the two difference channel amplifiers by the input to the sum channel amplifier.

In two channel monopulse receivers, separate AGCs could be used on each channel, the AGC could be derived from one channel and applied to both channels, or the AGC could be derived from a combination of the two channels, and applied to both channels.

A basic premise of the use of AGC in two channel monopulse receivers is that, during track, the $\Sigma$ signal has a much larger magnitude than the $\Delta$ signal, so that $|\Sigma \pm \Delta| \approx |\Sigma|$. This may not be a valid assumption during search or acquisition. This is another reason to disconnect the $\Delta$ signal from the receiver input during search and acquisition. Alternately, the AGC could be disabled during search.

AGC normalization is best suited for single target tracking radars because of the time required for the AGC servo loop to reach steady state. If it is used in multi-target track radars, or multi-function radars, part of the track dwell would need to be devoted to the AGC operation.

Yet another class of normalization methods is based on the use of a bandpass limiter [8, 24, 36]. In those approaches, the $\Sigma$ and $\Delta$ signals are multiplexed and sent through the same bandpass limiter. As with AGC normalization, an assumption is that the magnitude of the $\Sigma$ signal is much larger than the magnitude of the $\Delta$ signal so that the bandpass limiter effectively divides the $\Sigma$ and $\Delta$ signals by the magnitude of the $\Sigma$ signal. With this type of normalization, an extra step is needed to recover the phase between the $\Delta$ and $\Sigma$ to obtain the sign of the error signals sent to the angle track filters. Discussions of AGC, bandpass limiter, and other types of monopulse normalization techniques are contained in various texts [5, 8, 15, 16, 24, 31].

### 5.8.1 Exact Processor

The exact processor forms the ratio

$$e = \text{Re}\left[\frac{V_\Delta(\tau_{trk})}{V_\Sigma(\tau_{trk})}\right] S_D \tag{5.100}$$

or

$$e = \text{Im}\left[\frac{V_\Delta(\tau_{trk})}{V_\Sigma(\tau_{trk})}\right] S_D \tag{5.101}$$

depending on whether the radar is using amplitude or phase comparison monopulse. Equation (5.100) applies to radars that use amplitude comparison monopulse (radars with space fed phased arrays or radars with reflector antennas), and (5.101) applies to radars that use phase comparison monopulse (radars with constrained feed phase array antennas). In (5.100) and (5.101), $e$ is the error signal sent to the track filters, and $S_D$ is the scaling factor mentioned earlier. The radar will have two monopulse processors that produce $e_u$ and $e_v$. In general, the two processors will have different scaling factors, different $S_D$'s, because the antenna difference pattern will generally be different in the $u$ and $v$ planes. In (5.100) and (5.101), $V_\Delta(\tau_{trk})/V_\Sigma(\tau_{trk})$ is termed the *monopulse ratio* [5, 16].

The need for using the real or imaginary part of the monopulse ratio for amplitude and phase comparison monopulse derives from the form of the ratio. As an example, for amplitude comparison monopulse we can use (5.67), (5.68), (5.63) and (5.64) to write[23]

$$V_\Delta(\tau_{trk}) = e^{-j\pi(K-1)\frac{d}{\lambda}(s_T - s_{trk})} \left[ \frac{\sin\left(\pi K \frac{d}{\lambda}(s_T - s_{trk} - s_s)\right)}{\sin\left(\pi \frac{d}{\lambda}(s_T - s_{trk} - s_s)\right)} - \frac{\sin\left(\pi K \frac{d}{\lambda}(s_T - s_{trk} + s_s)\right)}{\sin\left(\pi \frac{d}{\lambda}(s_T - s_{trk} + s_s)\right)} \right] \tag{5.102}$$

and

[23] Recall we are using $s$ as the general representation of $u$ and $v$.

$$V_{\Sigma}(\tau_{trk}) = e^{-j\pi(K-1)\frac{d}{\lambda}(s_T - s_{trk})} \left[ \frac{\sin\left(\pi K \frac{d}{\lambda}(s_T - s_{trk} - s_s)\right)}{\sin\left(\pi \frac{d}{\lambda}(s_T - s_{trk} - s_s)\right)} + \frac{\sin\left(\pi K \frac{d}{\lambda}(s_T - s_{trk} + s_s)\right)}{\sin\left(\pi \frac{d}{\lambda}(s_T - s_{trk} + s_s)\right)} \right]. \tag{5.103}$$

From this we note the ratio of $V_{\Delta}(\tau_{trk})$ to $V_{\Sigma}(\tau_{trk})$ is a real number.

For phase comparison monopulse, we can use (5.30) and (5.33) to write

$$V_{\Delta}(\tau_{trk}) = j2e^{-j\pi(2K-1)\frac{d}{\lambda}(s_T - s_{trk})} \times \sin\left(\pi K \frac{d}{\lambda}(s_T - s_{trk})\right) \frac{\sin\left(\pi K \frac{d}{\lambda}(s_T - s_{trk})\right)}{\sin\left(\pi \frac{d}{\lambda}(s_T - s_{trk})\right)} \tag{5.104}$$

and

$$V_{\Sigma}(\tau_{trk}) = 2e^{-j\pi(2K-1)\frac{d}{\lambda}(s_T - s_{trk})} \times \cos\left(\pi K \frac{d}{\lambda}(s_T - s_{trk})\right) \frac{\sin\left(\pi K \frac{d}{\lambda}(s_T - s_{trk})\right)}{\sin\left(\pi \frac{d}{\lambda}(s_T - s_{trk})\right)} \tag{5.105}$$

In this case, the ratio of $V_{\Delta}(\tau_{trk})$ to $V_{\Sigma}(\tau_{trk})$ is a purely imaginary number. Thus, to form a real error signal, we must use its imaginary part.

As a note, in general, of $V_{\Delta}(\tau_{trk})$ and $V_{\Sigma}(\tau_{trk})$ will be of the form

$$V_{\Delta}(\tau_{trk}) = G_{\Delta} e^{j\phi_{\Delta}} \left[ A_{\Delta}(\tau_{trk}) e^{j\psi_{\Delta}(\tau_{trk})} V_{\Delta}(s_T - s_{trk}) + n_{\Delta}(\tau_{trk}) \right] \tag{5.106}$$

and

$$V_{\Sigma}(\tau_{trk}) = G_{\Sigma} e^{j\phi_{\Sigma}} \left[ A_{\Sigma}(\tau_{trk}) e^{j\psi_{\Sigma}(\tau_{trk})} V_{\Sigma}(s_T - s_{trk}) + n_{\Sigma}(\tau_{trk}) \right]. \tag{5.107}$$

In (5.106) and (5.107), $G_\Delta$, $G_\Sigma$, $\phi_\Delta$ and $\phi_\Sigma$ are the gains and phase shifts caused by the receivers. If the receivers are balanced, $G_\Delta = G_\Sigma$ and $\phi_\Delta = \phi_\Sigma$. $A_\Delta$, $A_\Sigma$, $\psi_\Delta$ and $\psi_\Sigma$ are amplitude and phase terms due to target Doppler frequency, the responses of the matched filter, and terms of the radar range equation, specifically, the target RCS fluctuation (e.g., Swerling fluctuation—[7]). If the matched filters are matched, $A_\Delta = A_\Sigma$ and $\psi_\Delta = \psi_\Sigma$. $n_\Delta$ and $n_\Sigma$ represent the receiver noise terms. They are complex, base band signals in the equation forms of (5.106) and (5.107).

The reason for including (5.106) and (5.107) was to point out that $V_\Delta(\tau_{trk})$ and $V_\Sigma(\tau_{trk})$ will not be purely real or purely imaginary. Instead, they will be complex numbers. However, from (5.102), (5.103), (5.104), and (5.105), we know we must use either the real or imaginary part of the ratio of $V_\Delta(\tau_{trk})$ to $V_\Sigma(\tau_{trk})$ to form a valid error signal.

5.8.1.1 Constrained Feed Array

If we consider the phase comparison monopulse case and form $e$ using (5.101), (5.104), and (5.105), we get (see Problem 11)

$$\begin{aligned} e_s\left(s_T - s_{trk}\right) &= \mathrm{Im}\left[\frac{V_\Delta\left(\tau_{trk}\right)}{V_\Sigma\left(\tau_{trk}\right)}\right] S_D = \tan\left(\pi K \frac{d}{\lambda}\left(s_T - s_{trk}\right)\right) S_D \\ &= \tan\left(\pi K \frac{d}{\lambda} \Delta s\right) S_D = e_s\left(\Delta s\right) \end{aligned} . \tag{5.108}$$

In many cases, we want $e_s(\Delta s) = \Delta s$ for $\Delta s$ near zero. This means we want to choose $S_D$ so that

$$\left.\frac{\partial e_s\left(\Delta s\right)}{\partial \Delta s}\right|_{\Delta s=0} = 1 = \left.\frac{\partial \tan\left(\pi K \frac{d}{\lambda} \Delta s\right)}{\partial \Delta s} S_D\right|_{\Delta s=0} = \pi K \frac{d}{\lambda} S_D \tag{5.109}$$

or

$$S_D = \frac{\lambda}{\pi K d} . \tag{5.110}$$

Figure 5.34 contains a plot of $e_s(\Delta s)$ for the 50-element, constrained feed uniformly illuminated, linear array example of Section 5.3. For that example, $K = 50$ and $d = \lambda/2$, so $S_D = 1/(25\pi)$ sine/sine. The plot is reasonably linear within ±1/2 beam width of $\Delta v = 0$. However, outside of that region, the slope rapidly increases as the curve heads toward infinity. This is an undesirable behavior that we will discuss later.

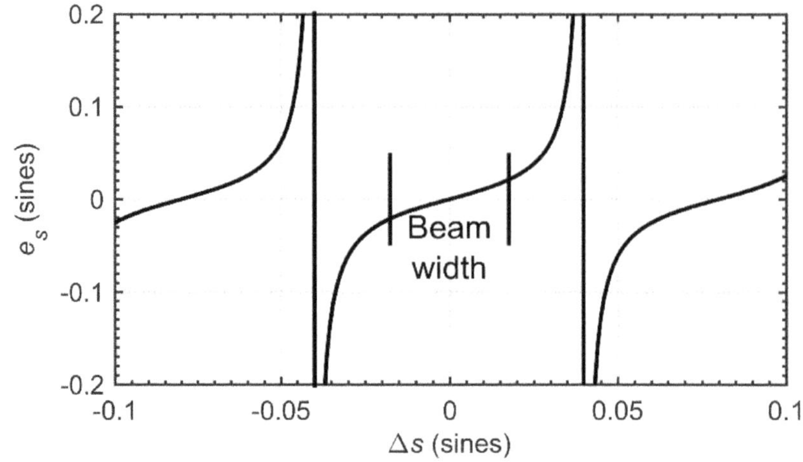


**Figure 5.34** Discriminator curve for a 50-element, constrained feed, linear array with half-wavelength element spacing.

### 5.8.1.2 Space-Fed Array

For the space-fed array, we form $e_s(\Delta s)$ using (5.100), (5.102), and (5.103). Unfortunately, it does not reduce to a simple form, as was the case for the constrained feed array. Figure 5.35 contains a plot of $e_s(\Delta s)$ for the space fed example of Section 5.4. For that example, $K = 50$, $d = \lambda/2$, and $s_s = 0.0136$ sines. In this case, $S_D$ was set to

$$S_D = \frac{2.5\lambda}{\pi K d} \tag{5.111}$$

to provide a slope of one near $\Delta s = 0$. The multiplier of 2.5 was determined experimentally. It is interesting that the discriminator curve of Figure 5.35 has the same tangent-looking shape as the discriminator curve of Figure 5.34.

The repeating nature of the discriminator curves of Figure 5.34 and Figure 5.35 are somewhat misleading. The portion of the curve in the center is due to the division of center portion of the difference pattern by the center portion of the sum pattern. The portions away from the center are due to division of the sidelobes of the difference patterns by the sidelobes of the sum pattern. In those regions, the sum and difference signals will be dominated by noise, and the discriminator curve will be the ratio of two, predominately noise signals. An example of this is indicated in Figure 5.36, which plot like Figure 5.34 where noise was added to the sum and difference signals. The noise level was chosen so the SNR at the peak of the sum pattern was 15 dB. The region in the center is somewhat smooth, but the region outside of about ±0.03 sines is predominated by the noise. The beam width for this antenna is about 0.035 sines (see the text near (5.66)), which means the discriminator curve is fairly smooth over the main beam region.

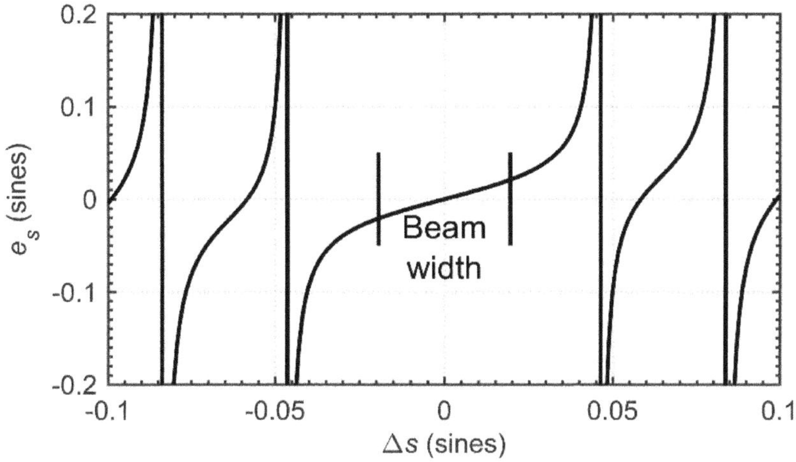


**Figure 5.35** Discriminator curve for a 50-element, space fed, linear array with half-wavelength element spacing and a squint angle of 0.0136 sines.

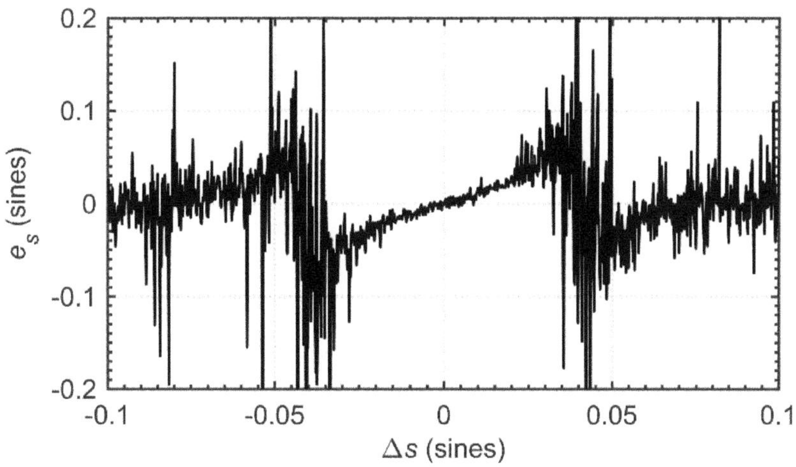


**Figure 5.36** Discriminator curve of Figure 5.33 with the influence of noise.

### 5.8.2 Modified Exact Processor

The fact that the slope of the discriminator curves of Figure 5.34 and Figure 5.35 rapidly increases as $\Delta s$ moves away from zero can cause angle tracking problems. Since the closed loop gain of the angle tracker is the slope of the discriminator curve, the fact that it increases can cause unexpected transient behavior of the angle tracker. A potential means of mitigating this behavior is to use a modified, exact processor. To derive that form, we start by multiplying the numerator and denominator of the $V_\Delta/V_\Sigma$ ratio in (5.100) and (5.101) by $V_\Sigma^*$. This does not change the form of (5.100) and (5.101) since we are essentially multiplying the ratio by unity, we add $|V_\Delta|^2$ to the denominator. The result of this is

$$e = \mathrm{Re}\left[\frac{V_\Delta(\tau_{trk})V_\Sigma^*(\tau_{trk})}{|V_\Sigma(\tau_{trk})|^2 + |V_\Delta(\tau_{trk})|^2}\right]S_D \tag{5.112}$$

and

$$e = \mathrm{Im}\left[\frac{V_\Delta(\tau_{trk})V_\Sigma^*(\tau_{trk})}{|V_\Sigma(\tau_{trk})|^2 + |V_\Delta(\tau_{trk})|^2}\right]S_D. \tag{5.113}$$

The addition of $|V_\Delta|^2$ prevents the denominator of the two equations from going to zero when $V_\Sigma$ goes through zero. If $V_\Sigma$ and $V_\Delta$ are zero at the same time, the denominator could be zero. However, in that case, the numerator would also be zero.

Figures 5.37 and 5.38 contain discriminator curves for the constrained and space fed antennas used to generate Figures 5.34 and 5.35. The net effect of including $|V_\Delta|^2$ in the denominator of (5.112) and (5.113) is to remove the large peaks of Figures 5.34 and 5.35. However, the shape of the discriminator curve in the main beam region is not affected. We will examine the effect of using the two types of processors when we consider example angle track loops.

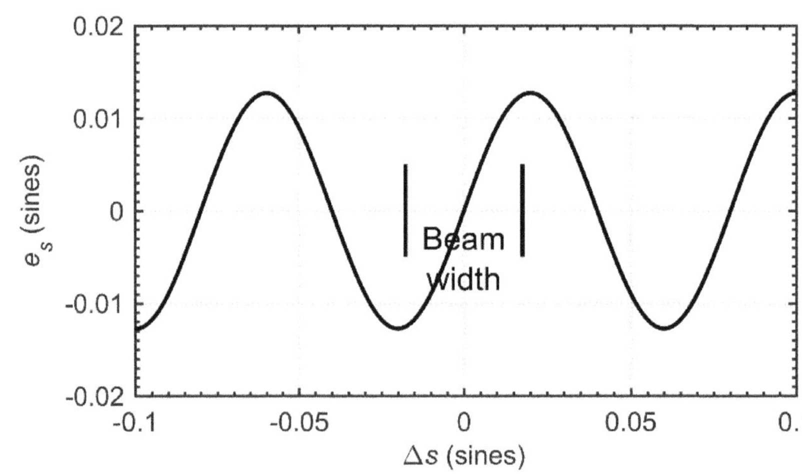


**Figure 5.37** Discriminator curve for a 50-element, constrained feed, linear array with half-wavelength element spacing—modified exact processor.

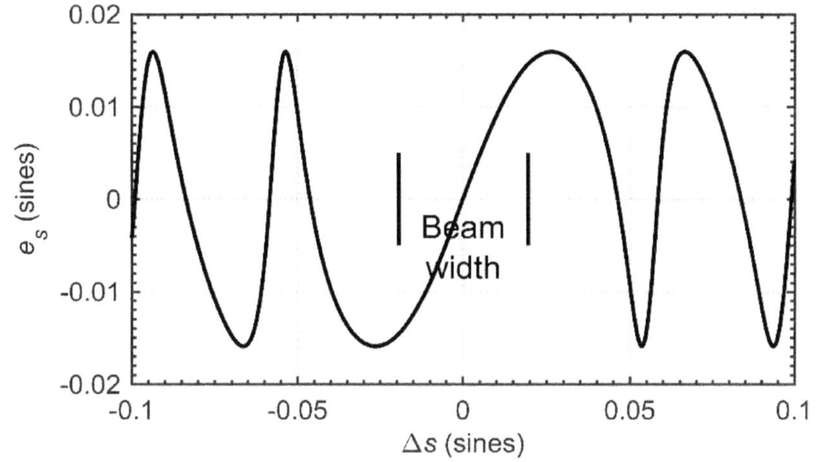


**Figure 5.38** Discriminator curve for a 50-element, space-fed, linear array with half-wavelength element spacing and a squint angle of 0.0136 sines—modified exact processor.

Figure 5.39 contains a plot like Figure 5.36 for the case where the modified exact processor is used. The vertical scales of the two figures were kept the same to illustrate that the modified exact processor also offers some noise reduction in the regions beyond the main beam. This should help reduce the noise in the angle tracker.

The addition of $|V_\Delta|^2$ to the denominator of the exact processor gives an additional degree of freedom we can use to help shape the discriminator curve. In particular, if we write (5.112) and (5.113) as

$$e = \mathrm{Re}\left[\frac{V_\Delta(\tau_{trk})V_\Sigma^*(\tau_{trk})}{|V_\Sigma(\tau_{trk})|^2 + k_\Delta |V_\Delta(\tau_{trk})|^2}\right] S_D \tag{5.114}$$

and

$$e = \mathrm{Im}\left[\frac{V_\Delta(\tau_{trk})V_\Sigma^*(\tau_{trk})}{|V_\Sigma(\tau_{trk})|^2 + k_\Delta |V_\Delta(\tau_{trk})|^2}\right] S_D \tag{5.115}$$

We can change the shape of the discriminator curve by selecting different values of $k_\Delta$. As an example, Figure 5.40 contains plots like Figure 5.37 with $k_\Delta$ values of 1.0 and 0.2. By changing $k_\Delta$ from 1.0 to 0.2, we extended the linear region of the discriminator curve.

We note that there is no theoretical basis for the form of the modified exact processor. However, its use does improve tracking behavior.

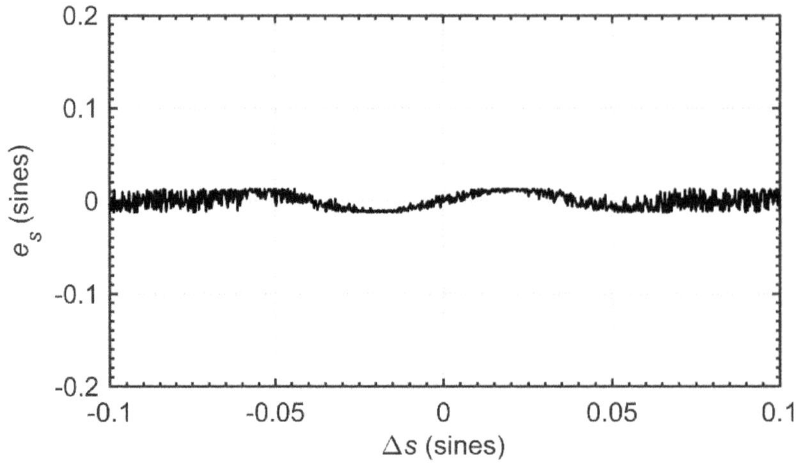


**Figure 5.39** Discriminator curve of Figure 5.36 with the influence of noise.

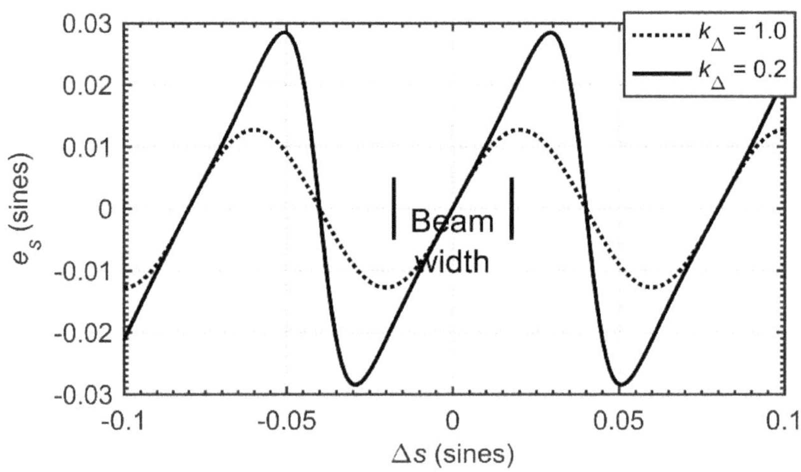


**Figure 5.40** Discriminator curve for a 50-element, constrained feed, linear array with half-wavelength element spacing—modified exact processor with different values of $k_\Delta$.

### 5.8.3 Log-Based Processor

Figure 5.41 contains a functional block diagram of a logarithm-based monopulse processor (log processor). The inputs are $V_-(\tau_{trk}) = V_\Sigma(\tau_{trk}) - V_\Delta(\tau_{trk})$ and $V_+(\tau_{trk}) = V_\Sigma(\tau_{trk}) + V_\Delta(\tau_{trk})$. They can be directly obtained at the outputs of a two-channel monopulse receiver/signal processor, or by combining appropriate signals from a three channel monopulse receiver/signal processor. A basic assumption of the log processor is that $V_\Sigma(\tau_{trk})$ and $V_\Delta(\tau_{trk})$ are in phase, or 180° out of phase. This occurs naturally in amplitude comparison monopulse. In phase comparison monopulse, the phase of $V_\Delta(\tau_{trk})$ must be shifted by 90°. In two-channel monopulse receivers, this is usually done when the $V_{RF\Sigma}(t) \pm V_{RF\Delta}(t)$ signals are formed (see Figures 5.22, 5.24, and 5.25). In three-channel monopulse receivers, it can be done at RF, or right before the $V_\Sigma(\tau_{trk}) \pm V_\Delta(\tau_{trk})$ signals are formed.

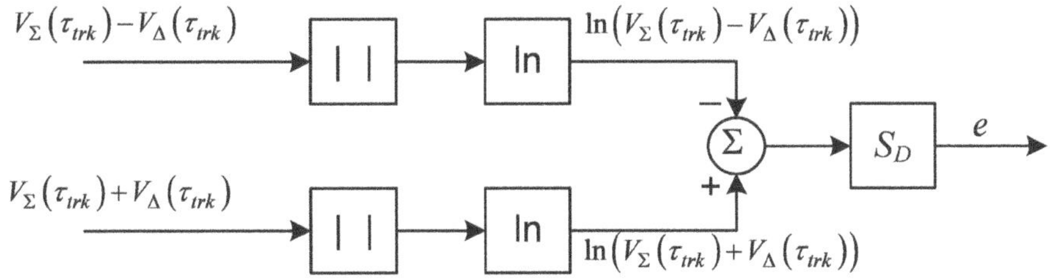


**Figure 5.41** Logarithm-based monopulse processor.

From the block diagram of Figure 5.40, we get

$$\begin{aligned} e &= \left[\ln\left|V_+(\tau_{trk})\right| - \ln\left|V_-(\tau_{trk})\right|\right] S_D \\ &= \left[\ln\left|V_\Sigma(\tau_{trk}) + V_\Delta(\tau_{trk})\right| - \ln\left|V_\Sigma(\tau_{trk}) - V_\Delta(\tau_{trk})\right|\right] S_D \end{aligned} . \quad (5.116)$$

Since the difference of logarithms is the ratio of their arguments, we can write

$$e = \ln\left(\frac{\left|V_\Sigma(\tau_{trk}) + V_\Delta(\tau_{trk})\right|}{\left|V_\Sigma(\tau_{trk}) - V_\Delta(\tau_{trk})\right|}\right) S_D . \quad (5.117)$$

If we factor $V_\Sigma(\tau_{trk})$ from the numerator and denominator we get

$$\begin{aligned} e &= \ln\left(\frac{\left|V_\Sigma(\tau_{trk})\left(1 + V_\Delta(\tau_{trk})/V_\Sigma(\tau_{trk})\right)\right|}{\left|V_\Sigma(\tau_{trk})\left(1 - V_\Delta(\tau_{trk})/V_\Sigma(\tau_{trk})\right)\right|}\right) S_D \\ &= \ln\left(\frac{\left|1 + V_\Delta(\tau_{trk})/V_\Sigma(\tau_{trk})\right|}{\left|1 - V_\Delta(\tau_{trk})/V_\Sigma(\tau_{trk})\right|}\right) S_D \end{aligned} . \quad (5.118)$$

If the radar is tracking the target, we have $|V_\Delta(\tau_{trk})| << |V_\Sigma(\tau_{trk})|$, which leads to $|V_\Delta(\tau_{trk})|/|V_\Sigma(\tau_{trk})|<<1$ and

$$\left|1 \pm V_\Delta(\tau_{trk})/V_\Sigma(\tau_{trk})\right| \approx 1 \pm V_\Delta(\tau_{trk})/V_\Sigma(\tau_{trk}) . \quad (5.119)$$

With this we get

$$
\begin{aligned}
e &= \ln\left(\frac{\left|1+V_\Delta(\tau_{trk})/V_\Sigma(\tau_{trk})\right|}{\left|1-V_\Delta(\tau_{trk})/V_\Sigma(\tau_{trk})\right|}\right)S_D \\
&= \ln\left(\frac{1+V_\Delta(\tau_{trk})/V_\Sigma(\tau_{trk})}{1-V_\Delta(\tau_{trk})/V_\Sigma(\tau_{trk})}\right)S_D \\
&= \left[\ln\left(1+V_\Delta(\tau_{trk})/V_\Sigma(\tau_{trk})\right)-\ln\left(1-V_\Delta(\tau_{trk})/V_\Sigma(\tau_{trk})\right)\right]S_D
\end{aligned} \quad . \tag{5.120}
$$

From the properties of logarithms, we know that if $|x|<<1$, $\ln(1\pm x) \approx \pm x$. We can use this to write

$$
\begin{aligned}
e &\approx \left[V_\Delta(\tau_{trk})/V_\Sigma(\tau_{trk})-\left(-V_\Delta(\tau_{trk})/V_\Sigma(\tau_{trk})\right)\right]S_D \\
&= 2\frac{V_\Delta(\tau_{trk})}{V_\Sigma(\tau_{trk})}S_D
\end{aligned} \tag{5.121}
$$

Equation (5.121) tells us that we can use the log processor to approximate the exact processor, albeit within a factor of two, which we can account for in $S_D$.

Figure 5.42 contains discriminator curves for the exact and logarithm-based monopulse processors and the linear array used to generate Figure 5.35. The shapes for the curves are similar in the main beam region. The discriminator curve for the log processor appears to head toward ±infinity, but never quite does. Examination of (5.116) indicates the curve would go to ±infinity if $V_\Sigma(\tau_{trk}) + V_\Delta(\tau_{trk})$ or $V_\Sigma(\tau_{trk}) - V_\Delta(\tau_{trk})$ became zero. Neither of these conditions apparently happen for the linear array used to generate Figure 5.42.

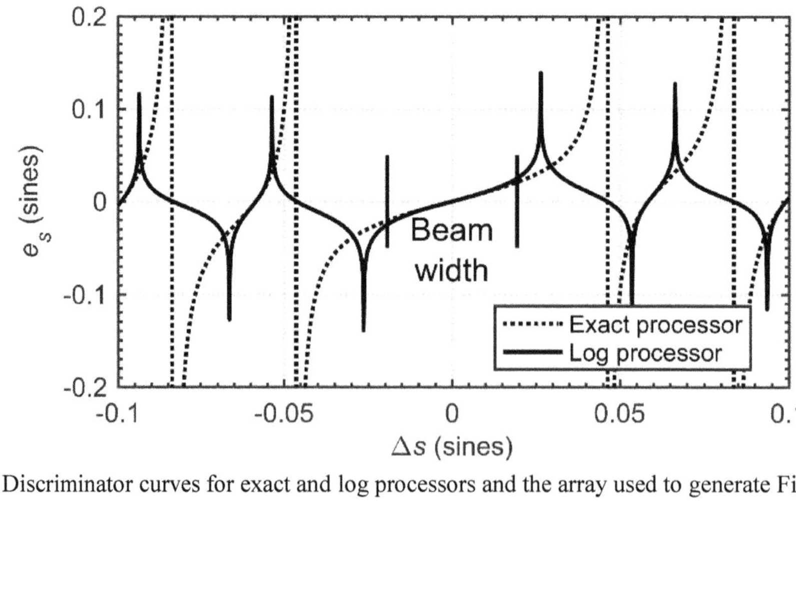


**Figure 5.42** Discriminator curves for exact and log processors and the array used to generate Figure 5.35.

## 5.9 EXAMPLE 1: ANGLE TRACKER WITH DIFFERENT MONOPULSE PROCESSORS

To close out this chapter, we will integrate the three types of monopulse processors (exact, modified exact, and logarithm), the uniformly illuminated, linear, space-fed array of Section 5.4, the Benedict-Bordner α-β track filter of Chapter 3, and a linear array equivalent of a three channel monopulse receiver into an angle tracking simulation. The monopulse receiver processes the sum ($V_{RF\Sigma}(t)$) and difference ($V_{RF\Delta}(t)$) signals from the monopulse combiner. That is why we term it an equivalent three channel monopulse receiver. Although not used in this example, the waveform is a 15-$\mu$s linear frequency modulated (LFM) pulse with a bandwidth of 1 MHz (this fact is used in the next example).

Figure 5.43 contains a block diagram of the model. The target model provides the range delay, $\tau_R$, the target RCS, and the sine space position of the target, $u_T$. The target RCS is used in the radar range equation to compute $P_S$, the target signal power, which is used to scale voltage levels out of the monopulse combiner. The model computes the sine-space angle error, $\Delta u$, which is subsequently used in the antenna model to compute normalized values of the complex voltage directivity of the two squinted beams of the space fed array. These normalized directivities are interpreted as normalized voltages, $V_a$ and $V_b$ (see (5.63) and (5.64)). $V_a$ and $V_b$ are subsequently combined, in the monopulse combiner, to produce normalized sum and difference voltages, $V_\Sigma$ and $V_\Delta$. These are subsequently scaled by $\sqrt{P_S}$.

It is assumed the matched filters are matched to the transmit LFM pulse and that $\tau_{trk} = \tau_R$, so that the outputs of the matched filters are always at their maximum value. The range gates, which are not explicitly shown, sample the matched filter output at $\tau_{trk}$. Noise is subsequently added to the matched filter outputs, and the sums are sent to the monopulse processor block which forms the error signal, $e_u$. The error signal from the monopulse processor is the input to the track filter. The track filter produces the tracked, sine-space, angle, $u_{trk}$, which is subsequently fed back to the antenna model. The antenna model acts as the controlled element.

As indicated above, all three monopulse processor types were included in the model. The exact processor is represented by (5.100), and the log processor is represented by (5.116). The modified exact processor is represented by (5.114) with $k_\Delta = 0.2$. $S_D$ is given by (5.111). For the log processor, $S_D$ was divided by 2, as indicated in Section 5.8.3.

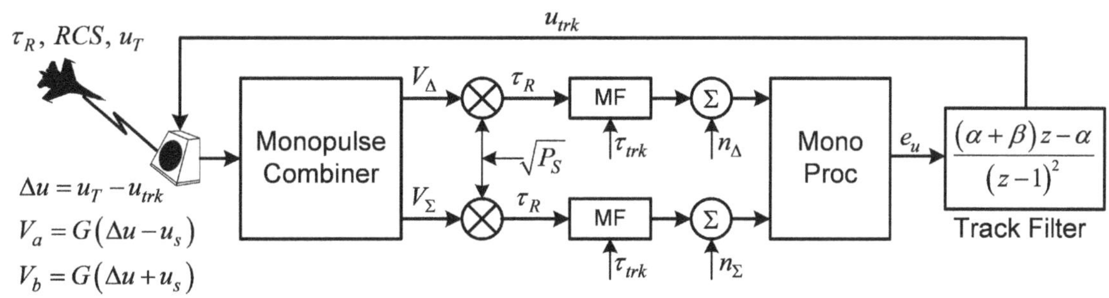


**Figure 5.43** Closed loop angle tracker model block diagram.

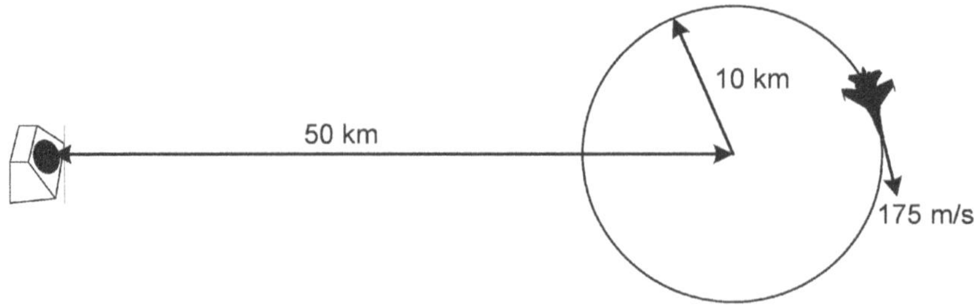


**Figure 5.44** Scenario used in the angle track simulations.

We used a track update period of $T = 0.1$ seconds, or a 10 Hz track rate. The α-β filter has a cutoff frequency of 1 Hz. This, the 10 Hz track rate, and the Benedict-Bordner configuration means the filter coefficients are (see Section 3.3)

$$\alpha = 0.323 \text{ and } \beta = 0.0622. \tag{5.122}$$

The scenario used in the simulation is the one used in Chapter 3, which is illustrated in Figure 5.44. In this scenario, the target is flying a circular path with a radial dimension of 10 km. The tangential velocity of the target is constant at approximately 175 m/s, or about Mach 0.5 [37]. The constant radial velocity of 175 m/s and radius of 10 km means the target is experiencing about 1/3 g of centripetal acceleration [38]. We assumed the target was at an altitude of 0 m, and the angle tracker was tracking in azimuth. Thus, in the notation of the previous sections, $\theta$ is azimuth angle, and the sine-space angle is $u$.

Tracking starts when the target is at the bottom of the circle and continues for 200 s, or until the target is slightly past the top of the circle. During the track, the azimuth angle varies from about −11.5° to about +11.5°.

The remaining simulation parameters are listed in Table 5.1. The reference SNR, $SNR_{ref}$, is the SNR at the start of tracking. The SNR starts at 10 dB and reaches a maximum of 14.2 dB before decreasing to 9 dB at the end of the simulation run. The position state of the track filter was initialized to the actual azimuth angle (−11.5°) plus an offset of ½ of the antenna beam width, or 1°. The initial rate state was set to zero.

The simulation procedure followed the same steps outlined in the range tracking example of Chapter 4. Specifically, the sequence of operations in the iterative section of the simulation are:

1. Compute the predicted track filter state, $X_p$, at the current time using (3.19).
2. Compute the predicted angle, $u_{trk}$, at the current time using (3.22).
3. Compute the true target range and angle, $R$ and $u_T$, at the current time.
4. Use $u_T$ and $u_{trk}$ to compute the sum and difference pattern directivities from (5.102) and (5.105). The equations must be normalized to a sum pattern directivity of 1.0 by dividing $V_\Sigma$ and $V_\Delta$ by the un-normalized peak of $V_\Sigma$. (This is necessary to assure proper SNR to Ps scaling.)
5. Compute the $SNR$ using (4.90).
6. Multiply $\sqrt{P_S}$ by the normalized sum and difference pattern directivities to get $v_\Sigma(\tau_{trk})$ and $v_\Delta(\tau_{trk})$, the interim values of $V_\Sigma(\tau_{trk})$ and $V_\Delta(\tau_{trk})$.

7. Draw four, independent, zero mean, unit variance, Gaussian random numbers and use them to form the noise samples $n_\Sigma = (n_1 + jn_2)/\sqrt{2}$ and $n_\Delta = (n_3 + jn_4)/\sqrt{2}$. We can assume independent noise samples because the noise in the sum and difference channels are from different sources.
8. Add the noise samples to $v_\Sigma(\tau_{trk})$ and $v_\Delta(\tau_{trk})$ to form $V_\Sigma(\tau_{trk})$ and $V_\Delta(\tau_{trk})$.
9. Process $V_\Sigma(\tau_{trk})$ and $V_\Delta(\tau_{trk})$ through the appropriate monopulse processor to get the error signal $e_u$.
10. Use $e_u$ to compute the smoothed state, $X_s$, using (3.20).
11. Save the outputs.
12. Update $t$.

Figures 5.45, 5.46, and 5.47 contain simulation output plots for each of the three monopulse processors. In the plots, sine-space angles were converted to angles in degrees using $\theta = \sin^{-1}u$. The top plot contains the actual angle, $\theta_T$, (dotted line) and tracked angel $\theta_{trk}$ (solid line). Since the tracker is doing a good job of tracking the target, the curves appear as a single curve. The center plot is error, $\Delta\theta = \theta_T - \theta_{trk}$ between the actual target angle and the predicted target angle from the tracker. The bottom plot is the error signal, $e_\theta$. It is the error signal from the monopulse processor that is fed to the α-β filter. The units of $e_\theta$ is degrees because the processor scaling factor, $S_D$, was chosen to make the slope of the discriminator curve unity.

All three curves were generated using the same noise sequence to remove the influence different noise sequences might have on the relative performance of the trackers.

As can be seen, the performances of the tracker with the three monopulse processors are essentially the same. In all three cases, the $\Delta\theta$ plot has a smaller amplitude and is somewhat smoother than the $e_\theta$ plot. This is because the $\Delta\theta$ plot reflects the filtering properties (i.e., noise reduction) of the track filter (the α-β filter), whereas the $e_\theta$ contains the effect of the unfiltered receiver noise on the monopulse processor. The peak-to-peak error signal ($e_\theta$) varies roughly between ±1/2 beam width. This rather large variation is due to the low value of SNR. When the simulation was run with an initial SNR ($SNR_{ref}$) of 15 or 20 dB, the variation was considerably smaller.

The only time differences between the three monopulse processors showed up was when the initial SNR was reduced. At an initial SNR of 8 dB, the tracker with the exact processor lost track in 8 of 50 Monte Carlo runs. This occurred near the end of the 200 second simulation period, when the SNR dropped to about 7 dB. The trackers with the modified exact processor and the log processor did not lose track on any of the 50 Monte Carlo runs.

**Table 5.1**

Simulation Parameters for Example 1

| *Parameter* | *Value* | *Comment* |
|---|---|---|
| $SNR_{ref}$ | 10 dB | The reference SNR |
| $R_{ref}$ | 51 km | The reference range |
| $\sigma_{ref}$ | 1 m$^2$ | The reference RCS |
| $\sigma_{tgt}$ | 1 m$^2$ | The target RCS |
| $u_{init}$ | $u_{actual}(0)$ + BW/2 | The initial $u$ for the track filter |
| BW | 2° | Antenna Beam Width |
| $f_c$ | 1 Hz | Closed loop bandwidth |
| $T$ | 1/10 s | 10 Hz track rate |

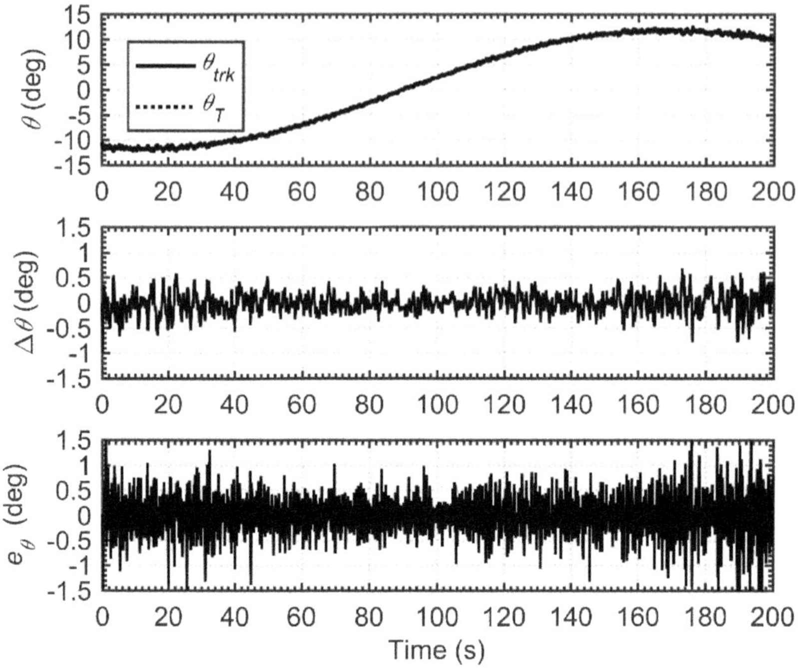


**Figure 5.45** Simulation output for the angle tracker with the exact processor.

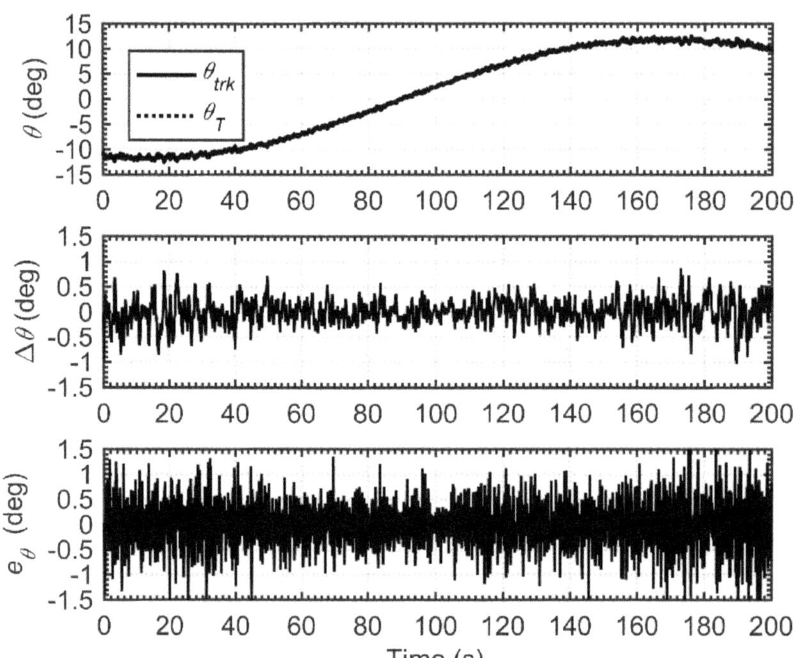


**Figure 5.46** Simulation output for the angle tracker with the modified exact processor.

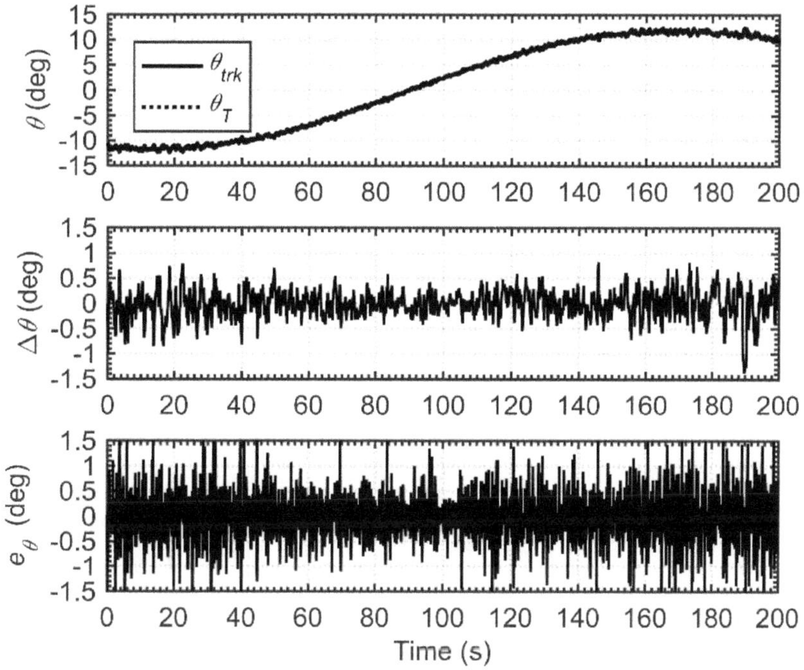


**Figure 5.47** Simulation output for the angle tracker with the log processor.

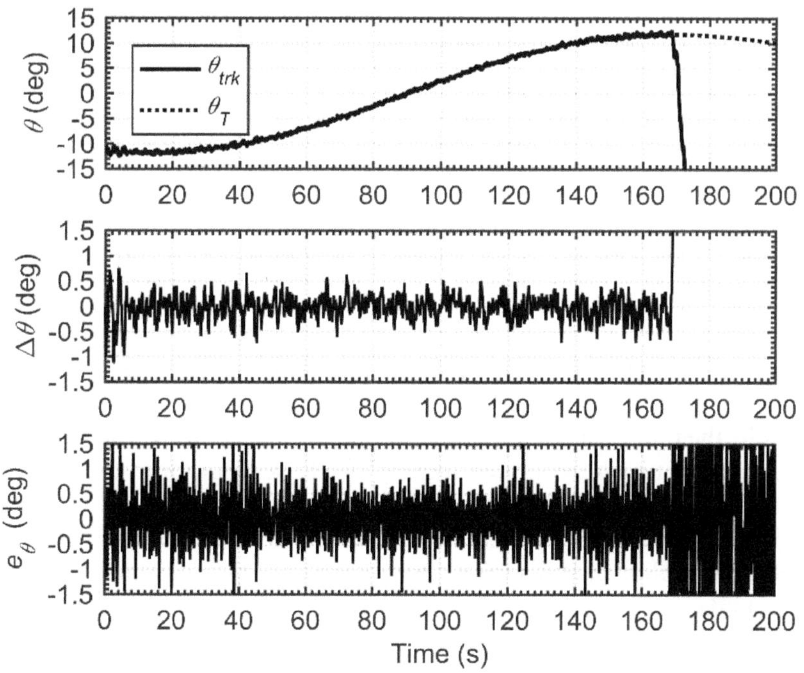


**Figure 5.48** Simulation output for the angle tracker with the exact processor—initial SNR of 8 dB.

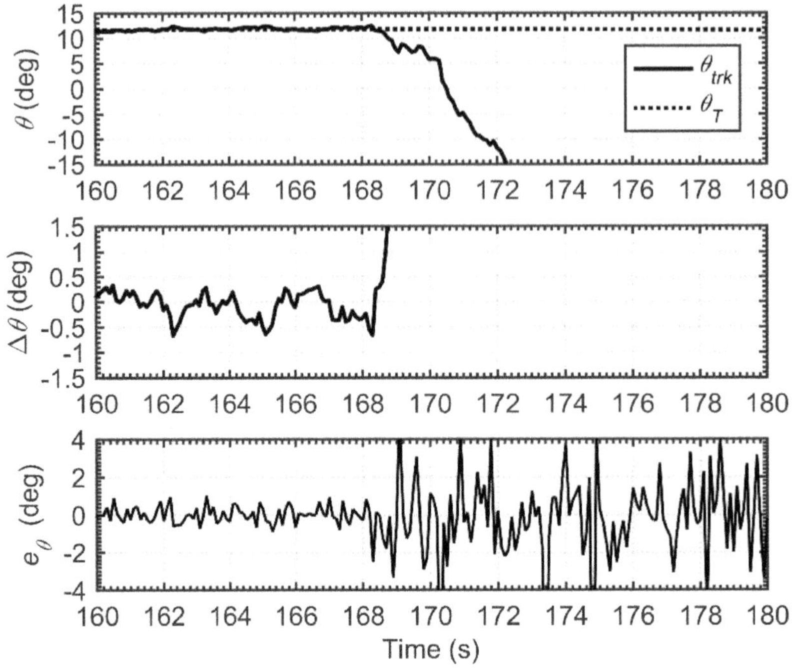


**Figure 5.49** Simulation output for the angle tracker with the exact processor—initial SNR of 8 dB—expanded time scale.

Figure 5.48 contains plots of the simulation output for one of the cases where the tracker lost track. An expanded version of the output is shown in Figure 5.49. Note that $\Delta\theta$ started drifting negative about 167 seconds. Shortly after 168 seconds, this drift and noise caused a succession of spikes in $e_\theta$, with amplitudes of about −2.5°, −2°, and −3°. These two events were most likely sufficient to drive $\Delta\theta$ out of the central region of the discriminator curve (see Figures 5.35 and 5.36) and resulted in $e_\theta$ being predominated by noise. Shortly after 169 seconds, the noise, and possibly a negative spike of the discriminator curve (see Figure 5.35), caused a pair of large positive spikes in $e_\theta$. This was apparently sufficient to drive the $\theta_{trk}$ so far away from $\theta_T$ that the angle tracker could not recover.

## 5.10 EXAMPLE 2: COMBINED ANGLE AND RANGE TRACKER

As a second example, the range tracker model of Chapter 4 was combined with the angle tracker model of this chapter to form an integrated tracker. A block diagram of the combined tracker is shown in Figure 5.50. The antenna model was the space fed model used in the angle model of Section 5.9. The modified exact processor was used as the monopulse processor. The range discriminator was the sampling gate discriminator, with $q = ½$ (see Chapter 4). The track filters for the range and angle track loops were the Benedict-Bordner α-β filters used in the range tracker model of Chapter 4 and the angle tracker model of Section 5.

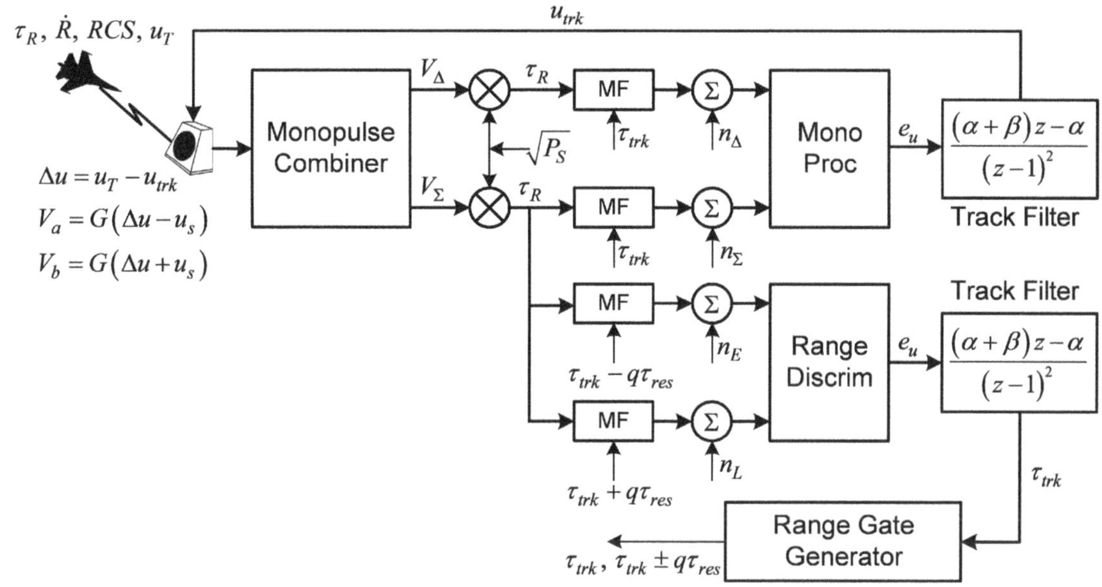


**Figure 5.50** Combined range and angle track model.

The range gates shown in other block diagrams are not specifically shown. As discussed earlier, in the actual radar, the sum and difference channels would each contain a single matched filter (and, possibly, a signal processor). Each of the difference channels would contain a single range gate that samples the matched filter output at the tracked range delay, $\tau_{trk}$. The sum channel would contain three range gates that sample the sum channel matched filter output at $\tau_{trk} - q\,\tau_{res}$, $\tau_{trk}$, and $\tau_{trk} + q\,\tau_{res}$. In the model, the sampling is modeled by computing the matched filter output at $\tau_{trk}$ in the difference channel (the model has only one difference channel) and the three values indicated in the previous sentence for the sum channel.

The controlled element in the range tracker is the range gate generator indicated in Figure 5.50. In the model, it uses the output of the range track filter to compute $\tau_{trk} - q\,\tau_{res}$, $\tau_{trk}$, and $\tau_{trk} + q\,\tau_{res}$. As discussed in the previous paragraph, they subsequently go to the matched filter model to compute their outputs at these times.

The target model is the same as that used in the example of Section 5.9. However, it was modified to provide range-rate, which was used to compute Doppler frequency, $f_d$. The Doppler frequency was used in the matched filter model. It was assumed the matched filter was matched to a Doppler frequency of $f_M$ of zero. The radar operating frequency, $f_c$, which is needed to compute $f_d$, was set to 5 GHz.

The sequence of operations in the model was a combination of the sequences of operation used in the range tracker model of Chapter 4 and the angle tracker model of Section 5.9. The range track filter was initialized to the true target range delay plus ½ of the range resolution of the LFM waveform, and the angle track filter was initialized to the true target angle plus ½ of the antenna beamwidth. The rate states of both filters were initialized to zero.

Figure 5.51 contains plots of the model output. The left column of plots is the output of the angle tracker, and the right column is the output of the range tracker. In general, both trackers work well. When the initial SNR was reduced to 7 or 8 dB, instead of the 10 dB used to generate

Figure 5.51, both the range and angle tracker sometimes lost track, or were never able to establish track.

Figure 5.52 contains plots of the simulation output for one of the cases where the radar lost track. An expanded version of the output is shown in Figure 5.53. It appears as if the radar first lost angle track and then range track. This conclusion is based on the observation that the $\Delta\theta$ plot deviated away from zero before the $\Delta R$ plot did. The exact reason the angle tracker lost track is not clear. There was an approximate $-1.8°$ spike in $e_\theta$ at about 188 seconds. This was followed by an approximate $+1.8°$ spike at about 189 seconds. It could have been these two spikes perturbed the angle track filter enough to cause $\Delta\theta$ to leave the central region of the discriminator curve. Examination of the $e_R$ plot indicates it had a negative spike of about $-60$ meters, followed by a positive spike of about $+60$ meters. It could have been this caused an oscillation in the range track filter, which subsequently caused the signals in the angle channel to move too far away from matched range. The counter to this argument is that the angle tracker appears to deviate before the range tracker.

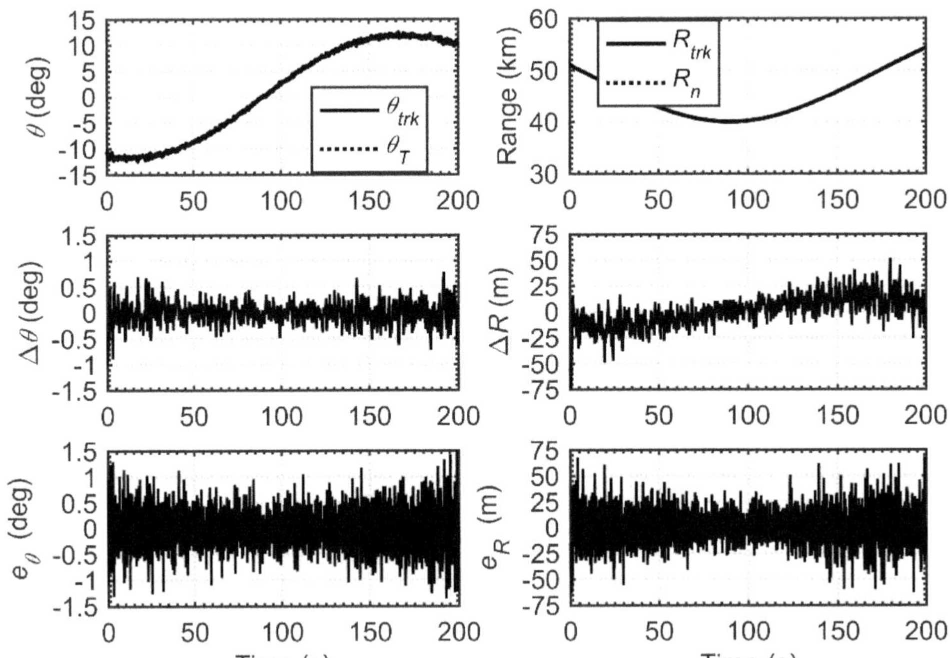


**Figure 5.51** Simulation output for combined range and angle trackers.

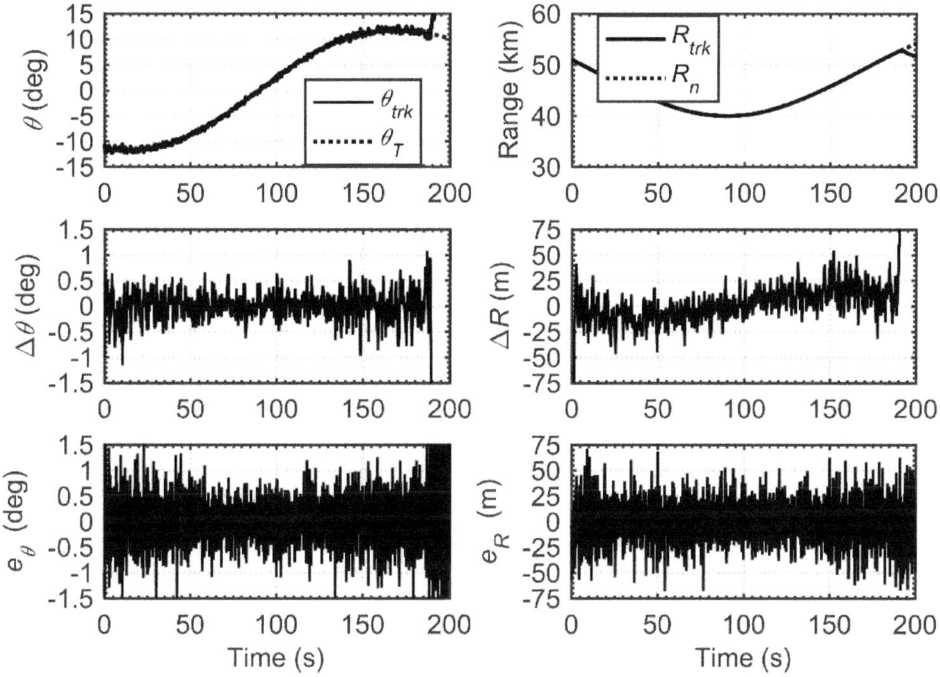


**Figure 5.52** Simulation output for combined range and angle trackers—8 dB initial SNR.

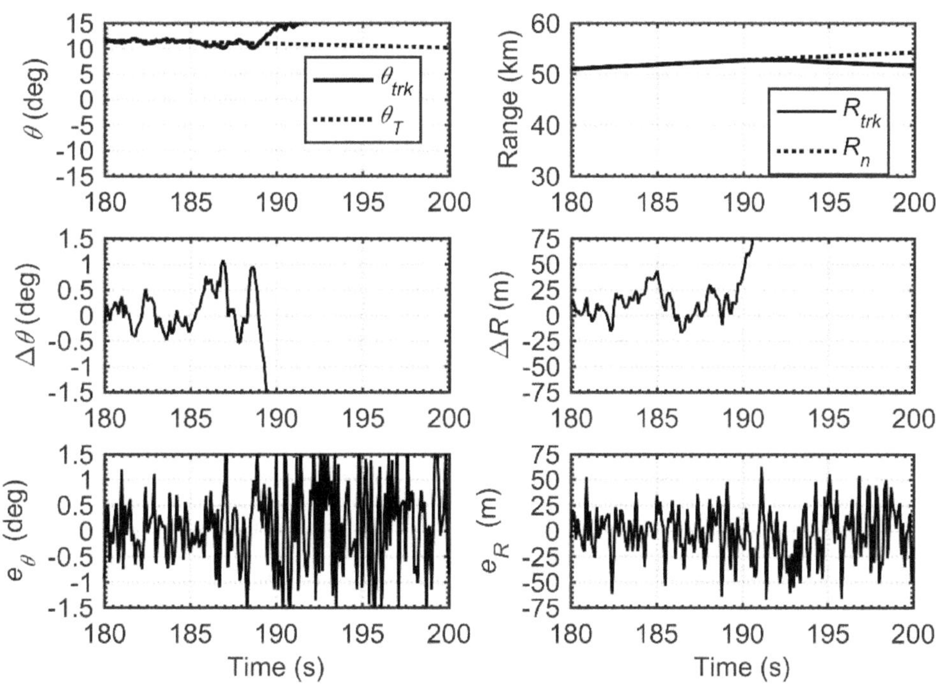


**Figure 5.53** Simulation output for combined range and angle trackers—8 dB initial SNR—expanded scale.

## 5.11 FUNCTIONAL LEVEL ERROR MODEL

In some preliminary or functional level tracking studies, development of an antenna, monopulse processor, and tracker model is not desirable. In such studies, the analyst simply wants an estimate of the expected angle measurement error as a function of SNR and antenna beam width. To that end, various authors provide equations of the form [24, p. 388; 32]

$$\sigma_{angmeas} = \frac{BW}{k_{ang}\sqrt{SNR}} \tag{5.123}$$

for the root-mean-square (RMS) angle measurement error. In (5.123), $k_{ang}$ is a scaling factor that depends on the normalized monopulse slope of the antenna [39, p. 136].

Barton [40, pp. 467–468] has a modified form of (5.123) that is applicable to lower values of *SNR*. His equation, in the form of (5.123) is

$$\sigma_{angmeas} = \frac{BW}{k_{ang}\sqrt{(1+SNR)}}. \tag{5.124}$$

To use the above equations, one would draw a zero-mean, Gaussian random number with a variance of $(\sigma_{angmeas})^2$ and add it to the true target angle.

Figure 5.54 contains plots of (5.123) and (5.124) with a $k_{ang}$ value of 3.2. The figure also contains a plot of an experimental $\sigma_{angmeas}$. The experimental plot was generated for the linear, uniformly illuminated, space fed antenna discussed in this chapter, and the modified exact monopulse processor. As can be seen, the theory equations and the experimental curves match well for SNR values above about 5 dB. The log processor yielded a similar plot. Also, similar results were obtained for the constrained feed array, with uniform weighting, when $k_{ang}$ was set to 2.8.

Figure 5.55 contains a plot for the exact processor. This plot indicates the actual error will be larger than the theoretical error for low values of SNR. This is due to the rapid increase in discriminator slope illustrated in Figures 5.34 and 5.35. This rapid increase in discriminator slope makes the monopulse processor output quite noisy, as was illustrated in Figure 5.36. As a note, generating the experimental curve for Figure 5.55 required 1,000,000 Monte Carlo tries. Even with that large number of Monte Carlo tries, the experimental curve changed somewhat from one set of 1,000,000 Monte Carlo tries to the next. However, the increasing error trend was consistent.

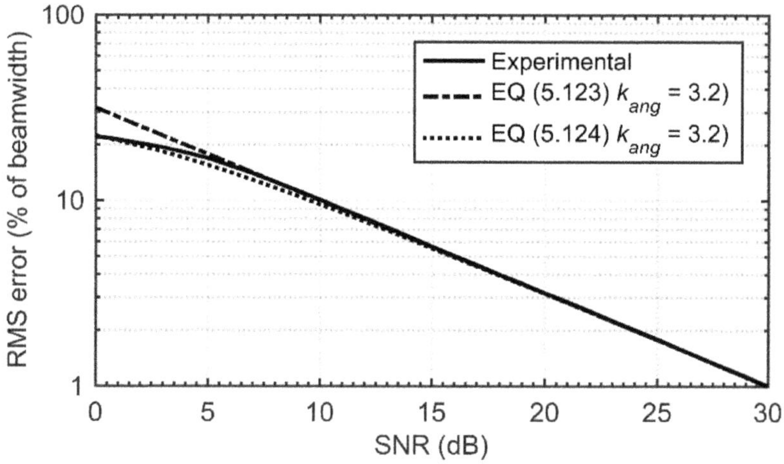


**Figure 5.54** RMS angle measurement error for a modified exact and log processors.

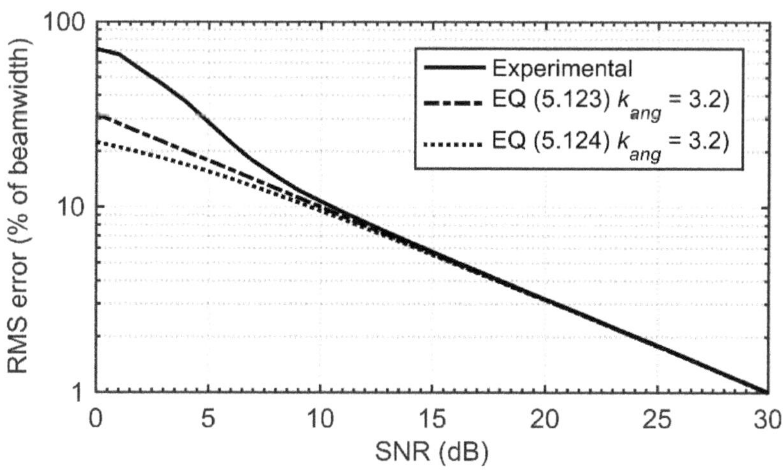


**Figure 5.55** RMS angle measurement error for the exact processor.

## 5.12 EXERCISES

1. Derive (5.29).
2. Derive (5.31).
3. Derive (5.32).
4. Derive (5.45)
5. Derive (5.50).
6. Derive (5.51).
7. Derive (5.58).
8. If the inputs are applied to ports 1 and 4 of the Magic Tee of Figure 5.15, show that the $\Sigma$ signal will appear at Port 3, and the $\Delta$ signal will appear at Port 2. Assume the E-field

direction in Port 4 is from left to right (the direction of the solid arrows in Figure 5.15), and the E-field direction in Port 1 is up. What happens if the E-field direction in Port 4 is from right to left (the direction of the dashed arrows)?

9. Provide the details of the math that lead to (5.75) and (5.76).
10. In footnote 2, we claim the operation of the rat race will be unaffected if all of the path lengths are increased by $\lambda_g/2$. Is this claim correct? Verify your answer.
11. If the input ports of the rat race of Figure 5.16 are 2 and 4, we claim that the $\Sigma$ and $\Delta$ signals will appear at ports 1 and 3. Is this correct? If so, which port will contain the $\Sigma$ signal, and which port will contain the $\Delta$ signal?
12. Derive (5.108).
13. Reproduce Figure 5.10.
14. Reproduce Figure 5.12.
15. Reproduce Figure 5.13.
16. Reproduce Figure 5.19.
17. Reproduce Figure 5.20.
18. Implement the conical scan scheme discussed in Section 5.7 and reproduce Figures 5.30, 5.31, 5.32, and 5.33.
19. Reproduce the discriminator curve of Figure 5.34. Apply the amplitude taper that led to Figure 5.19 and generate the modified discriminator curve.
20. Reproduce the discriminator curve of Figure 5.35. Apply the amplitude taper that led to Figure 5.20 and generate the modified discriminator curve.
21. Reproduce the discriminator curves of Figure 5.40. Generate the discriminator curves for the case where the amplitude taper used to generate Figure 5.19 is used.
22. Reproduce the log processor curve shown in Figure 5.42. Generate the discriminator curve for the case where the amplitude taper used to generate Figure 5.20 is used.
23. Implement the angle tracker of Section 5.9 (Example 1) and produce plots like Figures 5.45, 5.46, and 5.47.
24. Implement the combined range and angle tracker of Section 5.10 and produce plots like Figure 5.49.
25. Generate a plot like Figure 5.50 for the case of a space fed phased array that uses 30-dB, $\bar{n} = 6$, Taylor weighting.

## References


[1] P. J. Kahrilas, *Handbook, Electronic Scanning Radar Systems* (ESRS) Design, Norwood, MA: Artech House, 1976.

[2] W. H. Von Aulock, "Properties of Phased Arrays," *Proceedings of the IRE,* vol. 48, no. 10, pp. 1715-1727, Oct. 1960.

[3] R. J. Mailloux, *Phased Array Antenna Handbook*, 2nd ed., Norwood,, MA: Artech House, 2005.

[4] H. You, X. Jianjuan and G. Xin, *Radar Data Processing With Applications*, Singapore: John Wiley & Sons, 2016.

[5] S. M. Sherman and D. K. Barton, *Monopulse Principles and Techniques*, Second ed., Norwood, MA: Artech House, 2011.

[6] R. A. Fessenden, "Wireless Signaling". Patent 706740, 12 August 1902.

[7] M. C. Budge Jr. and S. R. *German, Basic Radar Analysis*, Norwood, MA: Artech House, 2015.

[8] M. I. Skolnik, Ed., Radar Handbook, 3rd ed., New York: McGraw-Hill, 2008.

[9] J. L. Eaves and E. K. Reedy, *Principles of Modern Radar*, New York: Van Nostrand Reinhold, 1987.

[10] R. Wallace and K. Andreasson, *Introduction to RF and Microwave Passive Components*, Norwood, MA: Artech House, 2015.

[11] W. T. Joines, W. D. Palmer and J. T. Bernhard, *Microwave Transmission Line Circuits*, Norwood, MA: Artech House, 2013.

[12] C. A. Balanis, *Antenna Theory Analysis and Design, Hoboken*, NJ: John Wiley & Sons, 2005.

[13] W. L. Stutzman and G. A. Theile, *Antenna Theory and Design*, Hoboken, NJ: John Wiley & Sons, Inc., 2012.

[14] I. S. Gradshteyn and I. M. Ryzhik, *Table of Integrals, Series, and Products*, Academic Press, 1980.

[15] D. K. Barton, Radars, Volume 1, *Monopulse Radar*, Dedham, MA: Artech House, 1975.

[16] A. I. Leonov and K. I. Fomichev, *Monopulse Radar*, Artech house, 1986.

[17] S. A. Maas, *Microwave Mixers*, 2nd ed., Norwood, MA: Artech House, 1993.

[18] D. M. Pozar, *Microwave Engineering*, New York: Addison-Wesley Publishing Company, Inc., 1990.

[19] P. W. Hannan, "Optimum feeds for all three modes of a monopulse antenna I: Theory," *IRE Transactions on Antennas and Propagation,* vol. 9, no. 5, pp. 444-454, September 1961.

[20] P. W. Hannan, "Optimum feeds for all three modes of a monopulse antenna II: Practice," *IRE Transactions on Antennas and Propagation,* vol. 9, no. 5, pp. 454-461, September 1691.

[21] D. K. Barton, "The 1993 Moscow Air Show," *Microwave Journal,* May 1994.

[22] M. Sarcione, J. Mulcahey, D. Schmidt, K. Chang, R. E. M. Russell, P. Rawlinson, W. Guzak, R. Howard and M. Mitchell, "The design, development and testing of the THAAD (Theater High Altitude Area Defense) solid state phased array (formerly ground based radar)," in *Proceedings of International Symposium on Phased Array Systems and Technology*, Boston, 1996.

[23] E. T. Bayliss, "Design of monopulse antenna difference patterns with low sidelobes," *The Bell System Technical Journal,* vol. 47, no. 5, pp. 623-650, May-June 1968.

[24] D. K. Barton, *Radar Systems Analysis and Modeling*, Boston, MA: Artech House, 2005.

[25] F. Le Chevalier, *Principles of Radar and Sonar Signal Processing*, Norwood, MA: Artech House, 2002.

[26] F. E. Nathanson, J. P. Reilly and M. N. Cohen, *Radar Design Priciples - Signal Processing and the Environment*, 2nd ed., New York: McGraw-Hill, 1991.

[27] R. Nitzberg, *Radar Signal Processing and Adaptive Systems*, Norwood, MA: Artech House, 1999.

[28] M. A. Richards, *Fundamentals of Radar Signal Processing*, Second ed., New York: McGraw-Hill Education, 2014.

[29] D. C. Schleher, *MTI and Pulsed Doppler Radar with MATLAB*, 2nd ed., Norwood: Artech House, 2010.

[30] M. O. Kolawole, *Radar Systems, Peak Detection and Tracking*, New York: Newnes, 2002.

[31] F. Neri, *Introduction to Electronic Defense Systems*, Norwood, MA: Artech House, 1991.

[32] M. I. Skolnik, *Introduction to Radar Systems*, 3rd ed., New York: McGraw-Hill, 2001.

[33] C. F. Chubb, B. L. Hulland and R. S. Noblit, "Simplified Monopulse Radar Receiver". Patent 3,239,836, 8 March 1966.

[34] L. A. Belov, S. M. Smolskiy and V. N. Kochemasov, *Handbook of RF, Microwave, and Millimeter-Wave Components*, Boston: Artech house, 2012.

[35] M. Abramowitz and I. S. Stegun, Eds., *Handbook of Mathematical Functions with Forumlas, Graphs, and Mathematical Tables*, New York: Dover Publications, Inc., 1965.

[36] ANRO Engineering, Inc., *Radar Evaluation Handbook*, D. K. Barton and P. Hamilton, Eds., Norwood, MA: Artech House, 1991.

[37] H. M. James, N. B. Nichols and R. S. Phillips, Eds., *Theory of Servomechanisms*, New York: McGraw-Hill Book Company, Inc., 1947.

[38] И. Н. Синицын (I. N. Sinitsyn), Фильтры Калмана и Пугачева (Kalman and Pugachev Filters), Москва: Логос (Moscow: Logos), 2006.

[39] S. M. Sherman, *Monopulse Principles and Techniques*, Norwood, MA: Artech House, 1984.

[40] D. K. Barton, *Modern Radar System Analysis*, Norwood, MA: Artech House, 1988.

[41] M. I. Skolnik, Ed., *Radar Handbook*, New York: McGraw-Hill, 1970.

# Chapter 6

# Closed Loop Doppler Tracking

## 6.1 INTRODUCTION

Figure 6.1 contains a block diagram of a closed loop Doppler[24] tracker. The input to the tracker is the signal returned from the target, which, for purposes of studying Doppler trackers, could take the form of a single pulse, a sequence, or burst, of pulses, or a continuous wave (CW) signal. The signal property of interest is its Doppler frequency, $f_d$. That is, the offset in the carrier frequency, $f_{RF}$, caused by motion of the target.

The returned signal is propagated through the receiver and on to the error sensor, which is termed a *Doppler discriminator*. The Doppler discriminator compares the Doppler frequency of the return signal to a predicted Doppler frequency, $f_{dtrk}$, and produces an error signal, $e_f$, that is proportional to $\Delta f = f_d - f_{dtrk}$, the error between the true Doppler frequency and the Doppler frequency predicted by the Doppler tracker. The structure of the Doppler discriminator depends on the characteristics of the transmit signal, the receiver, and the signal processor.

The error signal from the Doppler discriminator is sent to the track filter. The track filter shown in Figure 6.1 is an α-β filter of the type discussed in Chapter 3. In practice, it would not be uncommon to use a Type 1 servo, since $f_d$ generally varies slowly.

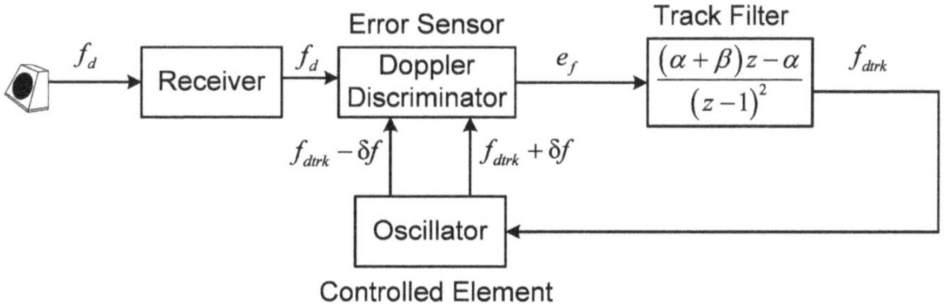


**Figure 6.1** Closed loop doppler frequency tracker.

[24] To be precise, we should use the phrase Doppler frequency. We will alternate between the word Doppler and Doppler frequency in this chapter.

Finally, the loop is closed through an oscillator that produces a pair of signals at frequencies of $f_{dtrk} - \delta f$ and $f_{dtrk} + \delta f$. These frequencies are analogous to the early and late gates used in range trackers, or the difference beams used in angle trackers. The oscillator closes the track loop by adjusting $f_{dtrk}$ to drive the error signal, $e_f$, to zero.

In this chapter, we discuss three Doppler discriminator configurations. The first is the type that might be used in a CW radar, or in a semi-active seeker that uses a CW illumination signal [1; 2, p. 346]. It consists of a pair of band pass filters (BPFs) followed by error signal formation logic.

The second structure is one that would be used in radars that employ pulsed Doppler waveforms and signal processors. It is similar to the CW radar configuration, but also includes pulse matched filters and range gates to accommodate the pulsed Doppler waveforms.

As an extension of the CW and pulsed Doppler configurations, we discuss a *direct Doppler measurement* technique where the Doppler discriminator is replaced by a bank of band pass filters whose outputs are used to provide an estimate of Doppler frequency, as opposed to an error signal.

The last configuration we consider would be applicable to a radar that derives track information from a single, long, modulated transmit pulse, as opposed to a burst of pulses or a CW signal. With this configuration, matched filters serve as the band pass filters of the Doppler discriminator.

## 6.2 CW DOPPLER DISCRIMINATOR

A representation of a normalized, transmit signal for a CW radar is

$$v_T(t) = e^{j2\pi f_{RF} t} \tag{6.1}$$

where $f_{RF}$ is the carrier frequency. $f_{RF}$ is also termed the *radio frequency*, or RF.

The associated, normalized, signal received from a target at a range of $R(t)$ is [3, p. 252]

$$v_R(t) = v_T\left(t - 2R(t)/c\right) = e^{j2\pi f_{RF}\left(t - 2R(t)/c\right)} \tag{6.2}$$

where $c = 3\times10^8$ m/s is the speed of light.

Radars that use Doppler trackers make use of the phase of the received signal

$$\phi(t) = -2\pi f_{RF}\left(2R(t)/c\right). \tag{6.3}$$

More specifically, they make use of the rate-of-change of the received signal phase because it provides a measurable quantity termed *Doppler frequency*, $f_d(t)$. The Doppler frequency of the returned signal is defined by [3, p. 11]

$$f_d(t) = \frac{1}{2\pi}\frac{d\phi(t)}{dt} = -\frac{2f_{RF}}{c}\frac{dR(t)}{dt} = -\frac{2\dot{R}(t)}{\lambda} \tag{6.4}$$

where $\lambda$ is the wavelength associated with $f_{RF}$ and is given by $\lambda = c/f_{RF}$.

. ($t$) is the *range-rate* of the target and depends on the target range, $R(t)$, and velocity. Specifically, if a target is located at $[x(t)\ y(t)\ z(t)]$ in a Cartesian coordinate system centered at the radar, the range to the target is given by

$$R(t) = \sqrt{x^2(t) + y^2(t) + z^2(t)}, \tag{6.5}$$

and its range-rate is given by

$$\dot{R}(t) = \frac{dR(t)}{dt} = \frac{x(t)v_x(t) + y(t)v_y(t) + z(t)v_z(t)}{R(t)} \tag{6.6}$$

where $[v_x(t)\ v_y(t)\ v_z(t)]$ is the velocity vector of the target (in the same coordinate system). $R(t)$, $\dot{R}(t)$, and $f_d(t)$ are termed instantaneous range, range-rate, and Doppler frequency, respectively.

Analogous to range trackers, closed loop Doppler trackers do not normally attempt to directly measure Doppler frequency. Instead, they measure the error between the actual target Doppler frequency, $f_d$, and the tracker's estimate of that Doppler frequency, $f_{dtrk}$. In range trackers, the error is determined by sampling the output of a matched filter at times earlier ($\tau_E$) and later ($\tau_L$) later than the expected range delay, $\tau_{trk}$. In a Doppler tracker, the signal out of the receiver is "sampled" at frequencies below ($f_{d-}$) and above ($f_{d+}$) the expected Doppler frequency, $f_{dtrk}$. The magnitudes of the signals at the two frequencies, denoted $V_-$ and $V_+$, are used to form an error signal, $e_f$.

A block diagram of a Doppler discriminator that performs these functions is shown in Figure 6.2. The blocks labeled BPF represent band pass filters centered at $f_{d-} = f_{dtrk} - \delta f$ and $f_{d+} = f_{dtrk} + \delta f$, where $\delta f$ is a frequency offset that is typically a function of the bandwidth of the BPFs. The blocks with | | in are envelope detectors that form the magnitude of the envelope of the BPF outputs to create $V_-$ and $V_+$.

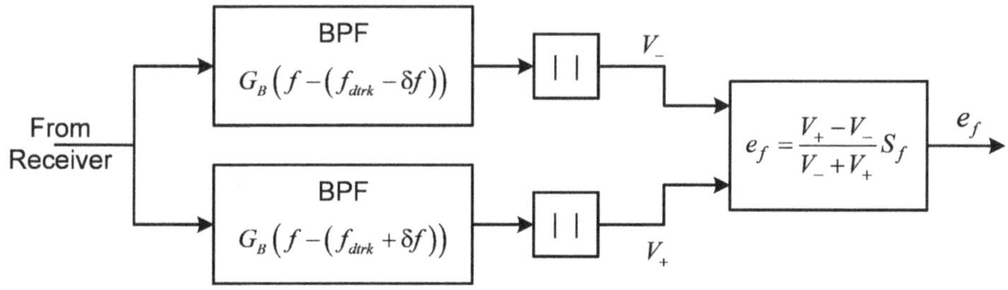


**Figure 6.2** Doppler discriminator block diagram.

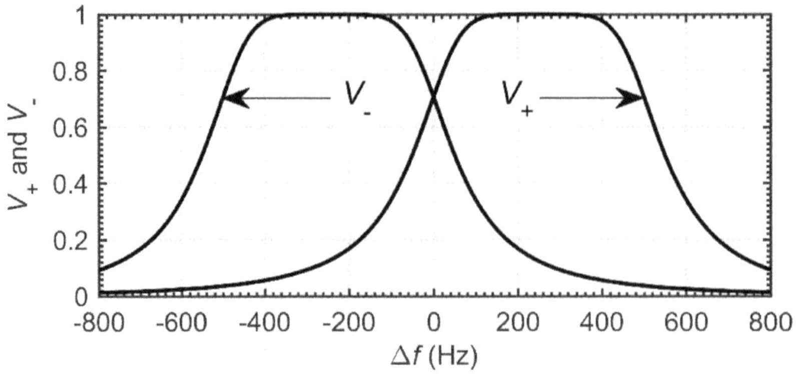


**Figure 6.3** Bandpass filter frequency responses.

Figure 6.3 contains plots of $V_- = |G_B(f_d-(f_{dtrk}-\delta f))| = |G_B(\Delta f+\delta f)|$ and $V_+ = |G_B(f_d-(f_{dtrk}+\delta f))| = |G_B(\Delta f-\delta f)|$ versus $\Delta f = f_d - f_{dtrk}$ for a representative pair of BPFs. The responses of the two BPFs cross at $\Delta f = 0$. At the crossing point, $|G_B(\delta f)| = |G_B(-\delta f)| = 0.707$. Thus, we say the filters cross at their 3-dB points.

Since $V_- = V_+$ when $\Delta f = f_d - f_{dtrk} = 0$, $V_+ - V_- = 0$. If $f_{dtrk} < f_d$, $\Delta f = f_d - f_{dtrk} > 0$, and $V_+ > V_-$. This means $V_+ - V_- > 0$. If $f_{dtrk} > f_d$, $\Delta f = f_d - f_{dtrk} < 0$, and $V_+ - V_- < 0$. This tells us we want to structure the error formation block such that $e_f$ is related to $V_+ - V_-$. If we were to simply use $V_+ - V_-$, we would find that $e_f$ depends on signal amplitude. To make $e_f$ independent of signal amplitude, we adopt the method used in range discriminators and divide by $V_+ + V_-$. Finally, we multiply the ratio $(V_+ - V_-)/(V_+ + V_-)$ by a scaling factor, $S_f$, to make the slope of the discriminator curve at $\Delta f = 0$ have a desired value. The resulting equation for the error formation block is

$$e_f = \frac{V_+ - V_-}{V_+ + V_-} S_f . \tag{6.7}$$

Figure 6.4 contains a plot of $e_f$ versus $\Delta f$ that corresponds to the BPF filter responses of Figure 6.3. In this case, $S_f$ was chosen to provide a discriminator slope of unity at $\Delta f = 0$ (more on this shortly).

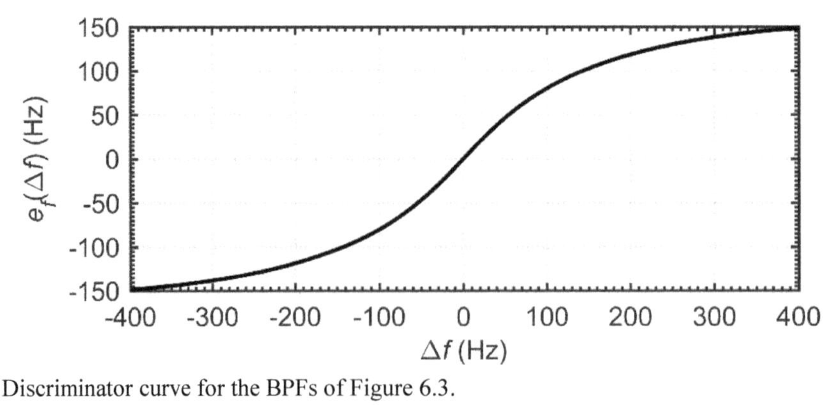


**Figure 6.4** Discriminator curve for the BPFs of Figure 6.3.

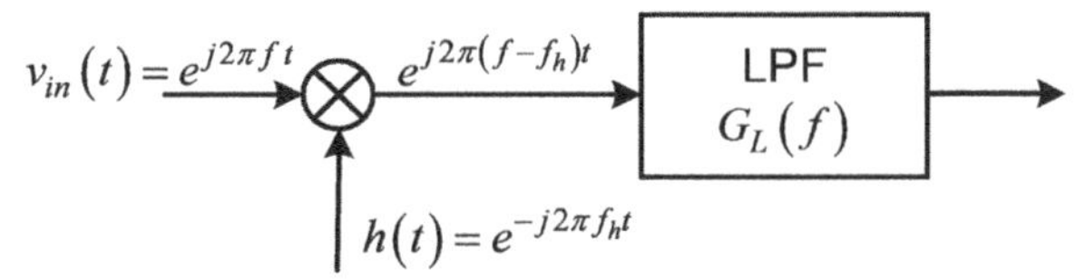


**Figure 6.5** Practical implementation of a tunable BPF.

An implication of the block diagram of Figure 6.2 is, the BPFs must be tunable. Since implementing such tunability by changing the coefficients of the BPF transfer functions is not practical, the alternate approach of Figure 6.5 is generally used.

The incoming signal, $v_{in}(t)$, is mixed with a signal, $h(t)$, at the desired center frequency to produce a low pass signal. The mixer output is then sent through a low pass filter (LPF) which preserves the desired low pass signal and rejects unwanted harmonics generated by the mixer. If the frequency, $f$, of the input signal is equal to $f_h$, the signal out of the mixer will be at a frequency of zero. Thus, the overall combination of the mixer and LPF will be a BPF tuned to $f_h$.

The implementation in Figure 6.5 uses baseband, complex signal notation. It would also represent the actual implementation if the radar uses I&Q processing in the discriminator. If the radar implements the discriminator at some IF, the inputs and output of the mixer would be real signals centered at an IF, and the LPF would be a BPF centered at a fixed IF. An example of this configuration is shown in Figure 6.6. In this case, the input signal is centered at $f_{IF1}$, and the $h(t)$ signal is centered at $f_{IF2} - f_{IF1} - f_h$. The signal out of the mixer is centered at $f_{IF2} + f_h$.

Figure 6.7 contains a block diagram of the discriminator of Figure 6.2 with the BPF blocks replaced by the diagram of Figure 6.5. The $f_{dtrk} - \delta f$ and $f_{dtrk} + \delta f$ inputs to the two mixers are the signals out of the oscillator (the controlled element), of Figure 6.1. We added two blocks labeled clutter filter, which are used to reject returns from clutter. The clutter filters must be placed before the mixers. If they are placed after the mixers, the mixers could move the clutter spectrum out of the clutter filter stop band, which would nullify their intended purpose.

The frequency response plots and discriminator curves of Figures 6.3 and 6.4 were generated for a third-order, Butterworth filter with a bandwidth of $f_B = 500$ Hz. The frequency offsets at the mixers are $+\delta f = f_B/2 = 250$ Hz and $-\delta f = -f_B/2 = -250$ Hz. The frequency response of the LPF is [4]

$$G_L(f) = \frac{1}{\sqrt{1+(f/f_C)^{2N}}} \text{ V/V} \tag{6.8}$$

where $f_C = f_B/2$ is the cutoff frequency of the filter, and $N$ is the number of poles in the filter.

To obtain the desired slope of 1.0 at $\Delta f = 0$, the discriminator scaling factor was computed (3 in this case) from

$$\left.\frac{de_f(\Delta f)}{d\Delta f}\right|_{\Delta f=0} = \frac{N}{\delta f\left[(\delta f/f_C)^{2N}+1\right](\delta f/f_C)^{2N}} S_f = 1. \tag{6.9}$$

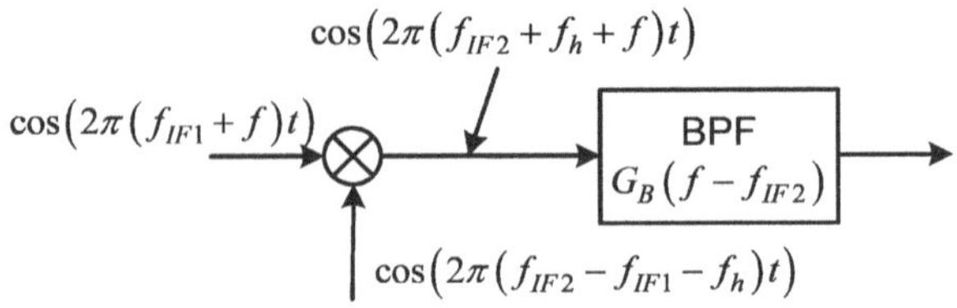


**Figure 6.6** Tunable IF bandpass filter.

Two other types of LPF that could be used are Chebychev Type I and Chebychev Type II LPFs [5, 6]. The frequency response for a Chebychev Type I filter is

$$G_L(f)=\frac{1}{\sqrt{1+\varepsilon^2\cos^2\left(N\cos^{-1}(\alpha f)\right)}} \text{ V/V} \tag{6.10}$$

where $\varepsilon$ is a function of the desired pass-band ripple, and $\alpha$ is a function of $\varepsilon$ and $f_C$, the filter cutoff frequency. As with the Butterworth filter, $f_C = f_B/2$ where $f_B$ is the desired bandwidth of the equivalent BPF. The equations for $\varepsilon$ and $\alpha$ are

$$\varepsilon=\sqrt{r-1} \tag{6.11}$$

and

$$\alpha=\frac{1}{f_C}\cosh\left[\frac{\cosh^{-1}(1/\varepsilon)}{N}\right]. \tag{6.12}$$

In these equations, $r = 10^{(\text{RdB}/10)}$ where RdB is the pass band ripple, in dB. RdB is typically 1 to 2 dB.

The frequency response for a Chebychev Type II filter is

$$G_L(f)=\frac{1}{\sqrt{1+\frac{1}{\varepsilon^2\cos^2\left(N\cos^{-1}(\alpha/f)\right)}}} \text{ V/V}. \tag{6.13}$$

In this case, $\varepsilon$ is a function of the desired stop-band ripple. As with the Type I filter, $\alpha$ is a function of $\varepsilon$ and $f_C$. The equations for $\varepsilon$ and $\alpha$ are

$$\varepsilon=\sqrt{\frac{r}{1-r}} \tag{6.14}$$

and

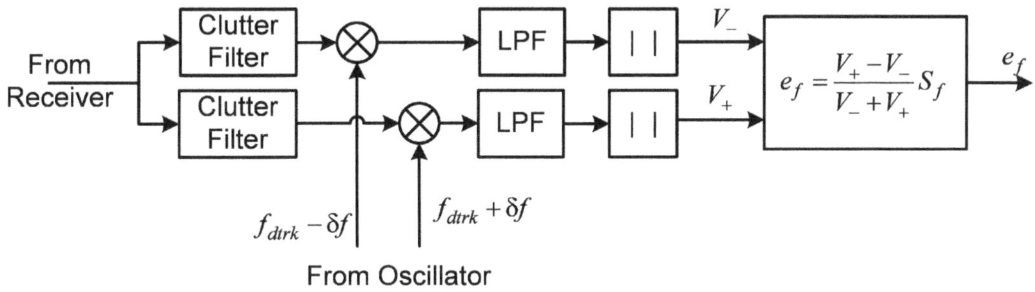


**Figure 6.7** Block diagram of a practical CW doppler discriminator.

$$\alpha = f_C \cosh\left[\frac{\cosh^{-1}(1/\varepsilon)}{N}\right]. \tag{6.15}$$

In this case, $\varepsilon$ is a function of the desired stop-band ripple. As with the Type I filter, $\alpha$ is a function of $\varepsilon$ and $f_C$. The equations for $\varepsilon$ and $\alpha$ are

$$\varepsilon = \sqrt{\frac{r}{1-r}} \tag{6.16}$$

and

$$\alpha = f_C \cosh\left[\frac{\cosh^{-1}(1/\varepsilon)}{N}\right]. \tag{6.17}$$

In (6.16) and (6.17), $r = 10^{(RdB/10)}$ where RdB is the stop band ripple, in dB. RdB is typically the peak value of $G_L(f)$ in the filter stop band. As an example, if a design objective for the filter is to have a greater than 30 dB attenuation in the stop band, RdB = –30.

Figure 6.8 contains plots of $G_L(f)$ for third order, Type I and Type II Chebychev filters with $f_C = 250$ Hz (an equivalent BPF bandwidth of 500 (N=3) Hz). The pass band ripple for the Type I filter was set to RdB = 1 dB, and the stop band ripple for the Type II filter was set to RdB = –30 dB. Figure 6.9 contains discriminator curves for discriminators that use the filters of Figure 6.8. $\delta f$ was set to $f_C = f_B/2 = 250$ Hz in both cases. In comparing Figures 6.8 and 6.9 to Figures 6.3 and 6.4, we note the frequency responses and discriminator curves for the Butterworth and Type II Chebychev filters are similar. However, the frequency response and discriminator curve for the Type I Chebychev filter is different from the other two filters. In particular, the linear region of the discriminator curve for the Type I Chebychev filter is much smaller than the linear region for the other two types of filters. This is due to the difference in steepness of the frequency response of the filters near the cutoff frequency (see Figure 6.8).

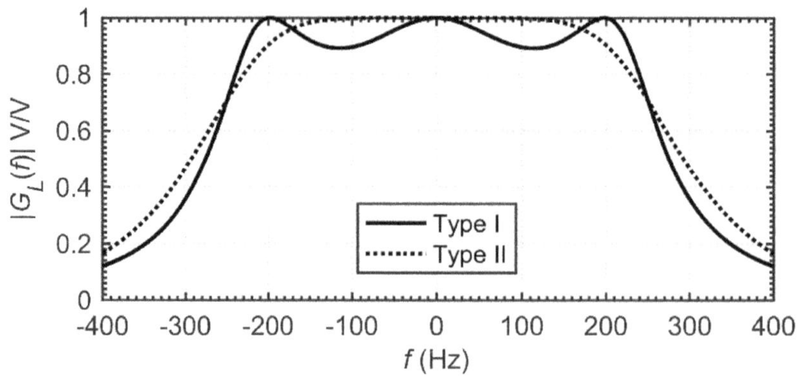


**Figure 6.8** Frequency response of 3rd order, Type I and Type II Chebychev filters with cutoff frequencies of 250 Hz.

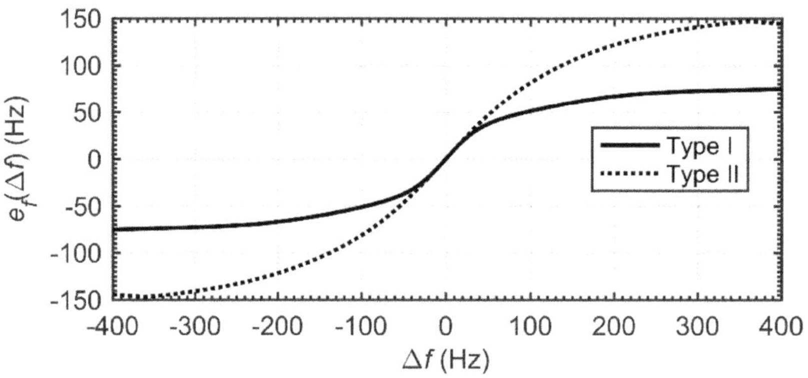


**Figure 6.9** Discriminator curves Type I and Type II Chebychev filters .

## 6.3 PULSED DOPPLER DISCRIMINATOR

Figure 6.10 contains a block diagram of a Doppler discriminator that might be used in a pulsed Doppler radar. It is basically the same as the CW radar Doppler discriminator except for the presence of the single pulse matched filter and range gate. The matched filter is matched to the pulses of the pulsed Doppler burst and the range gate samples, and holds, the matched filter output for a PRI. This is illustrated in Figure 6.11. The top plot of that figure represents the transmit signal, which consists of a burst of pulses with a constant PRI. The triangles in the bottom plot represent the matched filter output, and the dashed lines connecting the peaks of the triangles represent the output of the range gate. For the case shown, the radar is perfectly tracking the target in range so that the range gate is sampling the matched filter output at its peak. As a note, $\tau_{trk}$ is the ambiguous range delay, $\tau_{trkambig}$, and is related to the actual range delay by

$$\tau_{trkambig} = \tau_{trkactual}\big|_{\text{mod}_T} = \tau_{trkactual} - \left\lfloor \frac{\tau_{trkactual}}{T} \right\rfloor T . \tag{6.18}$$

In (6.18), the notation $\lfloor x \rfloor$ denotes the integer portion of $x$.

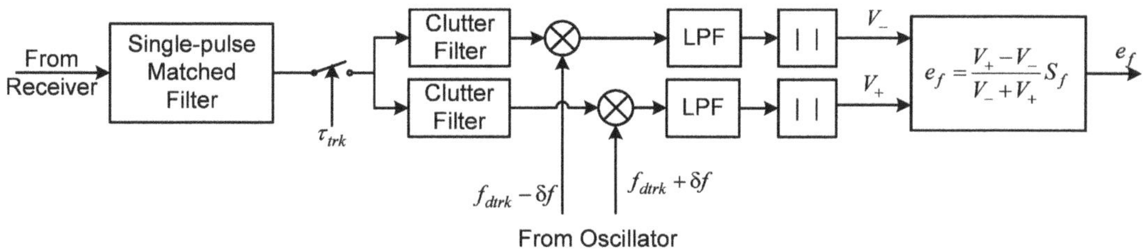


**Figure 6.10** Doppler discriminator for a pulsed doppler radar.

Since many pulsed Doppler radars use Doppler ambiguous waveforms, the signals from the oscillator, $f_{dtrk} \pm \delta f$ must also be Doppler ambiguous. This happens automatically because the output of the oscillator is sampled once per PRI. This sampling causes the Doppler-unambiguous oscillator output to be ambiguous, just as the sampling caused by the pulsed Doppler waveform causes the unambiguous target Doppler frequency to be ambiguous [3, p. 433].

The LPFs of Figure 6.10 could be Butterworth, Chebychev, or some other type of recursive filter. As an alternate, they could be implemented as finite impulse response (FIR) filters [7, p. 170]. The clutter filters would, generally, still need to be recursive because of their stringent stop band requirements.

A functional block diagram of a FIR filter is illustrated in Figure 6.12. It consists of a weight function followed by an integrator. The weight function multiplies the mixer outputs by a time varying weight, and the integrator sums, or integrates, the output of the weight function. In radars that use digital signal processing, the integrator is a summer. In radars that use analog signal processing, the integrator is a true integrator, or a LPF with a very small bandwidth.

The weight function is typically some type of standard amplitude weighting such as Chebychev, Taylor, Hamming, and so forth [5, 8, 9]. It is usually chosen so that $|G_L(f)|^2$ has a desired sidelobe level.

Figure 6.13 contains a block diagram of a combined Doppler, range, and angle tracker for a pulsed Doppler radar. It illustrates how the receiver outputs, range gates, and oscillator signals are distributed to the various trackers. For example, the Doppler tracker uses signals from the sum channel receiver that are sampled at $\tau_{trkrambig}$ and are mixed with the $f_{dtrk} \pm \delta f$ signals from the oscillator. The angle trackers use signals from the sum and difference channel receivers. These signals are sampled at $\tau_{trkambig}$ and mixed with the $f_{dtrk}$ signal from the oscillator. The range channel uses the signal from the sum channel receiver. This signal is sampled at $\tau_E$ and $\tau_L$, and mixed with the $f_{dtrk}$ signal from the oscillator.

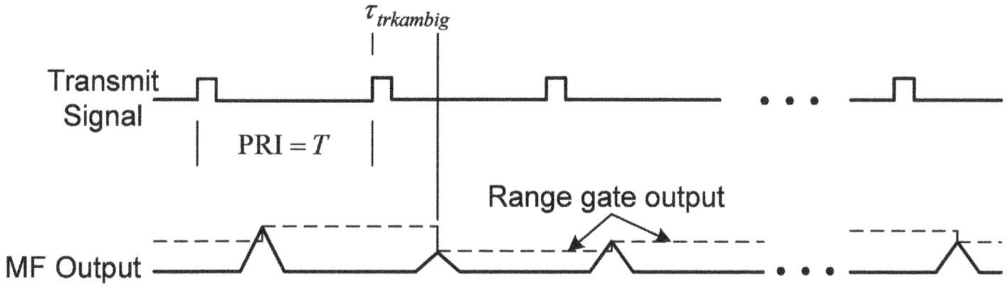


**Figure 6.11** Transmit signal, matched filter output and range gate output for a pulsed Doppler waveform.

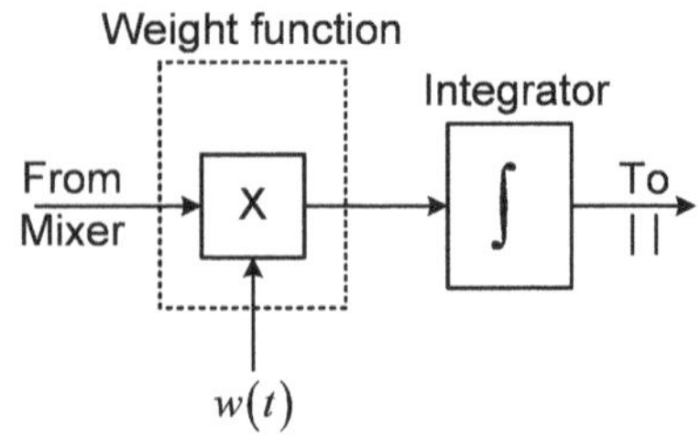


**Figure 6.12** FIR filter block diagram.

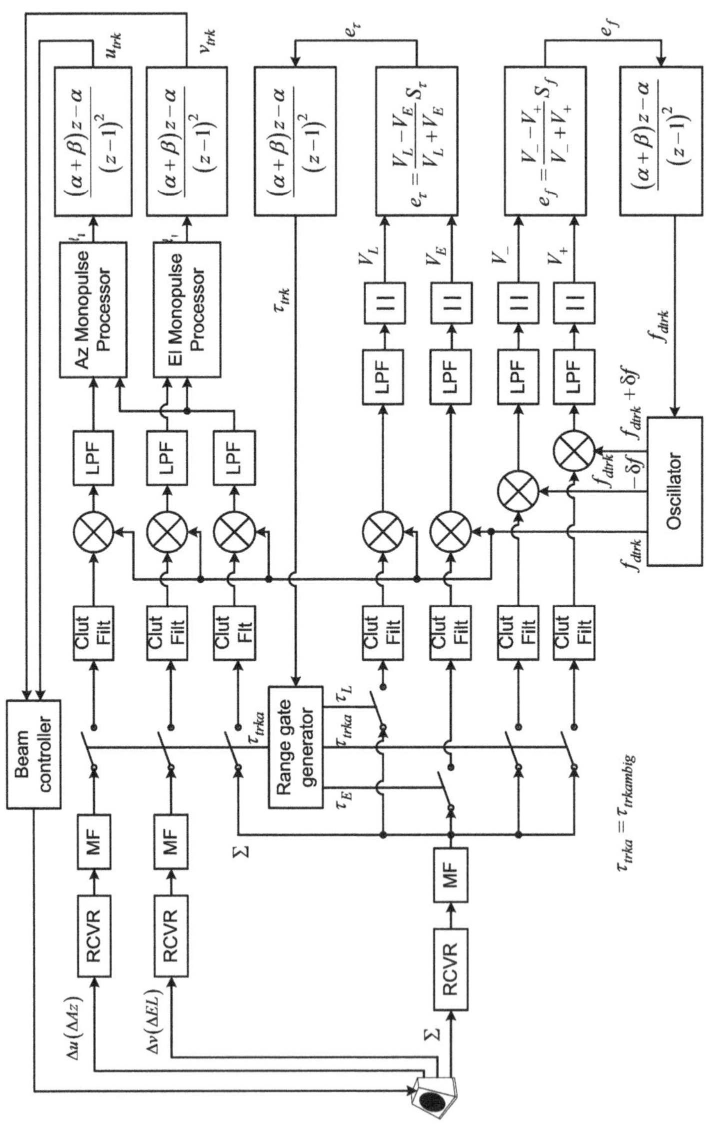


**Figure 6.13** Integrated Doppler, range and angle trackers for a pulsed Doppler radar.

## 6.4 DIRECT DOPPLER MEASUREMENT

An alternate to using a Doppler discriminator is direct Doppler frequency measurement. With this method, the pair of BPFs, envelope detectors, and error formation logic are replaced with a bank of BPFs whose combined bandwidth equal to the PRF (assuming a pulsed Doppler radar). In radars that use digital signal processing, such a bank of filters can be realized with a FFT [10, p. 116]. The taps of the FFT represent a bank of BPFs whose bandwidths are, nominally, the reciprocal of the coherent processing interval, $T_{CPI}$, and whose center frequencies are separated by $1/T_{CPI}$. Typically, the coherent processing interval (CPI) is about half the total duration of the pulsed Doppler waveform dwell, $T_{dwell}$, and occupies the last part of the dwell. The other half of the dwell is used to account for clutter, target, and clutter filter transients [3, p. 454]. A weighting function across the input taps of the FFT is typically used to reduce the sidelobes of the frequency response of the FFT.

With the direct measurement method, several BPF outputs (FFT output taps), centered on $f_{dtrkambig}$, are used in a weighted average equation to form the measured (ambiguous) Doppler frequency. Specifically, if $V(n)$, $n = 1, 2, \ldots, N_{BPF}$, are the envelope amplitudes of the $N_{BPF}$ BPF outputs centered on $f_{dtrkambig}$, and $f_d(n)$ are the center frequencies of those BPFs, the (ambiguous) measured Doppler frequency is given by

$$f_{dmeas} = \frac{\sum_{n=1}^{N_{BPF}} V(n) f_d(n)}{\sum_{n=1}^{N_{BPF}} V(n)}. \tag{6.19}$$

The value of $N_{BPF}$ depends on the anticipated error between $f_{dtrk}$ and the true Doppler frequency, $f_d$. A minimum value of $N_{BPF}$ would be three.

If direct Doppler measurement is used, the open loop track filter of Figures 6.10 and 6.13 is not used. Instead, the Doppler track loop is closed by a closed loop track filter. For example, an α-β filter that would be used with direct Doppler measurement has the form (see Chapter 3)

$$\begin{aligned}
&X_p(n) = F(n-1)X_s(n-1) \\
&X_s(n) = X_p(n) + K(n)\left(f_{dmeas}(n) - f_{fdtrkambig}(n)\right) \\
&F(n-1) = \begin{bmatrix} 1 & T_{n-1} \\ 0 & 1 \end{bmatrix} \quad X_p(n) = \begin{bmatrix} f_{dtrk}(n) \\ \dot{f}_{dtrk}(n) \end{bmatrix} \\
&K(n) = \begin{bmatrix} \alpha \\ \beta/T_{n-1} \end{bmatrix} \quad X_s(n) = \begin{bmatrix} f_{ds}(n) \\ \dot{f}_{ds}(n) \end{bmatrix}
\end{aligned} \tag{6.20}$$

In (6.20), $f_{dtrk}(n)$ and $f_{ds}(n)$ are the *unambiguous* tracked and smoothed Doppler frequencies, respectively, and $\dot{f}_{dtrk}(n)$ and $\dot{f}_{ds}(n)$ are their respective rates. Since the Doppler measurement will be the ambiguous Doppler, the predicted Doppler used in the second line of (6.20), $f_{dtrkambig}$, is the *ambiguous* predicted Doppler frequency. We will discuss ambiguous and unambiguous Doppler shortly.

The equivalent α-β track filter z-transfer function is (see Chapter 3)

$$G_{CL}(z) = \frac{f_{dtrkambig}(z)}{f_{dmeas}(z)} = \frac{(\alpha+\beta)z-\alpha}{z^2-(2-\alpha-\beta)z+1-\alpha}. \tag{6.21}$$

Since the Doppler track loop is closed through the track filter, the oscillator (controlled element) is not needed in the Doppler tracker. However, it may be needed in the range and angle trackers, if those track loops use the BPFs.

Ambiguous Doppler operation needs to be considered when the direct Doppler measurement technique is used. This stems from the fact that the bank of BPFs (FFT) measures ambiguous Doppler frequency, not actual Doppler frequency. This means the logic used to select the BPFs (FFT taps) used to compute $f_{dmeas}$ must convert $f_{dtrk}$, the true tracked Doppler, to its ambiguous equivalent, $f_{dtrkambig}$. That way, the Doppler frequencies used to form the Doppler error ($f_{dmeas}(n) - f_{dtrkambig}(n)$) in (6.20) will be compatible. As a note, ambiguous Doppler frequency is related to actual Doppler frequency by

$$f_{dambig} = f_{dactual}\big|_{\mathrm{mod}\,PRF} = f_{dactual} - \left\lfloor \frac{f_{dactual}}{PRF} \right\rfloor PRF \tag{6.22}$$

where $\lfloor x \rfloor$ denotes the integer portion of *x*, and *PRF* is the pulse repetition frequency of the pulsed Doppler waveform.

In pulsed Doppler radars that use very high PRFs (greater than 100 kHz or so) it may not be necessary to consider ambiguous Doppler operations. This would be the case if the waveform PRF is greater than twice the highest expected target Doppler frequency. If consideration of ambiguous Doppler operation is not necessary, $f_{dmeas}$ and $f_{dtrkambig}$ becomes simply their unambiguous values, and there is no need to consider the conversion to ambiguous Doppler frequency.

## 6.5 DOPPLER TRACKING IN LOW PRF PULSED RADARS

We don't normally think of a Doppler tracker being necessary in low PRF pulsed radars. In most low PRF pulsed radars, this is the case. However, in radars that use pulse widths of several hundred μs to a few ms, Doppler tracking may be needed to keep the matched filter "tuned" to the target Doppler frequency. This is especially true for radars that use long, phase coded pulses, such as pseudo random noise (PRN) coding, generalized Barker coding, or minimum peak sidelobe coding [3, 11]. If the matched filter is not tuned to the target Doppler frequency, the output of the matched filter could be attenuated to the point where it cannot be used to determine range.

Doppler tuning does not cause as much of a problem in radars that use LFM pulses. However, if the matched filter is not tuned to the target Doppler frequency, the matched filter could cause a bias error in the track filter.

To illustrate the effect of Doppler tuning on matched filter response, we consider the two plots of Figure 6.14. These plots are top-down views (contour plots) of the ambiguity function [3, 12] of a 1023 $\mu s$ pulse. The plot on the right is for a 1 MHz bandwidth, LFM pulse, and the plot on the left is for PRN coded pulses with a chip width of 1 $\mu s$ (an equivalent bandwidth of 1 MHz). The dark regions represent high amplitude while the light regions represent low amplitude.

The horizontal axis is $\tau_R - \tau_{meas}$, and the vertical axis is $f_d - f_{dtune}$. Thus, when $\tau_R - \tau_{meas} = 0$, the measured range delay, $\tau_{meas}$, will equal the actual range delay, $\tau_R$. When $f_d - f_{dtune} = 0$, we say the matched filter is tuned to the target Doppler frequency, $f_d$. The condition $(\tau_R - \tau_{meas}, f_d - f_{dtune}) = (0,0)$ is termed matched range and Doppler condition. The ambiguity function will attain its maximum value at this point.

The horizontal lines on the plots of Figure 6.14 correspond to $f_d - f_{dtune}$ values of 0 and 2 kHz. When $f_d - f_{dtune} = 0$, the matched Doppler condition, the line passes through a high amplitude region near $\tau_R - \tau_{meas} = 0$. A more quantitative illustration of this is contained in Figure 6.15, which contains plots of the matched filter output as $\tau_R - \tau_{meas}$ moves along the line at $f_d - f_{dtune} = 0$. Figure 6.15 is termed a matched Doppler range cut because it is a plot of the matched filter output when it is tuned, or matched, to the target Doppler frequency. Both plots peak at $\tau_R - \tau_{meas} = 0$, which means $\tau_{meas}$ provides a good indication of the actual range delay, $\tau_R$.

The line at $f_d - f_{dtune} = 2$ kHz on Figure 6.14 represents the case where the matched filter is not tuned to the target Doppler frequency. In this case, the line does not pass through a dark, or high amplitude, region of the ambiguity function plot for the PRN coded pulse. It does pass through a high amplitude region for the LFM pulse. However, the high amplitude region does not occur at $\tau_R - \tau_{meas} = 0$.

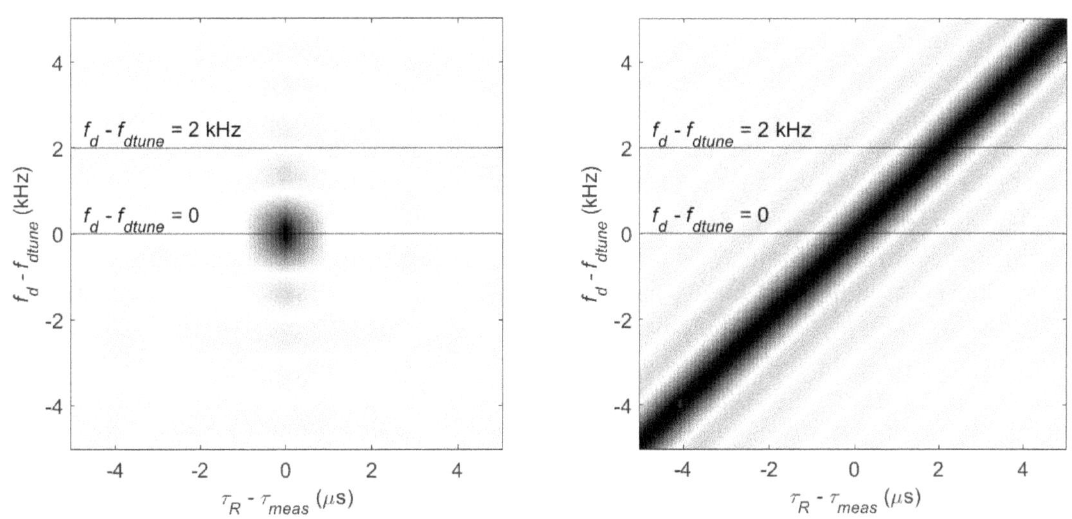


**Figure 6.14** Ambiguity function plots of a 1023 $\mu s$, 1023-chip, PRN coded pulse (left) and a 1023 $\mu s$, 1 MHz, LFM pulse (right).

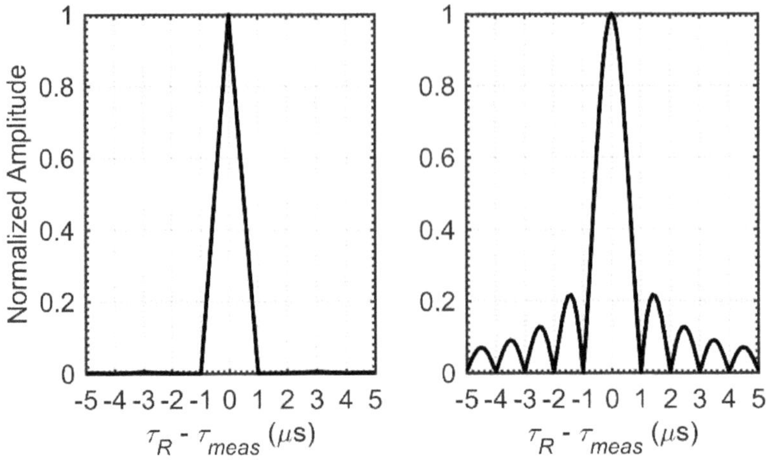


**Figure 6.15** Matched Doppler range cuts for a PRN coded pulse (left) and a LFM pulse (right).

Figure 6.16 contains plots of the amplitude versus $\tau_R - \tau_{meas}$ for this case. The plot for the LFM pulse has a peak, however, it occurs at $\tau_R - \tau_{meas} = 2$ $\mu$s, which means the measured range delay will differ from the actual range delay by 2 $\mu$s. That is, there will be a 2-$\mu$s error, or bias, in the range delay measurement.

For the PRN coded pulse, the plot of amplitude versus $\tau_R - \tau_{meas}$ is very small, and most likely close to, or below, the receiver noise level. This means it would not be possible to obtain a reliable range delay measurement from the signal out of the matched filter.

A Doppler mistuning error of 2 kHz translates to a range-rate mistuning error of $\Delta\dot{R} = \lambda\Delta f_d/2 = 30$ m/s at X-band. This means the radar must have a reasonably accurate estimate of the target range-rate to keep the matched filter tuned to the target Doppler frequency. If the matched filter can be initially tuned during acquisition or track initiation, its tuning could be updated by using the range-rate output of the range tracker.

An alternative, is to include a Doppler tracker as part of the range tracker. An example of such a configuration is shown in Figure 6.17. It is a block diagram of an integrated Doppler, range, and angle tracker that might be used in a low PRF radar that uses a sampling gate range tracker (see Chapter 4) and a three channel monopulse angle tracker (see Chapter 5).

The Doppler tracker part of the overall tracker consists of a Doppler discriminator, a track filter, and a controlled element, as in the CW and pulsed Doppler examples. The difference between the configuration of Figure 6.17 and the previously discussed Doppler trackers is that the LPFs of the Doppler discriminator are replaced by matched filters.

As was implied in the discussion of Figures 6.14, 6.15, and 6.16, the matched filters simultaneously act as "time" filters and, with the mixers, BPFs. They act as time filters because they isolate target return signals in time, or range delay. They act as BPFs because they isolate target return signals in frequency, or Doppler.

The basic matched filter response peaks at a frequency of zero, which is why we term it a LPF. The mixers that precede the matched filters cause the combination to act as a BPF, as was the case with the previously discussed tunable BPFs.

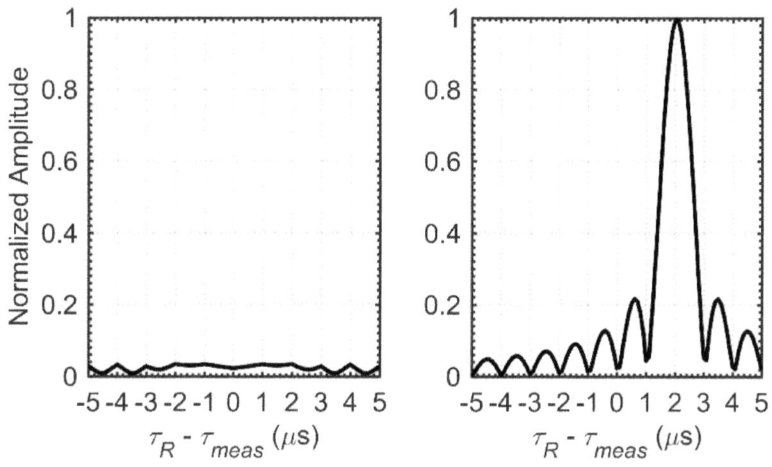


**Figure 6.16** Range cuts of a PRN coded pulse (left) and a LFM pulse (right), and a 2-kHz Doppler mismatch.

The mixers in the range and angle trackers tune the matched filters to $f_{dtrk}$, the tracked Doppler frequency. If the Doppler tracker is tracking the target, $f_{dtrk} = f_d$, which means the matched filters are tuned to the target Doppler frequency. The two mixers of the Doppler tracker tune the matched filters to $f_{dtrk} + \delta f$ and $f_{dtrk} - \delta f$, as in the previously discussed Doppler trackers.

To better understand how the Doppler discriminator of Figure 6.17 works, we will develop some equations that will lead to a discriminator curve for the Doppler discriminator. We start by considering a normalized transmit pulse of the form

$$v_T(t) = e^{j2\pi f_{RF} t} e^{j\phi(t)} \text{rect}\left[\frac{t}{\tau_p}\right] \tag{6.23}$$

where $f_{RF}$ is the transmit RF, and $\tau_p$ is the pulse width. The rect[ ] function is defined by

$$\text{rect}\left[\frac{t}{\tau_p}\right] = \begin{cases} 1 & |t| \le \tau_p/2 \\ 0 & |t| > \tau_p/2 \end{cases}. \tag{6.24}$$

The term $\phi(t)$ is the phase modulation on the pulse. For an LFM pulse, $\phi(t) = \pi\alpha t^2$ where $|\alpha| = B/\tau_p$ is the LFM slope, and $B$ is the LFM bandwidth. For a pulse with discrete phase coding, $\phi(t)$ is a piecewise continuous phase function. That is, the phase is constant over a period of $\tau_{chip}$, but can change every $\tau_{chip}$. $\tau_{chip}$ is termed the chip, or subpulse, width and is such that $N_{chip}\tau_{chip} = \tau_p$, where $N_{chip}$ is an integer. $1/\tau_{chip}$ is the waveform bandwidth, $B$, and the range resolution of the pulse is $\tau_{chip}$. $N_{chip}$ is also termed the bandwidth-time (BT) product of the waveform, because $N_{chip} = \tau_p/\tau_{chip} = B\tau_p$. The phase assigned to each chip is determined by some phase coding algorithm. Examples of phase coding algorithms are the PRN, generalized Barker, and minimum peak sidelobe codes mentioned earlier.

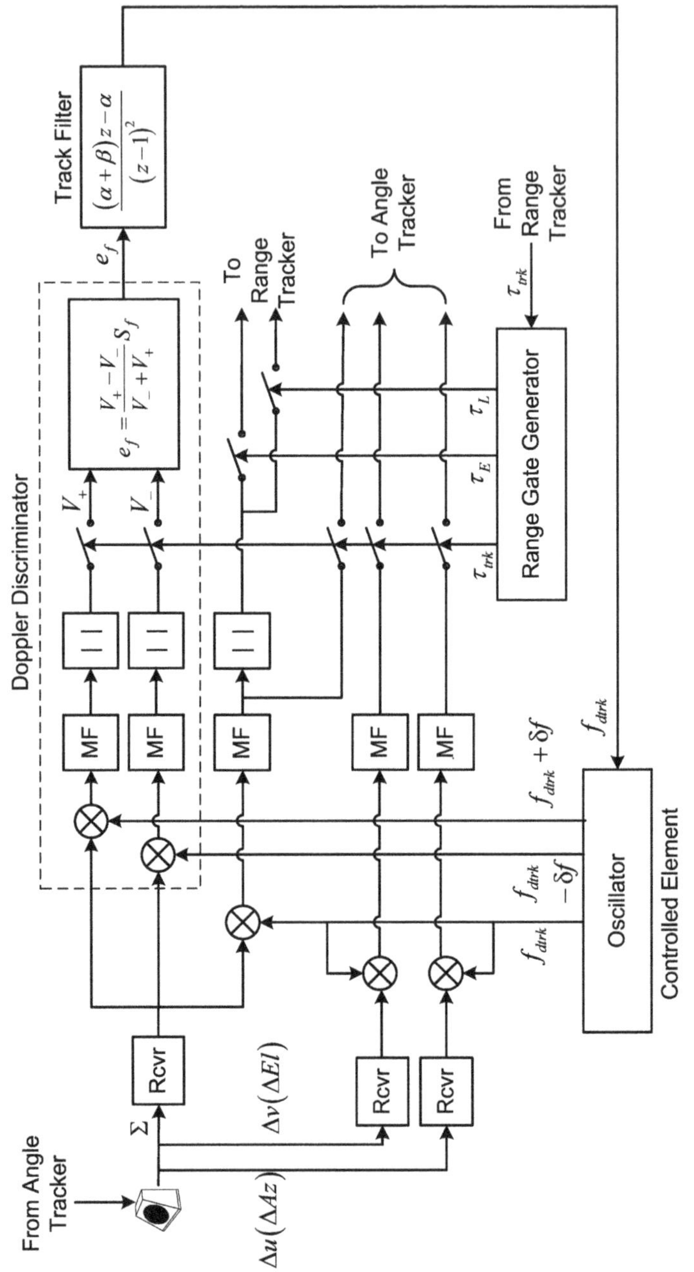


**Figure 6.17** Integrated Doppler, range and angle tracker for a low PRF radar.

The ideal, normalized signal returned from a point target that is at a range of $R$, and has a range rate of $\dot{R}$, is [3, p. 15]

$$v_R(t) = e^{j2\pi(f_{RF}-f_d)t} e^{j\phi(t-\tau_R)} \text{rect}\left[\frac{t-\tau_R}{\tau_p}\right] \quad (6.25)$$

where $f_d = -2\dot{R}/\lambda$ is the Doppler frequency of the return signal, and $\tau_R = 2R/c$ is the range delay to the target.

The ideal, normalized, baseband signal at the output of the sum ($\Sigma$) channel receiver can be written as

$$v_\Sigma(t) = e^{-j2\pi f_d t} e^{j\phi(t-\tau_R)} \text{rect}\left[\frac{t-\tau_R}{\tau_p}\right]. \quad (6.26)$$

If the frequency into the mixers of Figure 6.17 is $f_{dtune}$ (e.g., $f_{dtrk}$, $f_{dtrk} + \delta f$, $f_{dtrk} - \delta f$), we can write the output of the absolute value following the matched filter as

$$v_{MF}(\tau, f) = v_{MF}(t - \tau_R, f - f_{dtune}) \quad (6.27)$$

where $|MF(t,f)| = \chi(t,f)$ is the ambiguity function of $v_T(t)$ [3, p. 254].

In general, $MF(\tau - \tau_R, f - f_{dtune})$ is a complicated function of $\tau$ and $f$. However, for the special case where $\tau = \tau_R$, $MF(\tau - \tau_R, f - f_{dtune}) = MF(0, f - f_{dtune})$ has the form [3, p. 257]

$$MF(0, f - f_{dtune}) = e^{j\theta} \text{sinc}\left(\tau_p (f - f_{dtune})\right) \quad (6.28)$$

where $\theta$ is a phase that depends on $f - f_{dtune}$. This form is also approximately valid when $\tau$ is not exactly equal to $\tau_R$, but is close to $\tau_R$. Specifically, (6.28) holds, with reduced peak amplitude, when $|\tau - \tau_R|$ is less than about one-half of the range resolution of the transmit pulse.

Figure 6.18 contains a plot of $|MF(0, f - f_{dtune})|$ versus $f$. The plot exhibits a bandpass characteristic that is centered at $f_{dtune}$ and has a 3-dB bandwidth of approximately $0.88/\tau_p$.

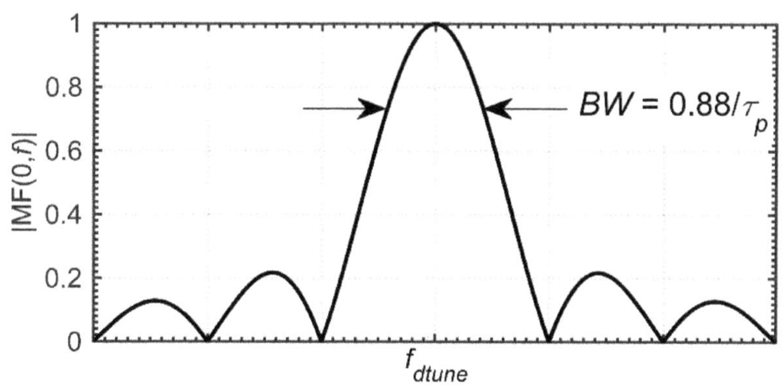


**Figure 6.18** Plot of the frequency response of a matched filter tuned to $f_{dtune}$.

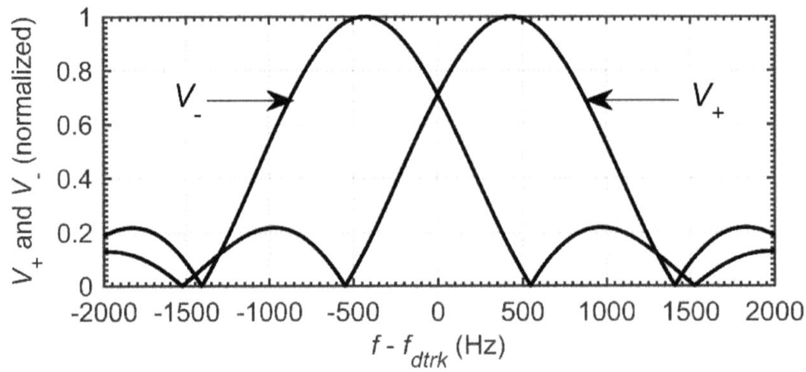


**Figure 6.19** Frequency responses of the matched filters of Figure 6.17.

In the range and angle trackers, $f_{dtune} = f_{dtrk}$. Thus, the matched filter (actually the mixer-MF combination) is a BPF tuned to the tracked Doppler frequency. In the Doppler discriminator, the two matched filters are tuned to $f_{dtrk} + \delta f$ and $f_{dtrk} - \delta f$. An example of the two matched filter frequency responses are contained in Figure 6.19 for the 1023 $\mu$s pulse examples previously discussed. The offset frequency, $\delta f$, was chosen as $|\delta f| = 0.44/\tau_p = 0.44/1023\ \mu\text{s} \approx 430$ Hz. With this value of $\delta f$, the two responses cross at their 3-dB points.

The switches (range gates) that follow the absolute value blocks sample the absolute values of the matched filter outputs at $\tau_{trk}$, the tracked range, to form $V_+$ and $V_-$. These are sent to the error formation block to form $e_f$. Finally, the Doppler track loop is closed through the track filter and the controlled element.

If we use (6.28), we can write the equation for $e_f$ as

$$e_f(\Delta f) = \frac{V_+ - V_-}{V_- + V_+} S_f = \frac{\left|\text{sinc}\left((\Delta f - \delta f)\tau_p\right)\right| - \left|\text{sinc}\left((\Delta f + \delta f)\tau_p\right)\right|}{\left|\text{sinc}\left((\Delta f + \delta f)\tau_p\right)\right| + \left|\text{sinc}\left((\Delta f - \delta f)\tau_p\right)\right|} S_f \tag{6.29}$$

where $\Delta f = f_d - f_{dtrk}$ is the Doppler frequency error. Figure 6.20 contains a plot of $e_f(\Delta f)$ versus $\Delta f = f_d - f_{dtrk}$ for the example conditions used to generate Figure 6.19. This plot is the *Doppler discriminator curve*. As before, the scaling factor was chosen so the slope of the discriminator curve was 1.0 at $\Delta f = 0$. With some manipulation, it can be shown that $S_f$ can be determined from

$$\left.\frac{\partial e(\Delta f)}{\partial \Delta f}\right|_{\Delta f = 0} = 1 = \left(\frac{1}{\delta f} - \pi\tau_p \cot\left(\pi\tau_p \delta f\right)\right) S_f . \tag{6.30}$$

The slope of the discriminator curve changes sign at the nulls of $V_+$ and $V_-$. From Figure 6.19, the first of these nulls occurs at $\pm(1 - 0.44)/\tau_p \approx 547$ Hz. If we change the frequency offset, $\delta f$, so that the $V_+$ and $V_-$ plots of Figure 6.18 overlap at their 2- or 4-dB points, rather than their 3-dB points, we get the discriminator curves of Figure 6.21. As expected, the central.

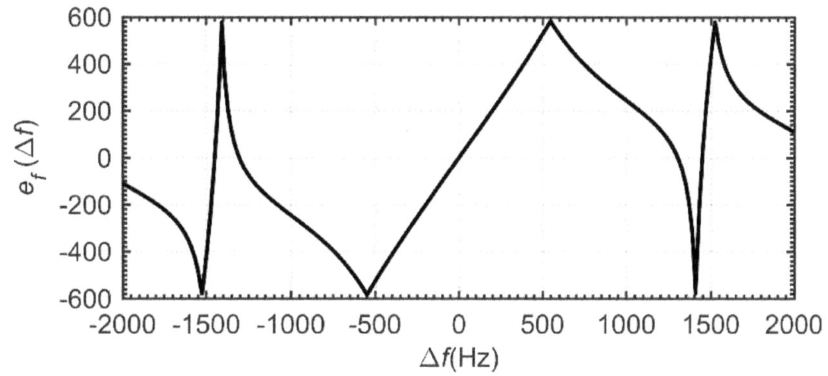


**Figure 6.20** Doppler discriminator curve for a matched filter-based, Doppler discriminator.

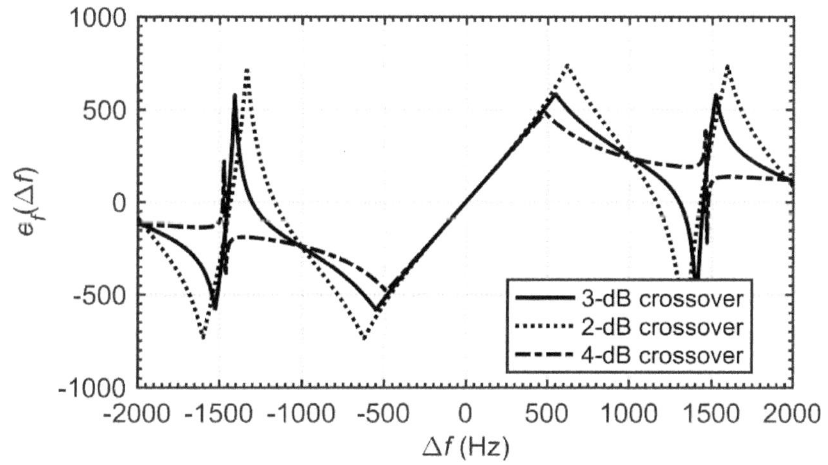


**Figure 6.21** Discriminator curves for $V_+$ and $V_-$ crossover points of 2, 3 and 4 dB.

region of the discriminator curve decreases in width as the crossover points vary from 2 to 3 to 4 dB

Equation (6.29) and the discriminator curves of Figures 6.20 and 6.21 will be the same for all phase coded pulses where the matched filter is matched to the transmit waveform. However, LFM pulses can use a mismatched filter to reduce the range sidelobes. In that case, the matched range, Doppler cut of the *mismatched*, or *cross*, ambiguity function, $|\chi_{mm}(0,f)|^2$ will not be a sinc-squared function. This will cause the discriminator curve to have a slightly different shape. Derivation of such a discriminator curve is left as a homework assignment.

Simulation experiments indicate the integrated range and Doppler tracker works well for phase coded pulses but not for LFM pulses. The reason it doesn't work well for LFM pulses appears to be the range-Doppler coupling property associated with the ambiguity function of LFM pulses. Specifically, an error in $f_{dtrk}$ causes $\tau_{meas}$ to be in error, and an error in $\tau_{trk}$ causes $f_{dmeas}$ to be in error. These measurement errors cause errors in $\tau_{trk}$ and $f_{dtrk}$, respectively, which cause further error in $f_{dmeas}$ and $\tau_{meas}$, which cause further error in $\tau_{trk}$ and $f_{dtrtk}$, and so forth.

Phase coded waveforms of the type discussed above do not exhibit range-Doppler coupling. Thus, range track errors do not directly cause Doppler measurement errors, and Doppler track errors do not directly cause range measurement errors. Range and Doppler track errors do indirectly cause measurement errors in cross domains because of the reduction in SNR due to not being matched in range and Doppler.

As a note, phase coded pulses such as Frank polyphase, Zadoff, and Chu [3, 11] do exhibit range-Doppler coupling. Thus, it is expected the integrated tracker would also not work well for these pulses.

Another possible combined range and Doppler tracker would be to use direct range and Doppler measurement in an open loop track configuration.

To close out this chapter, we consider two examples of Doppler trackers. The first is a CW tracker using the Doppler discriminator discussed in Section 6.2. The second is the Doppler and range tracker portions of the integrated low PRF tracker discussed in this section.

## 6.6 EXAMPLE 1: CW DOPPLER TRACKER

Figure 6.22 contains a block diagram of the CW tracker model we consider in this example. The Doppler discriminator is basically the same as in Figure 6.7, without the clutter filters. We added summers to accommodate the modeling of noise.

In an actual radar, noise enters the tracker in the RF portion of the receiver. However, for simulation purposes, we take advantage of superposition and inject noise after the LPFs of the discriminator (see the discussion of Section 4.5). Injecting noise after the LPFs also makes scaling of relative signal and noise levels easier, because SNR is specified at the output of the LPFs.

In an actual radar, the two branches of the discriminator are driven by the same noise. This means the noises out of the LPFs will be correlated with each other. Because of this, the noises we use in the simulation will need to be correlated. From our studies in Chapters 4 and 5, if $r_{12}$ is the correlation coefficient between random numbers $n_1$ and $n_2$, we would form properly correlated random numbers by first drawing two independent, zero-mean, unit variance, complex, Gaussian random numbers, $n_a$ and $n_b$. We would then use these to compute $n_1$ and $n_2$ from

$$\begin{aligned} n_1 &= n_a \\ n_2 &= r_{12} n_a + \left(1 - r_{12}\right) n_b \end{aligned} . \quad (6.31)$$

The random numbers $n_1$ and $n_2$ will be zero-mean, unit variance, complex, and Gaussian. However, they will not be independent. Instead, they will have a correlation coefficient of $r_{12}$. The proof of this is left as a homework assignment.

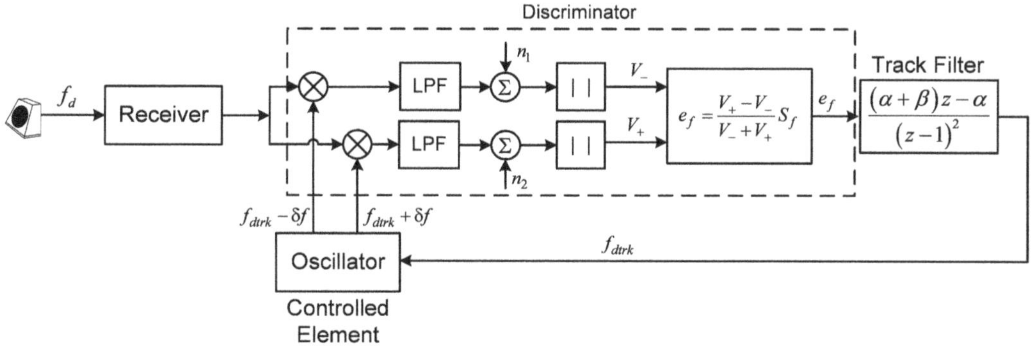


**Figure 6.22** CW tracker example.

We compute the correlation coefficient from [13]

$$r_{12} = \frac{\int_{-\infty}^{\infty} L(f - \delta f) L^*(f + \delta f) df}{\int_{-\infty}^{\infty} |L(f)|^2 df}. \tag{6.32}$$

The derivation of (6.32) is contained in the appendix. $L(f)$ is the transfer function of the LPFs in Figure 6.22. For this example, we use the 3rd order Butterworth filter discussed in Section 6.2. From (6.8), we have

$$|L(f)|^2 = \frac{1}{1 + (2f/f_B)^{2N}} \tag{6.33}$$

where $f_B$ is the bandwidth of the equivalent BPF and $N = 3$. From Butterworth filter theory [14,15, 16] we can write the transfer function of an $N^{\text{th}}$ order Butterworth, LPF as

$$L(f) = \prod_{k=1}^{N} \frac{1}{j2f/f_b - \exp\left[j(2k + N - 1)\pi/(2N)\right]}. \tag{6.34}$$

We assume outputs of the LPFs are sampled every $T = 0.01$ s and update the track filter, at a 100 Hz rate. Consistent with Section 6.2, we assume the bandwidth of the LPFs is 500 Hz (i.e., the cutoff frequency is $f_C = 250$ Hz). This combined with the 100 Hz sample rate means we can assume the noises out of the LPFs will be uncorrelated from sample-to-sample. Said another way, $n_1(k)$ and $n_1(k+1)$ are uncorrelated, and $n_2(k)$ and $n_2(k+1)$ are uncorrelated. Since they are also Gaussian, they are independent. This is what allowed us to choose $n_a$ and $n_b$ as independent random numbers.

We model the signal into the two mixers as a complex sinusoid of the form

$$v_{in}(k) = \sqrt{P_S}e^{j2\pi f_d kT} \tag{6.35}$$

where $f_d$ is the target Doppler frequency and $T = 0.01$ s is the track update period. We choose the scaling factor, $P_S$, to provide a specified SNR at the output of a mixer-LPF combination (i.e., BPF) that is tuned to $f_d$.

The LPF transfer function was chosen so that its gain is 0 dB at $f = 0$. This means the signal power out of the mixer-LPF combination tuned to $f_d$ will equal the signal power into the mixers. Given this, and the fact that $n_1$ and $n_2$ are unit variance, $P_S$ is equal to the SNR at the output of the mixer-LPF combination tuned to $f_d$. The topic of signal and noise scaling is discussed more fully in Chapter 4. With this, we can compute $P_S$ from

$$\frac{P_S}{P_N} = \frac{P_S}{\sigma_n^2} = SNR \text{ or } P_S = (SNR)\sigma_n^2 = SNR. \tag{6.36}$$

As in previous simulation examples, we avoid having to work directly with the radar range equation parameters by specifying a reference SNR, $SNR_{ref}$, for a reference RCS target, $\sigma_{ref}$, at a reference range, $R_{ref}$. Specifically, we use

$$SNR_{ref} = \frac{P_T G_T G_R \lambda^2 \sigma_{ref} \tau_p}{(4\pi)^3 R_{ref}^4 kT_s L} = \frac{K_{SNR}\sigma_{ref}}{R_{ref}^4} \tag{6.37}$$

which leads to

$$K_{SNR} = \frac{SNR_{ref} R_{ref}^4}{\sigma_{ref}}. \tag{6.38}$$

With this, the SNR for any range and RCS can be computed from

$$SNR = \frac{K_{SNR}\sigma}{R^4}. \tag{6.39}$$

The frequency offsets, $\delta f$, of the BPFs was set to $\delta f = f_C = 250$ Hz. With this, the correlation coefficient was computed from (6.32) as $r_{12} = -0.134 - j0.274$. The discriminator scaling factor, $S_f$, was computed from (6.9) as 166.7.

We assume the track filter is a Benedict-Bordner type of $\alpha$-$\beta$ filter with a cutoff frequency of 1Hz [17]. Thus, we use (3.50) to determine $\alpha$, and (3.43) to determine $\beta$. The results of those calculations are

$$\alpha = 0.0417 \text{ and } \beta = 8.9 \times 10^{-4}. \tag{6.40}$$

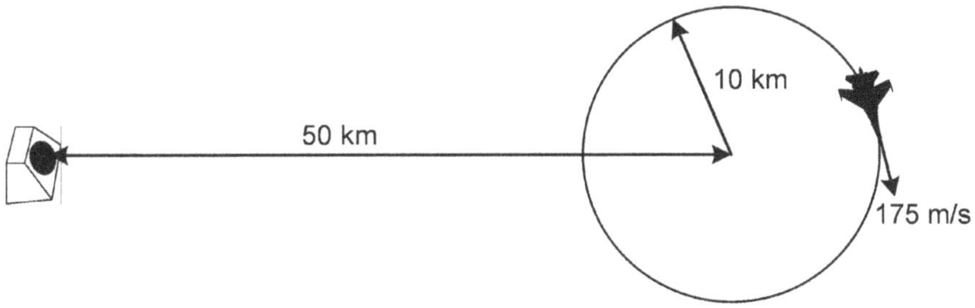


**Figure 6.23** Scenario use in the Doppler tracking simulation.

We will use a scenario that was used in Chapters 3 and 5, which is illustrated in Figure 6.23. In this scenario, the target is flying a circular trajectory with a radial dimension of 10 km. The tangential velocity of the target is constant at approximately 175 m/s, or about Mach 0.5. We assume the target altitude is (unrealistically) 0 m for convenience. Tracking starts when the target is at the bottom of the circle and continues for 200 s, or until the target is slightly past the top of the circle. Assuming an RF of $f_{RF}$ = 10 GHz, the Doppler frequency starts at

$$f_d(0) = -\frac{2\dot{R}(0)}{\lambda} = -\frac{2(-175)}{0.03} \approx 11.7 \text{ kHz} \tag{6.41}$$

and ends at a Doppler frequency of approximately −11.7 kHz.

The remaining simulation parameters are listed in Table 6.1. The reference SNR, $SNR_{ref}$, is the SNR at the start of tracking. The SNR starts at 10 dB and reaches a maximum of 14.2 dB before decreasing to 9 dB at the end of the simulation run. $f_{dtrk}$ was initialized to $f_d(0)+\delta f/2$. The initial Doppler frequency rate was set to zero.

**Table 6.1**

Simulation Parameters for CW Tracker Example

| *Parameter* | *Value* | *Comment* |
|---|---|---|
| $SNR_{ref}$ | 10 dB | The reference SNR |
| $R_{ref}$ | 51 km | The reference range |
| $\sigma_{ref}$ | 1 m$^2$ | The reference RCS |
| $f_{RF}$ | 10 GHz | The radar RF |
| $\sigma$ | 1 m$^2$ | The target RCS |
| $f_{dinit}$ | $f_d(0) + \delta f/2$ | The initial Doppler for the track filter |
| $f_B$ | 500 Hz | Doppler filter bandwidth |
| $\delta f$ | 250 Hz | BPF offsets |
| $S_f$ | 166.7 | Discriminator Scale Factor |
| $r_{12}$ | −0.134−j0.274 | Noise correlation coefficient |
| $f_c$ | 1 Hz | Closed loop track bandwidth |
| $T$ | 1/100 s | 100 Hz track rate |

The sequence of operations in the simulation is

1. Compute $f_d$ and the target range, $R$, from the target geometry and RF (10 GHz).
2. Compute $P_S$ from (6.36).
3. Compute the Doppler frequency error, $\Delta f = f_d - f_{dtrk}$.
4. Compute the outputs of the LPF as

$$v_{L1} = \sqrt{P_S}\,L(\Delta f - \delta f)$$
$$v_{L2} = \sqrt{P_S}\,L(\Delta f + \delta f).$$

5. Compute the noise samples, $n_1$ and $n_2$ from (6.31) and form

$$V_+ = |v_{L1} + n_1|$$
$$V_- = |v_{L2} + n_2|.$$

6. Compute $e_f$ from (6.7).
7. Update the track filter to produce $f_{dtrk}$.
8. Advance time by $T$ and go to step 1.

Figure 6.24 contains results of one of the simulation runs. The top graph contains plots of actual Doppler frequency, $f_d$, and the Doppler frequency predicted by the tracker, $f_{dtrk}$. Because of the scale, and the fact that the tracker is working very well, the two curves appear as one. The bottom graph is a plot of the error signal at the output of the Doppler discriminator, $e_f$, and the middle graph is a plot of $\Delta f = f_d - f_{dtrk}$, the error between the actual Doppler frequency and the tracker's estimate of it. While the $e_f$ is somewhat large, $\Delta f$ is quite small. The mean squared value of the measurement error, $E\{e_f^2\}$, is about 1,260 Hz$^2$ while the mean squared value of the tracked error, $E\{\Delta f^2\}$, is about 63 Hz$^2$. Thus, the tracker provides about a 20-to-1 reduction in mean squared values of the errors. Said another way, the Doppler tracker is providing about 13 dB on noncoherent integration gain. This is not unexpected given that the ratio of the track update rate (100 Hz) and the one-sided track loop bandwidth is 1 Hz.

As an experiment, the initial SNR was reduced until the track loop had difficulty maintaining Doppler track on the target. This occurred at an initial SNR of about −1 dB. This is illustrated in Figure 6.25, which is the output from one of the simulation runs. In this case, the $\Delta f$ error is somewhat large at the beginning and the end of the track interval. This was consistent over several simulation runs. In some cases, the errors were a little larger than shown in Figure 6.25, and in other cases it was a little smaller. The SNR at the end of the simulation run was about −2 dB, versus −1 dB at the beginning of the run. As a result, the error was generally larger at the end of the simulation run than at the beginning. The SNR near the middle of the run, when the target was closest to the radar, was about 3 dB, and the error was correspondingly smaller than at the beginning or the end of the simulation run.

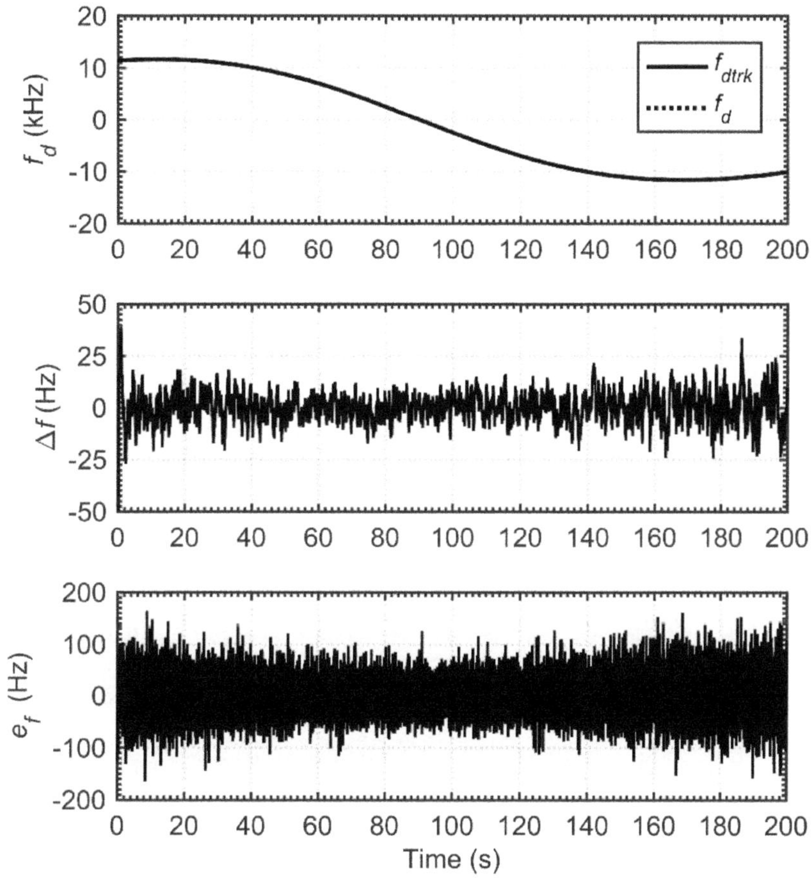


**Figure 6.24** Simulation output for the CW Doppler tracker and a 10 dB initial SNR.

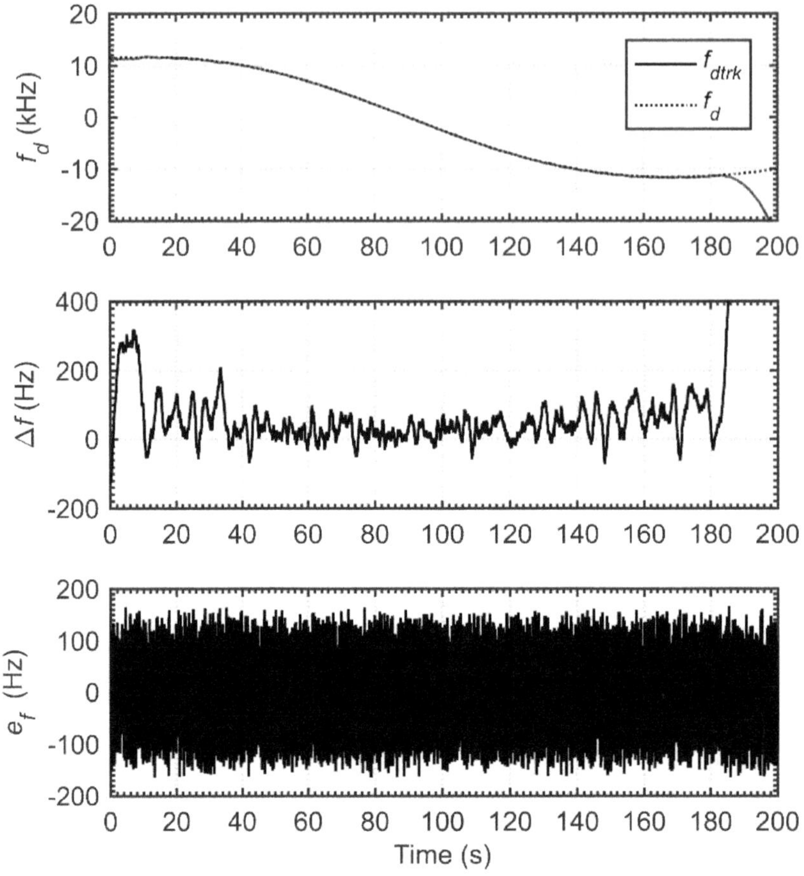


**Figure 6.25** Simulation output for the CW Doppler tracker and a −1 dB initial SNR.

At an initial SNR of −2 dB, the tracker lost track about 50% of the time, usually near the end of the simulation run, when the SNR dropped below −2 dB. At an initial SNR of −3 dB, the tracker consistently lost track. This usually occurred at the beginning of the simulation run and was attributed to a combination of the initial Doppler error of 125 Hz and noise. When the initial Doppler error was reduced to zero, the loop was sometimes able to maintain track until near the end of the simulation run, where the SNR dropped below −3 dB.

## 6.7 EXAMPLE 2: LOW PRF TRACKER

For a second example, we consider a combined range and Doppler tracker of the type discussed in Section 6.4. We use the target model from Example 1. The other simulations parameters are similar to Example 1 and listed in Table 6.2.

We assume the transmit pulse uses a 1023 chip, PRN code with a chip width of $\tau_{chip} = 1\ \mu s$. This means the transmit pulse width is $\tau_p = 1023\ \mu s$. We assume the matched filters of Figure 6.17 are matched to the transmit pulse.

To avoid having to develop a detailed matched filter model, which is quite complicated, we model the matched filter output in only the central region of range-Doppler space; specifically, the region between approximately $\pm\tau_{chip}$ in range delay, and $\pm 1/\tau_p$ in Doppler. It can be shown (see Exercise 8) that an approximate equation for the matched filter output in this region is

$$MF(\tau, f) = e^{j\psi} e^{j2\pi f\tau} \frac{\tau_p}{\tau_{chip}} \left(\tau_{chip} - |\tau|\right) \text{sinc}\left(f\left(\tau_p - |\tau|\right)\right) \text{rect}\left[\frac{\tau}{2\tau_{chip}}\right] \tag{6.42}$$

where $\psi$ is a phase due to range delay and target radar cross section properties.

The model of (6.42) does not account for the range side lobes of the matched filter output. However, the average range side lobe level for a 1023 chip, PRN coded pulses is about 30 dB below the peak of the matched filter output. This means the range side lobes will be well below the noise floor for SNR values below 20 to 30 dB.

We assume the track update period is $T = 0.1$ s (10 Hz track rate). We assume both track filters (range and Doppler) are of the Benedict-Bordner type with a cutoff frequency of 1 Hz. Thus, we use (3.67) to determine $\alpha$, and (3.60) to determine $\beta$. The results of these calculations are

$$\alpha = 0.3230 \text{ and } \beta = 0.0622\,. \tag{6.43}$$

We use a value of $q = ½$ to compute the early and late gate values for the range tracker (see Chapter 4). Thus, the scaling factor for the range discriminator is

$$S_\tau = (1-q)\tau_{res} = (1-q)\tau_{chip} = \tau_{chip}/2\,. \tag{6.44}$$

We use frequency offsets of $\delta f = 0.44/\tau_p$ and use (6.30) to compute the Doppler discriminator scaling factor.

Since $\tau_E$ and $\tau_L$ are separated by $\tau_{chip}$, we can assume the noise on the early and late gate signals are uncorrelated, and thus independent since we also assume they are Gaussian.

We use (6.32), along with (6.42) to compute the correlation coefficient between the noises out of the two matched filters of the Doppler discriminator (see Figure 6.17) as $r_{12} = 0.1342$. We then use (6.31) to compute the noise samples, $n_1$ and $n_2$. In (6.32), we used $L(f) = MF(0,f)$.

Strictly speaking, we should consider the correlation between all of the noises in the range and Doppler channels. However, we save that development for another time.

**Table 6.2**
Simulation Parameters for Low PRF Tracker Example

| *Parameter* | *Value* | *Comment* |
|---|---|---|
| $SNR_{ref}$ | 10 dB | The reference SNR |
| $R_{ref}$ | 51 km | The reference range |
| $\sigma_{ref}$ | 1 m$^2$ | The reference RCS |
| $f_{RF}$ | 5 GHz | The radar RF |
| $\sigma$ | 1 m$^2$ | The target RCS |
| $f_{dinit}$ | $f_d(0) + \delta f/2$ | The initial Doppler for the track filter |
| $\tau_{init}$ | $\tau_R(0) + \tau_{res}/2$ | The initial range delay for the track filter |
| $f_B$ | $1/\tau_p = 977$ Hz | Doppler filter bandwidth |
| $\delta f$ | $0.44/\tau_p$ | Doppler filter offsets |
| $q$ | ½ | Range gate offset |
| $S_f$ | 584.1 | Doppler discriminator scale factor |
| $S_\tau$ | $(1-q)\tau_{chip}$ | Range discriminator scale factor |
| $r_{12}$ | −0.134−j0.274 | Noise correlation coefficient |
| $f_c$ | 1 Hz | Closed loop track bandwidth |
| $T$ | 1/10 s | 10 Hz track rate |

The sequence of operations for this experiment is:

1. Compute $f_d$ and $\tau_R$ from the target geometry and RF.
2. Compute $P_S$ from (6.36).
3. Compute, $\tau_E = \tau_{trk} - q\tau_{chip}$, $\tau_L = \tau_{trk} + q\tau_{chip}$, and $\Delta f = f_d - f_{dtrk}$.
4. Compute the matched filter outputs as

$$
\begin{aligned}
v_+ &= \sqrt{P_S}MF\left(\tau_{trk}, \Delta f - \delta f\right) \\
v_- &= \sqrt{P_S}MF\left(\tau_{trk}, \Delta f + \delta f\right) \\
v_E &= \sqrt{P_S}MF\left(\tau_E, \Delta f\right) \\
v_L &= \sqrt{P_S}MF\left(\tau_L, \Delta f\right)
\end{aligned}.
$$

5. Compute the noise samples, $n_E$, $n_L$, $n_1$, and $n_2$ and form

$$
\begin{aligned}
V_+ &= \left|v_+ + n_1\right| \\
V_- &= \left|v_- + n_2\right| \\
V_E &= \left|v_E + n_E\right| \\
V_L &= \left|v_L + n_L\right|
\end{aligned}.
$$

6. Compute $e_f$ from (6.7) and $e_\tau$ from

$$e_\tau = \frac{V_E - V_L}{V_E + V_L} S_\tau .$$

7. Update the track filter to produce $f_{dtrk}$ and $\tau_{trk}$.
8. Advance time by $T$ and go to step 1.

Figure 6.26 contains plots of one of the simulation runs. The left three plots apply to the range tracker, and the right three plots apply to the Doppler tracker. The top plots depict actual and predicted values. Because of the scale, and the fact that the tracker is working very well, the two curves on each plot appear as one. The bottom plots depict the error signals at the output of the discriminators, and the middle plots depict the errors between the actual range and Doppler, and the trackers' estimate of them. In both cases, the trackers provide a reduction of about 3.5 between the variances of error signals out of the discriminator ($e_\tau$ and $e_f$) and the variances of $\Delta R$ and $\Delta f$. This is somewhat consistent with the ratio of the 10-Hz track rate and the 2-Hz track filter one-sided bandwidth (the two-sided track filter bandwidth is twice the cutoff frequency of 100 Hz).

As an experiment, the initial SNR was reduced until the track loop had difficulty maintaining track on the target. At an initial SNR of about 5 dB, the tracker lost track in about 50% of the simulation runs. When it lost track, half the time it did so near the end of the run, and half the time it did so near the beginning or middle of the run. When the initial SNR was reduced to 4 dB, the tracker lost track in about 90% of the simulation runs. Half the time it lost track near the end of the run, and half the time it lost track near the beginning or middle of the run.

Figure 6.27 contains a sample simulation output for the case where the initial SNR was 5 dB. In this case, the tracker lost track near the end of the simulation run. It is not clear whether the range or Doppler tracker was the first to lose track. However, based on other simulation runs, it is expected the Doppler tracker lost track before the range tracker.

We indicated earlier that an alternate to using a Doppler tracker to keep the matched filter tuned to the target Doppler was to use the range rate estimate from the range track filter. An example of a simulation run where that approach was used is illustrated in Figure 6.28. The parameters for the simulation run were the same as for the runs that generated Figures 6.26 and 6.27. However, the initial SNR was increased to 20 dB. If the initial SNR was allowed to be lower than 20 dB, the trackers usually lost track. Even at an SNR of 20 dB, the trackers frequently lost track. The plots of Figure 6.28 indicate the range tracker initially had difficulty tracking the target, but eventually settled down. However, near the end of the simulation run, the tracker lost track.

While the tracker was tracking the target, the range error, $\Delta R$, and the discriminator output, $e_R$, are quite small. However, this is expected since the SNR is quite high.

We previously noted that the use of a Doppler tracker to tune the matched filters for an LFM waveform did not work. However, the use of the range rate estimate from the range tracker did seem to work, somewhat. An example of a simulation output for this case is contained in Figure 6.29. In this particular case, the transmit waveform was a 1023-$\mu$s LFM pulse with bandwidth of 1 MHz. Thus, it is compatible with the PRN coded waveform used in the previous examples. The initial SNR for the simulation run was 20 dB, which is compatible with the simulation used to generate Figure 6.28. In this case, both $\Delta R$ and $e_R$ are significantly larger than in Figure 6.28. This is attributed to the range-Doppler coupling property of LFM pulses. Basically, errors in the range rate estimate caused errors in the tuning frequency of the matched filter. Because of range-Doppler coupling, the Doppler errors caused range errors, and these range errors were larger than the range errors cause by the noise.

A possible alternative for LFM waveforms is to take advantage of the range-Doppler coupling property and not attempt to tune the matched filter. Instead, the tracker would be allowed to develop a bias error. A smoothed version of the range rate estimate from the track filter could then be used to remove the bias error. Figure 6.30 contains a sample simulation output for this case. The waveform and other simulation conditions were the same as those used to generate Figure 6.29. However, in this case, the matched filter was tuned to a Doppler frequency of zero. Also, the track filter was initialized at the range of the target plus the bias error expected because of the Doppler frequency of the target. In practice, it should be possible to initialize the track filter thusly if the biased search range is used.

As can be seen, $\Delta R$ is very large, which is due to range-Doppler coupling. However, $e_R$ is quite small. This is expected because $e_R$ is affected by only the receiver noise, which is small relative to the signal power.

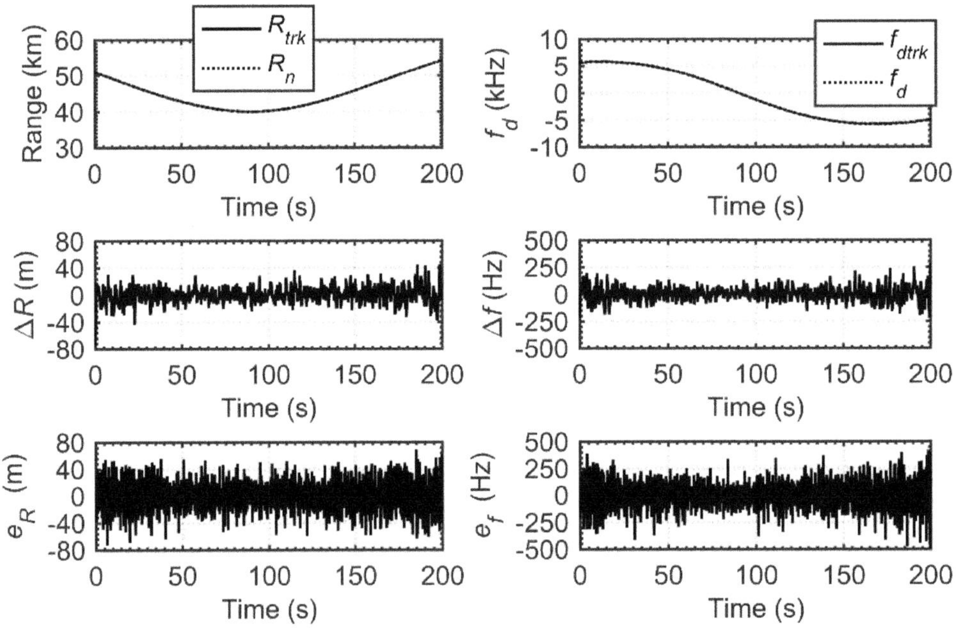


**Figure 6.26** Simulation output for the low PRF tracker and a 10-dB initial SNR.

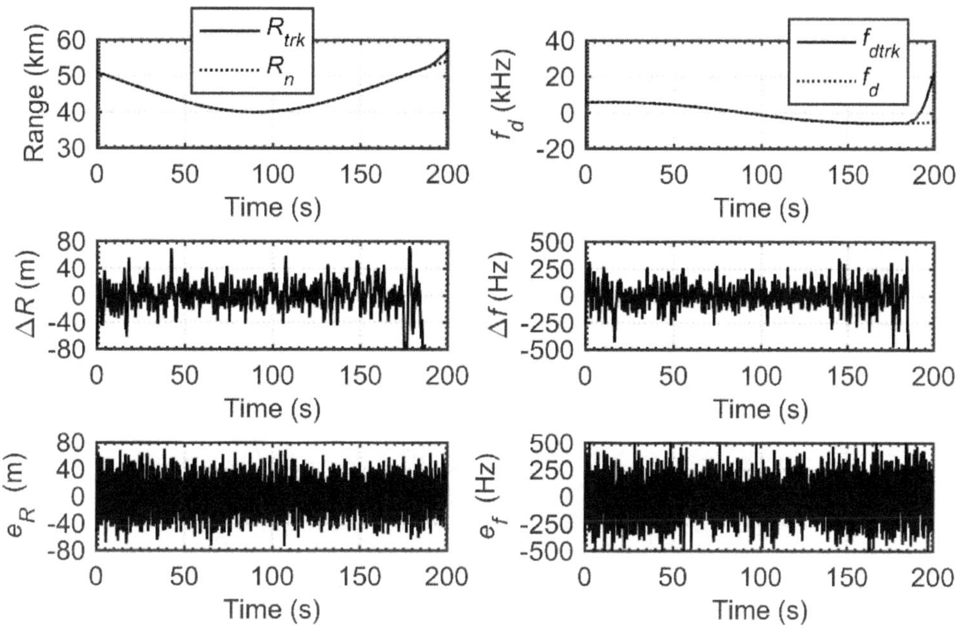


**Figure 6.27** Simulation output for the low PRF tracker and a 5-dB initial SNR.

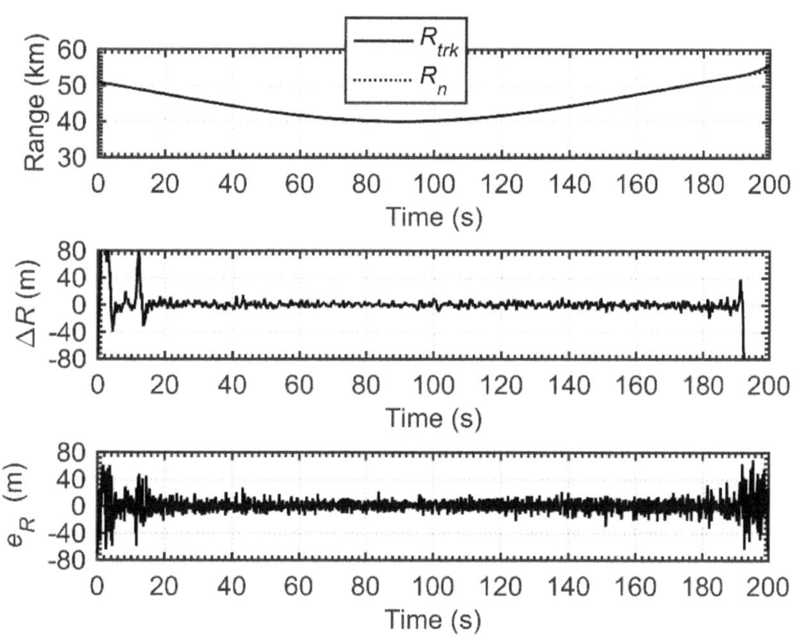


**Figure 6.28** Simulation output for the low PRF tracker where the range track filter output is used to tune the matched filter, 20-dB initial SNR.

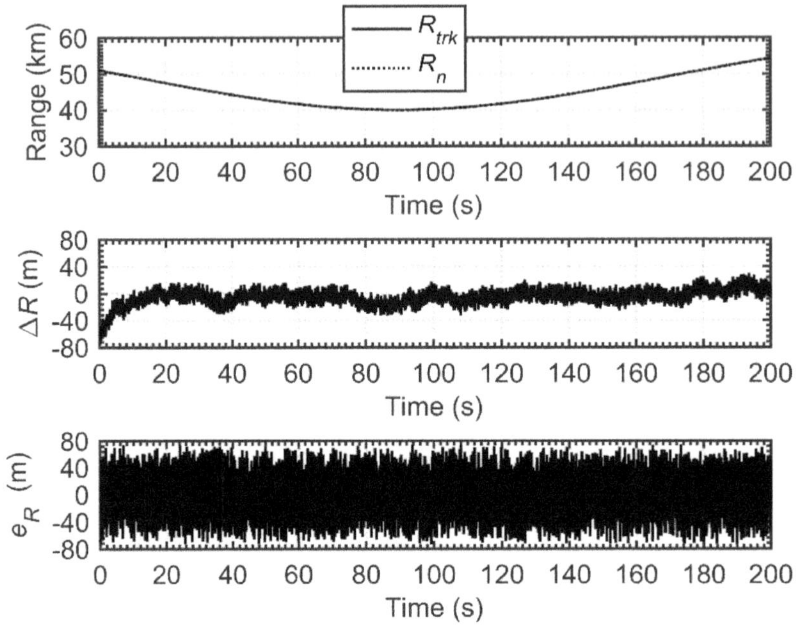


**Figure 6.29** Simulation output for the low PRF tracker; range track filter output used to tune the matched filter; 1023-μs, 1-MHz bandwidth LFM pulse; 20-dB initial SNR.

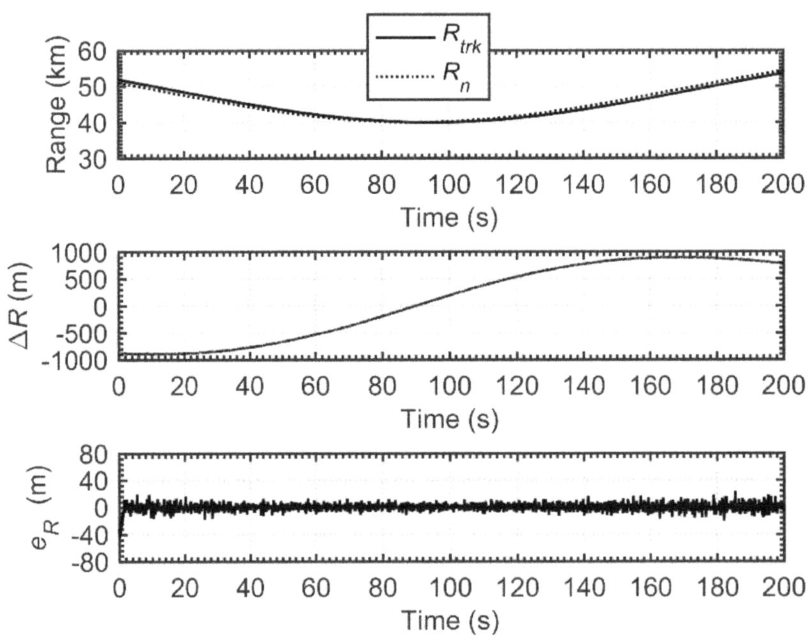


**Figure 6.30** Simulation output for the low PRF tracker; matched filter tuned to zero Doppler; 1023-μs, 1-MHz bandwidth LFM pulse; 20-dB initial SNR.

## 6.8 FUNCTIONAL LEVEL ERROR MODEL

In some preliminary or functional level tracking studies, development of a discriminator and tracker model is not desirable. In such studies, the analyst simply wants an estimate of the expected Doppler measurement error as a function of SNR and Doppler filter bandwidth, or Doppler resolution. To that end, various authors provide the equation [18, 19]

$$\sigma_{fdmeas} = \frac{(f_B/2)}{k_f\sqrt{SNR}} \tag{6.45}$$

for the root-mean-square (RMS) Doppler measurement error. To use the equation, one would draw a zero-mean, Gaussian random number with a variance of $(\sigma_{fdmeas})^2$ and add it to the true Doppler frequency.

Figure 6.31 contains a plot of (6.45) with $k_f = 2.24$, and a plot of $\sigma_{Rmeas}$ for the discriminator of (6.7), and a third-order, Butterworth filter. As can be seen, (6.45) and the experiment results match reasonably well for SNR values above about 8 dB. For SNR values below 8 dB, the two curves deviate. Unfortunately, experiments indicate the value of $k_f$ depends on filter order, and most likely the filter type. Thus, it is questionable whether (6.45) would be universally applicable

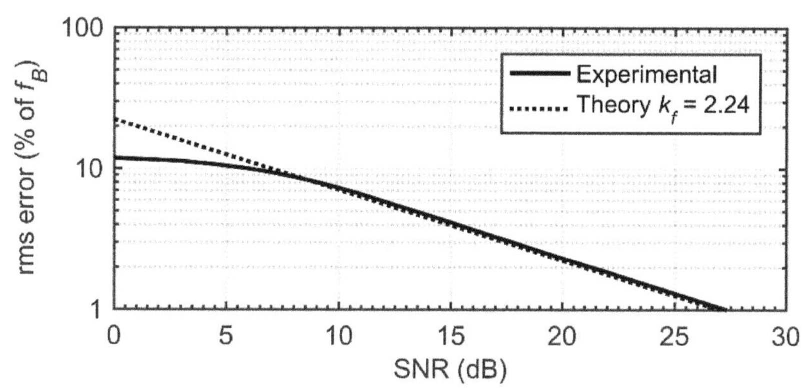


**Figure 6.31** RMS Doppler measurement error for a third-order Butterworth filter.

## 6.9 EXERCISES

1. Generate plots like Figures 6.3 and 6.4 for a fourth-order Butterworth filter with a bandwidth of $f_B = 500$ Hz and $\delta f = f_B/2$.
2. Derive (6.9).
3. Recreate Figure 6.8 and Figure 6.9.
4. Derive (6.30).
5. Reproduce Figure 6.21.
6. Plot the discriminator curve for an LFM pulse where weighting is used to reduce range sidelobes. Use a 1023 $\mu$s pulse with an LFM bandwidth of 1 MHz. Assume the weighting is 30 dB, $\bar{n} = 6$, Taylor weighting.
7. Implement the simulation discussed in Section 6.5 and generate plots like Figures 6.24 and 6.25.
8. Show that (6.42) provides a reasonable representation of the center portion of a phase coded pulse.
9. Implement the first simulation discussed in Section 6.5 and produce plots like Figures 6.26 and 6.27.
10. Modify the simulation from problem 9 as indicated in the discussion above Figure 6.28 and generate a plot like Figure 6.28.
11. Modify the simulation from problem 10 as indicated in the discussion above Figure 6.29 and generate a plot like Figure 6.29.
12. Modify the simulation from problem 11 as indicated in the discussion above Figure 6.30 and generate a plot like Figure 6.30.
13. Determine suitable values of $k_f$ for 2$^{nd}$, 4$^{th}$, and 5$^{th}$ order Butterworth filters (see Section 6.7). See if you can develop a general expression for $k_f$ versus filter order.
14. Repeat problem 13 for a Type I Chebychev filter.
15. Repeat problem 13 for a Type II Chebychev filter.
16. Repeat problem 13 for a $\tau_p$ $\mu$s, phase coded pulse. Use a separation of $\delta f = 0.44/\tau_p$.
17. Show that $n_1$ and $n_2$, as defined by (6.31), are zero mean and have a variance of 1.0.

### References


[1] D. K. Barton, Radars, Volume 7, CW and Doppler Radar, Dedham, MA: Artech House, 1978.

[2] S. M. Sherman and D. K. Barton, Monopulse Principles and Techniques, Second ed., Norwood, MA: Artech House, 2011.

[3] M. C. Budge Jr. and S. R. German, Basic Radar Analysis, Norwood, MA: Artech House, 2015.

[4] S. Butterworth, "On the Theory of Filter Amplifiers," *Experimental Wireless & the Wireless Engineer,* vol. 7, pp. 536-541, October 1930.

[5] C. L. Dolph, "A Current Distribution for Broadside Arrays Which Optimizes the Relationship Between Beam Width and Side-Lobe Level," *Proceedings IRE,* vol. 34, no. 6, pp. 335-348, June 1946.

[6] H. J. Riblet, "Discussion on Dolph's Paper," *Proc. IRE,* vol. 35, no. 5, pp. 489-492, May 1947.

[7] R. G. Lyons, Understanding Digital Signal Processing, 3rd ed., New York: Prentice Hall, 2011.

[8] R. W. Hamming, Digital Filters, 2nd ed., Englewood Cliffs, NJ: Prentice-Hall, Inc., 1983.

[9] T. T. Taylor, "Design of Line-Source Antennas for Narrow Beamwidths and Low Sidelobes," *IRE Transactions on Antennas and Propagation,* Vols. AP-3, pp. 16-28, January 1955.

[10] E. O. Brigham, The Fast Fourier Transmform and Its Applications, Englewood Cliffs, NJ: Preintice-Hall, Inc., 1988.

[11] N. Levanon and E. Mozeson, Radar Signals, Hoboken, New Jersey: Wiley-Interscience, 2004.

[12] P. M. Woodward, Probability and Information Theory with Applications to Radar, 2nd ed., New York: Pergamon Press, 1953, Artech House, 1980.

[13] A. Papoulis, Probability, Random Vairables, and Stochastic Processes, 3rd Edition ed., New York: McGraw-Hill, Inc., 1991.

[14] H. J. Blinchikoff and A. I. Zverev, Filtering in the Time and Frequency Domains, New York: John Wiley & Sons, Inc., 1976.

[15] A. I. Zverev, Handbook of Filter Synthesis, John Wiley & Sons, 1967.

[16] R. Schaumann and . M. E. V. Valkenburg, Design of Analog Filters, New York: Oxford University Press, 2001.

[17] T. Benedict and G. Bordner, "Synthesis of an optimal set of radar track-while-scan smoothing equations," *IRE Transactions on Automatic Control,* vol. 7, no. 4, pp. 27-32, 1962.

[18] D. K. Barton and H. R. Ward, Handbook of Radar Measurement, Englewood Cliffs, New Jersey: Prentice Hall, 1969.

[19] M. I. Skolnik, Introduction to Radar Systems, 3rd Edition ed., New York: McGraw-Hill, 2001.

## APPENDIX 6A: DERIVATION OF THE CORRELATION COEFFICIENT OF THE BPF OUTPUTS

In this appendix, we derive an equation for the correlation between the noises out of the BPF of the Doppler discriminator. Specifically, we want to derive (6.32), which is

$$r_{12} = \frac{\int_{-\infty}^{\infty} L(f-\delta f) L^*(f+\delta f) df}{\int_{-\infty}^{\infty} |L(f)|^2 df}. \tag{6A.1}$$

Let $n_1(t)$ and $n_2(t)$ be the noises out of the two BPFs. From random processes theory [13], the correlation between $n_1(t)$ and $n_2(t)$ is

$$R_{12}(\tau)\Big|_{\tau=0} = E\left\{n_1(t+\tau)n_2^*(t)\right\}\Big|_{\tau=0} . \tag{6A.2}$$

If we assume $n_1(t)$ and $n_2(t)$ are zero mean, which we do, (6A.2) is also the covariance between $n_1(t)$ and $n_2(t)$, $C_{12}(\tau)$. From random processes theory, we know that $R_{12}(\tau)$ is equal to the inverse Fourier transform of the cross, power spectrum between $n_1(t)$ and $n_2(t)$. That is

$$R_{12}(\tau) = \int_{-\infty}^{\infty} S_{12}(f)e^{j2\pi f\tau}df . \tag{6A.3}$$

From this we have

$$R_{12}(0) = \int_{-\infty}^{\infty} S_{12}(f)df . \tag{6A.4}$$

Thus, to find $R_{12}(0)$ we need $S_{12}(f)$.

Again drawing on random processes theory, if we have the configuration shown in Figure A.1, we can write $S_{12}(f)$ as

$$S_{12}(f) = S_n(f)B_1(f)B_2^*(f). \tag{6A.5}$$

In our case, $B_1(f)$ and $B_2(f)$ are shifted versions of the same LPF filter response. That is

$$\begin{aligned} B_1(f) &= L(f-\delta f) \\ B_2(f) &= L(f+\delta f) \end{aligned} . \tag{6A.6}$$

We assume the noise into the filters is white so that

$$S_n(f) = N_0 \tag{6A.7}$$

where $N_0$ is the power spectral density of the input noise, $n(t)$. With this

$$S_{12}(f) = N_0 L(f-\delta f)L^*(f+\delta f) \tag{6A.8}$$

and

$$R_{12}(0) = N_0 \int_{-\infty}^{\infty} L(f-\delta f)L^*(f+\delta f)df . \tag{6A.9}$$

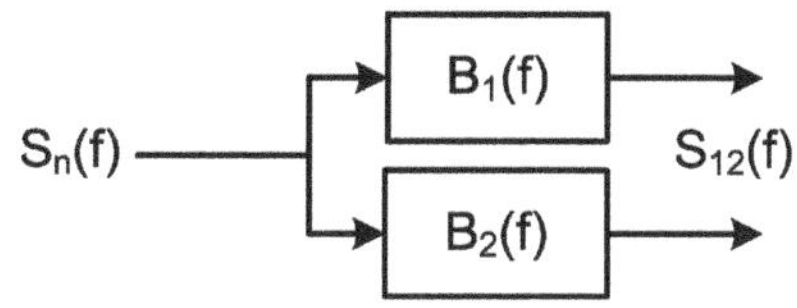


**Figure A.1** Filter Configuration Used to Determine $S_{12}(f)$

Once we have $R_{12}(0)$ we can write the correlation coefficient as

$$r_{12} = \frac{R_{12}(0)}{\sqrt{R_1(0)R_2(0)}}. \tag{6A.10}$$

We can find $R_1(0)$ and $R_2(0)$ from (6A.2), (6A.4), (6A.5), and (6A.9) by letting $n_1(t) = n_2(t)$. Thus,

$$\begin{aligned} R_1(0) &= \int_{-\infty}^{\infty} S_1(f)df = N_0 \int_{-\infty}^{\infty} B_1(f)B_1^*(f)df \\ &= N_0 \int_{-\infty}^{\infty} |B_1(f)|^2 df = N_0 \int_{-\infty}^{\infty} |L(f-\delta f)|^2 df \end{aligned} \tag{6A.11}$$

and

$$R_2(0) = N_0 \int_{-\infty}^{\infty} |L(f+\delta f)|^2 df. \tag{6A.12}$$

Since $L(f-\delta f)$, $L(f+\delta f)$, and $L(f)$ have the same shape, and the integrals of (6A.11) and (6A.12) are from $-\infty$ to $\infty$,

$$R_1(0) = R_2(0) = N_0 \int_{-\infty}^{\infty} |L(f)|^2 df = R(0). \tag{6A.13}$$

With this we get

$$r_{12} = \frac{R_{12}(0)}{\sqrt{R_1(0)R_2(0)}} = \frac{R_{12}(0)}{R(0)} = \frac{\int_{-\infty}^{\infty} L(f-\delta f)L^*(f+\delta f)df}{\int_{-\infty}^{\infty} |L(f)|^2 df} \tag{6A.14}$$

which is (6.32).

# Chapter 7

# Simulation Examples

## 7.1 INTRODUCTION

In this chapter, we combine some of the trackers developed in previous chapters into simulations and use the simulations to demonstrate some theoretical predictions related to range and angle tracking. In particular, we examine the impact of radar cross section (RCS) fluctuation on tracking, how trackers behave in the presence of two, interfering targets, and tracking in the presence of specular multipath.

## 7.2 RCS FLUCTUATION

In the examples of Chapters 4, 5, and 6, we assumed the RCS of the target was constant. For practical targets, this is not the case. Practical targets consist of a large number of surfaces that capture energy from the radar and reradiate it back to the radar [1, p. 65; 2-5]. The properties of this reradiated energy depend on several factors that include size, composition, complexity, orientation of the target, operating frequency (RF), and target speed. Generally, the signal returned from a target looks like a random signal [1, 4-6]. That is, its amplitude and phase vary with time in an apparent random fashion. To account for this, Peter Swerling developed four random process target models that almost all radar analysts use to analyze detection performance [7]. The Swerling RCS models are commonly termed Swerling 1, 2, 3, and 4, or SW1, SW2, SW3, and SW4. Constant RCS is often referred to as SW0 or SW5.

With the SW1 and SW2 models, the RCS is governed by the exponential, or Chi-squared, two-degree-of-freedom, density function

$$f(\sigma) = \frac{1}{\sigma_{AV}} e^{-\sigma/\sigma_{AV}} U(\sigma). \tag{7.1}$$

with SW3 and SW4 models, the RCS is governed by the Chi-squared, four-degree-of-freedom, density function

$$f(\sigma) = \frac{4\sigma}{\sigma_{AV}^2} e^{-2\sigma/\sigma_{AV}} U(\sigma). \tag{7.2}$$

In (7.1) and (7.2), $\sigma$ is the RCS, and $\sigma_{AV}$ is the average RCS, both in m$^2$. $\sigma_{AV}$ is the RCS used in the radar range equation [1, 8–10]. $U(\sigma)$ is the unit step function and is defined by

$$U(\sigma)=\begin{cases}1 & \sigma\geq 0\\ 0 & \sigma<0\end{cases}. \tag{7.3}$$

The difference between the SWodd (SW1, SW3) and SWeven (SW2, SW4) models is how the RCS varies with time. With SWodd models, the RCS changes from "scan to scan". With the SWeven models, the RCS changes from pulse to pulse. The two terms, scan to scan and pulse to pulse, are a carryover from rotating search radars. In those radars, a scan is one rotation of the antenna. Thus, with a SWodd target, the RCS is assumed to remain constant during the time the radar beam is scanning by the target, on any one rotation. However, it changes randomly from rotation to rotation, or scan to scan. We can think of SWodd being associated with targets whose RCS varies slowly and SWeven being associated with targets whose RCS varies rapidly [1, 4, 6, 8].

SW1 and SW2 RCS models are normally associated with complex targets such as aircraft and ships. SW3 and SW4 RCS models are normally associated with simple targets such as artillery shells and reentry vehicles.

The Swerling RCS models can be considered limiting cases of actual RCS fluctuation. Because of this, they are difficult to use in tracking simulations. Also, the Swerling models represent only amplitude fluctuation. They don't accommodate the phase fluctuation exhibited by actual targets [1, pp. 66–70].

In this book, we adopt one of the "Swerling-like" fluctuation models discussed in [1]. With that model, the RCS amplitude is governed by (7.1), and the phase of the return signal is governed by the uniform density

$$f(\psi)=\frac{1}{2\pi}\text{rect}\left[\frac{\psi}{2\pi}\right] \tag{7.4}$$

where the rect[$x$] function is defined by

$$\text{rect}[x]=\begin{cases}1 & |x|\leq 1/2\\ 0 & |x|>1/2\end{cases}. \tag{7.5}$$

The temporal variation of both RCS ($\sigma$) and phase ($\psi$) is determined by a low pass filter (LPF) with a prescribed bandwidth. The bandwidth of the LPF depends on the RF of the radar, the size of the target, and the speed of the target. Representative bandwidth values are 0.5 to 6 Hz. A block diagram of the method proposed in [1], and used here, is shown in Figure 7.1.

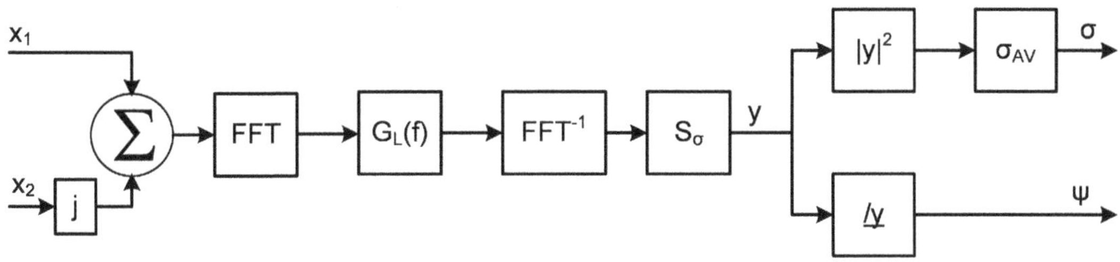


**Figure 7.1** Fluctuating RCS generation method.

$x_1$ and $x_2$ are independent sequences of zero-mean, unit-variance, Gaussian random numbers. They are combined to form the complex random sequence, $x = x_1 + jx_2$. It is assumed $x$ is a discrete-time, random process with a sample period equal to the track update period, $T$. The FFT-$G_L(f)$-FFT$^{-1}$ blocks provide an implementation of the LPF, and the block labeled $S_\sigma$ scales the FFT$^{-1}$ output so that $y$ is a unit variance, complex, random process. Since $x_1$ and $x_2$ are zero-mean and Gaussian, $y$ is also zero mean and Gaussian. Because of this, $|y|^2$ has an exponential density [11–13]. Scaling $|y|^2$ by $\sigma_{AV}$ yields a sequence of random numbers with an average value of $\sigma_{AV}$. Further, $\psi$ is a sequence of uniform random numbers with a density governed by (7.4) [11–3].

The FFT length, $N_{FFT}$, is chosen to be larger than the number of samples used in the simulation. For example, if the simulation is to run for $T_{sim}$ seconds, $N_{FFT}$ should be chosen to satisfy

$$N_{FFT} > T_{sim}/T \ .$$

$G_L(f)$ is a desired low pass filter function. An example might be the Butterworth filter function discussed in Chapter 6. The scaling factor, $S_\sigma$, is best determined experimentally, because the theoretical equation for it can be somewhat complicated.

### 7.2.1 Example 1

As a first example, we examine the effect of RCS fluctuation on tracking error. For this example, we use the range tracker discussed in Section 4.7.1. That tracker used a sampling gate discriminator and a 1 $\mu$s, unmodulated pulse. Since we are interested in characterizing only the effect of RCS fluctuation, we use a stationary target at the reference range of 50 km. Also, we initialize the tracker to the true target range and true target range rate of 0 m/s.

Figure 7.2 contains a plot of the simulation output for a SNR of 10 dB. The top two plots are of the error, $\Delta R = R - R_{trk}$, between the true range and the tracked range. The bottom two plots are of the error signal out of the range discriminator, $e_\tau$, converted to meters. The left two plots apply to the case where the target RCS was held constant at $\sigma = 1$ m$^2$, and the right two plots apply to the case where the target RCS fluctuation model discussed above is used. In that case, the average RCS, $\sigma_{AV}$, was set to 1 m$^2$. A third order, Butterworth filter function was used in the RCS generation model, and the bandwidth of the filter was set to 0.5 Hz. For this case, the main effect of the fluctuating RCS is to cause some spikes in the plot of $\Delta R$. The plots of $e_\tau$ are similar and is indicative of the fact that the error signal out of the discriminator is driven mainly by noise (because of the somewhat low value of SNR).

Further examples with the 10 dB SNR case indicate that the plots of Figure 7.2 are somewhat unusual. In most cases, the fluctuating RCS caused the SNR at any one time to drop so low that the tracker lost track. An example of this is illustrated in Figure 7.3, which is a plot like Figure 7.2 later in the simulation run. In this case, $\Delta R$ became larger, until the tracker finally lost track at about the 630 second mark.

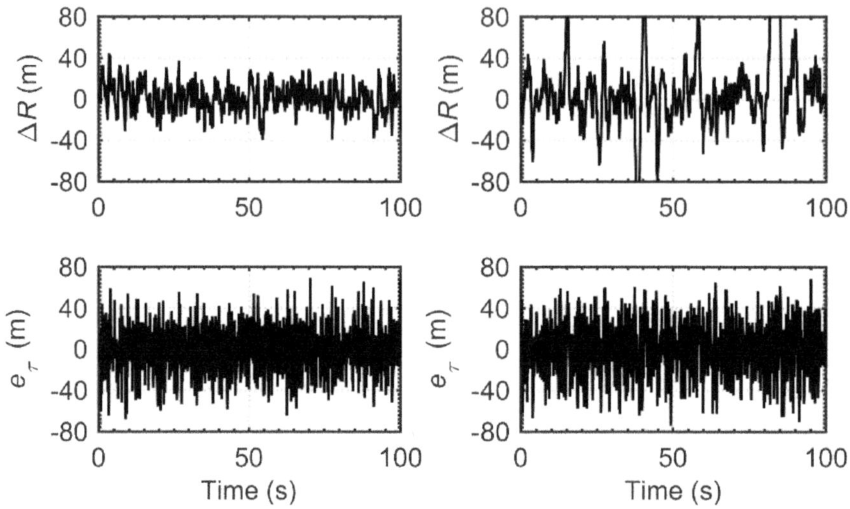


**Figure 7.2** Results of fluctuating target example for a 10 dB SNR—left plots: constant RCS—right plots: fluctuating RCS.

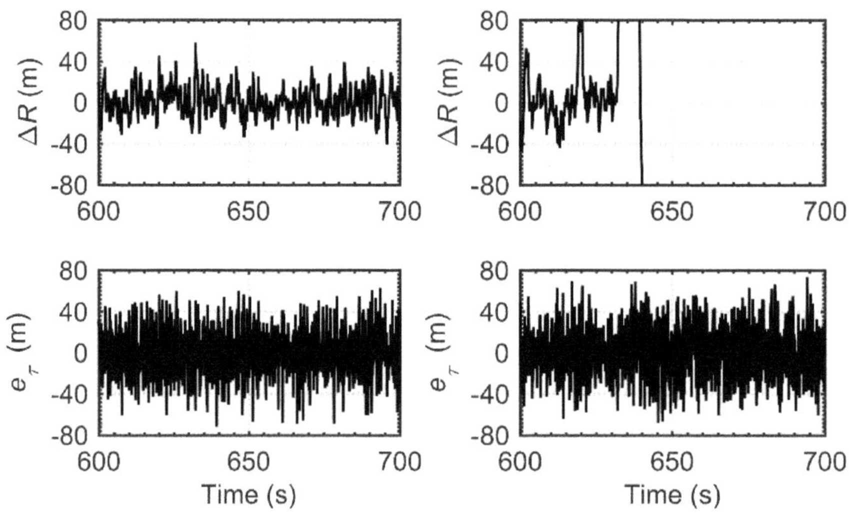


**Figure 7.3** Results of fluctuating target example for a 10 dB SNR—left plots: constant RCS—right plots: fluctuating RCS—at a later time.

Even at larger SNR values, the fluctuating RCS affected the tracking error, but to a lesser extent. This is illustrated in Figure 7.4 which is a plot like Figures 7.2, but at a SNR of 20 dB. In this case, the $\Delta R$ is significantly reduced, however, the plot for the fluctuating RCS case still has some rather large spikes. Also, the fluctuating RCS has more of an influence on the output of the discriminator. This is due to the fact that the RCS fluctuation allows the SNR to momentarily decrease. When this happens, the impact of noise on the discriminator output increases. For the constant RCS case (the left plots) the SNR never varies so the overall influence of noise on $e_\tau$ is fairly constant.

In a 1990 paper [14], Swerling and Peterman developed an equation that provides an estimate of the impact of RCS fluctuation on root-mean-squared (rms) tracking error. The equation also accounts for the effect of the non-coherent integration provided by the tracker. Their equation, in a form consistent with (4.100) of Chapter 4, is

$$\sigma_{Rfluct} = \frac{\rho \tau_{res}}{k_r \sqrt{n(SNR)}} \tag{7.6}$$

where $\rho$ is a parameter that depends on a detection threshold relative to the $SNR$ corresponding to the average target RCS, and $n$ is the effective number of pulses noncoherently integrated by the tracker. Since the track rate in the simulations used to generate Figure 7.4 is 10 Hz ($T = 0.1$ s), and the two-sided bandwidth is $B = 2f_c = 2$ Hz, an appropriate value of $n$ would be 5. However, a value of 3 seemed to work better.

Determination of $\rho$ is not straightforward. In the simulation used to generate Figure 7.5, no detection threshold was used. Because of this, it is not clear what value should be used for $\rho$. A value of 1.75 was suggested by the reviewer and used here. Values of 2 and 150 m were used for $k_r$ and $\tau_{res}$, respectively.

Figure 7.5 contains plots of (7.6) and the rms values of $\Delta R$ and $e_\tau$ versus SNR for the constant and fluctuating RCS cases. For the constant RCS case, $\rho$ was set to 1.0. As expected, RCS fluctuation increases the rms values of both $\Delta R$ and $e_\tau$. The simulation results match the theory quite well for the constant RCS case, and reasonably well for the fluctuating RCS case. Although not evident from the plots of Figure 7.6, the relative effect of the fluctuating RCS increases with SNR. This is due to the fact that, for smaller SNRs, the noise has more of an effect on $\Delta R$ and $e_\tau$ than does signal level. Verification of this is left as an exercise.

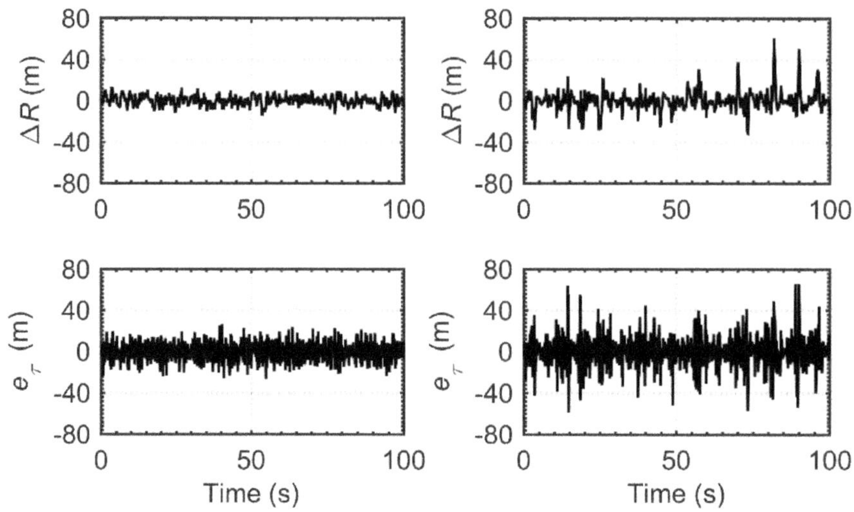


Figure 7.4 Results of fluctuating target example for a 20 dB SNR—left plots: constant RCS—right plots: fluctuating RCS.

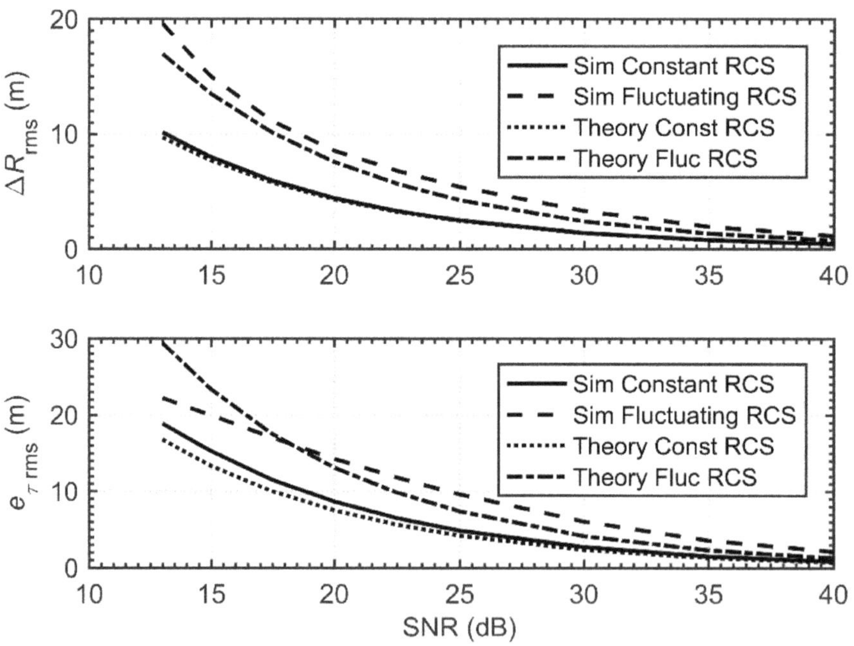


Figure 7.5 Plots of rms error versus SNR for constant and fluctuating RCS targets—same simulation conditions as Figure 7.2.

## 7.3 DUAL TARGET TRACKING

### 7.3.1 Background

In the next set of examples, we investigate how the integrated, range and angle tracker of Section 5.10 behaves in the presence of two targets, when the spacing between the targets is less than the range and/or angle resolution of the radar. The track filter perspective of this situation is discussed extensively in text books and papers on multiple target tracking [15–19]. It is also discussed somewhat from the perspective of discriminators [20–26]. In this chapter, we briefly outline the dual target tracking problem and consider some examples using an integrated, range and angle tracker.

Theory states that, for two, constant RCS (point) targets that have exactly the same Doppler frequency and phase, and are separated by less than the beam width of the antenna, the angle tracker will tend to follow the voltage centroid of the two targets. In terms of equations, if the targets are located at $u_1$ and $u_2$, in sine space, and if the signal amplitudes at the matched filter output are $S_1$ and $S_2$, the tracked angle will be

$$u_{trk} = \frac{u_1 S_1 + u_2 S_2}{S_1 + S_2}. \tag{7.7}$$

If the targets have different Doppler frequencies, the average behavior of the tracker will depend on how the monopulse normalization is performed. If the monopulse normalization is performed by an automatic gain control circuit (AGC), the value of $u_{trk}$, relative to the ratio, $R_S = S_2/S_1$, will depend on the value of the difference in Doppler frequencies, $\Delta f_d$, relative to the AGC bandwidth, $B_{AGC}$. Specifically, if $\Delta f_d \ll B_{AGC}$, $u_{trk}$ tends to abruptly switch between $u_1$ and $u_2$ as $R_S$ transitions above and below 1.0. If $\Delta f_d \gg B_{AGC}$, $u_{trk}$ follows what is termed an elliptic centroid between $u_1$ and $u_2$. Limiting cases of the above were treated by Kliger, Olenberger, Kanter, Andrews, and others [1–4]. In particular, they show that, if $B_{AGC} = \infty$, utrk is governed by the equation

$$u_{trk} = \begin{cases} u_1 & R_S < 1 \\ u_2 & R_S > 1 \\ \dfrac{u_1 + u_2}{2} & R_S = 1 \end{cases} \tag{7.8}$$

If $B_{AGC} = 0$, $u_{trk}$ is governed by the equation

$$u_{trk} = \frac{1}{\pi\sqrt{1 + R_S^2}}\left[(u_1 - u_2)(1 - R_S)K(r) + (u_1 + u_2)(1 + R_S)E(r)\right] \tag{7.9}$$

where $K(k)$ and $E(k)$ are elliptic integrals of the first and second kind, respectively, and

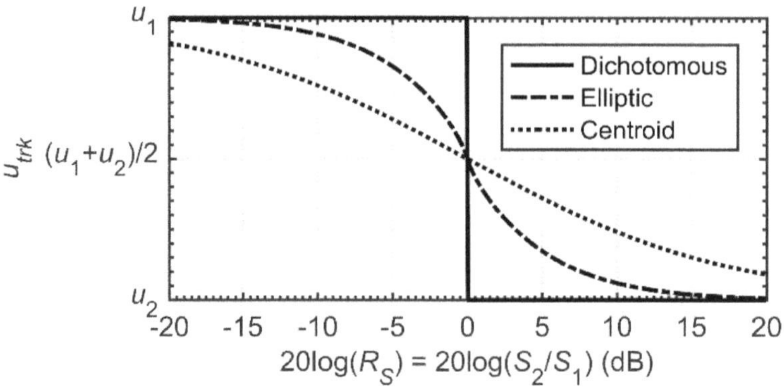


**Figure 7.6** Illustration of dichotomous, elliptic centroid, and voltage centroid tracking.

$$k = \frac{4R_S}{\left(1+R_S\right)^2}. \tag{7.10}$$

These conditions are illustrated in Figure 7.6. The concept of $u_{trk}$ abruptly switching between $u_1$ and $u_2$ is termed *dichotomous tracking*. Andrews took the development a step further and demonstrated that $u_{trk}$ is characterized by a transition that lies between those described by (7.8) and (7.9) when the AGC bandwidth is finite and non-zero [1].

In arriving at (7.8) and (7.9), it is assumed the tracker bandwidth is zero. That is, it acts as a perfect integrator, or "averager". As will be shown in some of the examples to follow, the behavior of $u_{trk}$ can be quite interesting when a non-zero bandwidth tracker is considered.

The infinite bandwidth AGC condition (i.e., $B_{AGC} = \infty$) is termed *instantaneous normalization*. For that case, the difference channel signal is normalized by the magnitude of the sum channel signal. This is the normalization method used in the exact discriminator discussed in Chapter 5. Thus, we would expect the angle tracker with the exact discriminator will exhibit dichotomous behavior. In a similar fashion, the range error signal is normalized by dividing by $|V_E| + |V_L|$, which means it also uses instantaneous normalization.

### 7.3.2 Example 2: Ideal Angle Tracker—Demonstration of Dichotomous Tracking

For this example, we use an idealized angle tracker to demonstrate the dichotomous tracking behavior predicted by (7.8). For the discriminator, we use the exact monopulse discriminator discussed in Sections 5.8.1 and 5.8.2. In particular, we use the error signal equation of (5.114) with $k_\Delta = 0$. That is, we form the error signal using

$$e_u = \text{Re}\left[\frac{V_\Delta\left(\tau_{trk}\right)V_\Sigma^*\left(\tau_{trk}\right)}{\left|V_\Sigma\left(\tau_{trk}\right)\right|^2}\right]S_u. \tag{7.11}$$

Since this discriminator effectively normalizes $V_\Delta$ by $|V_\Sigma|$, it is implementing instantaneous normalization.

The form of (7.11) implies the antenna uses amplitude comparison monopulse. This stems from the fact that $G_\Delta(u)$ and $G_\Sigma(u)$, as defined by (7.12) and (7.13) (below), are in-phase or 180°out of phase, which is the situation that occurs in antennas that use amplitude comparison monopulse. Had we assumed the antenna used phase comparison monopulse, we would have had to shift $G_\Delta(u)$ by 90° and use the imaginary operator in (7.11).

In keeping with the developments that lead to (7.8) [1–6], we need to remove the effects of the shapes of the antenna sum and difference directivity patterns. We do this by assuming ideal sum and difference voltage directivity patterns. That is, we use

$$G_\Sigma(u) = 1.0 \text{ V/V} \tag{7.12}$$

and

$$G_\Delta(u) = u \text{ V/V}. \tag{7.13}$$

We assume the targets are at a fixed range of $R$ and that the range tracker is perfectly tracking the target. This means $\tau_{trk} = \tau_R$ and that we are sampling the output of the sum and difference angle channel matched filters at matched range (see Chapters 4 and 5). We assume the RCSs of the two targets are constant and have values of $\sigma_1$ and $\sigma_2$. We assume Target 1 has a Doppler frequency of $f_{d1} = 0$, and Target 2 has a Doppler frequency of $f_{d2} = \Delta f$. Finally, we assume Target 1 is located at an angle of $\Delta u$, and Target 2 is located at an angle of $-\Delta u$. A block diagram of the angle track portion of the tracker used in the example is illustrated in Figure 7.7.

We can write $V_\Sigma$ and $V_\Delta$ (see Figure 7.7) as

$$V_\Sigma = \sqrt{\frac{K_{ref}\sigma_1}{R^4}} e^{j0} G_\Sigma(\Delta u - u_{trk}) + \sqrt{\frac{K_{ref}\sigma_2}{R^4}} e^{j2\pi\Delta f t} G_\Sigma(-\Delta u - u_{trk}) \tag{7.14}$$

and

$$V_\Delta = \sqrt{\frac{K_{ref}\sigma_1}{R^4}} e^{j0} G_\Delta(\Delta u - u_{trk}) + \sqrt{\frac{K_{ref}\sigma_2}{R^4}} e^{j2\pi\Delta f t} G_\Delta(-\Delta u - u_{trk}), \tag{7.15}$$

or, with (7.12) and (7.13)

$$V_\Sigma = K_V\sqrt{\sigma_1} + K_V\sqrt{\sigma_2} e^{j2\pi\Delta f t} \tag{7.16}$$

and

$$V_\Delta = K_V\sqrt{\sigma_1}(\Delta u - u_{trk}) + K_V\sqrt{\sigma_2} e^{j2\pi\Delta f t}(-\Delta u - u_{trk}), \tag{7.17}$$

where we used

$$K_V = \sqrt{\frac{K_{ref}}{R^4}} . \tag{7.18}$$

Since the slope of $G_\Delta(u)$ is 1.0, and $G_\Sigma(u) = 1.0$, it can be shown that $S_u = 1.0$ (see Section 5.8).

Another of the assumptions in the derivation of (7.8) is that the closed loop tracker provides significant averaging. To model this, we change the cutoff frequency of the tracker to $f_c = 0.01$ Hz, instead of the 1 Hz we have used in the previous chapters. We make $\Delta f = 0.1$ Hz, so it will be large relative to $f_c$. We again use a Benedict and Bordner type of α-β track filter, and a track update rate of 10 Hz ($T = 0.1$ s) [7]. With these conditions, we have

$$\alpha = 0.0043 \text{ and } \beta = 9.21 \times 10^{-6} . \tag{7.19}$$

To be consistent with Figure 7.6, we fix $\sigma_1$ at 1 $m^2$ (0 dBsm) and vary $\sigma_2$ from 0.01 $m^2$ to 100 $m^2$ (-20 dBsm to 20 dBsm). To be sure we allow enough time for the tracker to settle to some average $u_{trk}$, we allow the simulation to run for 1000 seconds before recording $u_{trk}$. Finally, we initialize the angle tracker to $u_{trk} = 0$ and $\dot{u}_{trk} = 0$. Since we are only interested in the performance of the tracker as a function of relative target sizes, we set the noise terms to zero.

Figure 7.8 contains plots of the results of this example. The dotted curve corresponds to the case where we let $\Delta f = 0$. As predicted by theory, it is the same as the dotted curve of Figure 7.6, the centroid track. The circles represent the response when $\Delta f = 0.1$ Hz. Again, it is consistent with theoretical predictions. That is, it tracks the larger target. When the target sizes are the same, it tracks the centroid, as indicated by the circle in the center of the plot.

We note that the tracker of this example is not a "practical" tracker. The antenna is ideal, the range track is ideal, the update rate is unrealistically large (for the multi-function track radars we have been considering in this book), and the target parameters are not realistic. This example is included only to corroborate the theoretical prediction concerning dichotomous track. A somewhat more realistic example is discussed in the next section.

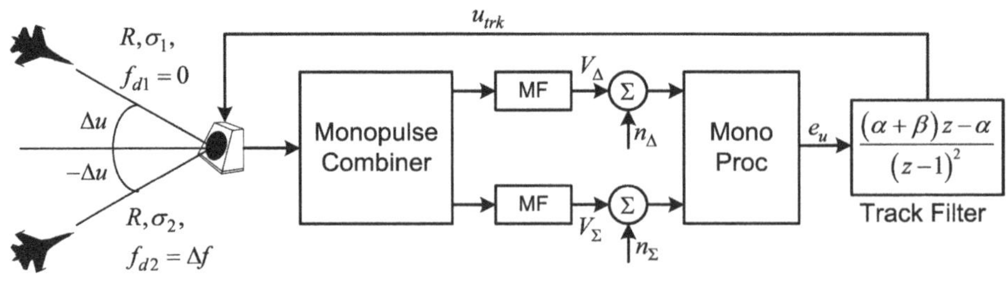


**Figure 7.7** Block diagram of angle tracker used in Example 2.

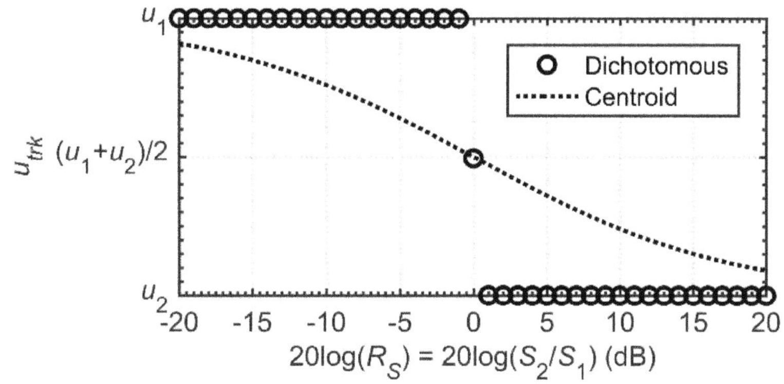


**Figure 7.8** Results of Example 2, illustrating dichotomous and centroid tracking.

### 7.3.3 Example 3: A Variation on Example 2

Figure 7.9 contains a plot of the simulation output for Example 2 with the track filter cutoff frequency set back $f_c = 1$ Hz. $\Delta f$ was also set to 0.5 Hz.[1] With these values of $f_c$ and $\Delta f$, $\Delta f$ is less than the track loop cutoff frequency of $f_c$. We again set the noises to zero, use the ideal antenna of Example 2, and assume the tracker has perfect range track. We initialize the angle tracker to $[u_{trk}, \dot{u}_{trk}] = [0, 0]$ and begin the simulation with Target 1 one dB larger than Target 2. After 10 seconds, the target sizes were made equal and, after another 10 seconds, Target 2 was made one dB larger than Target 1.

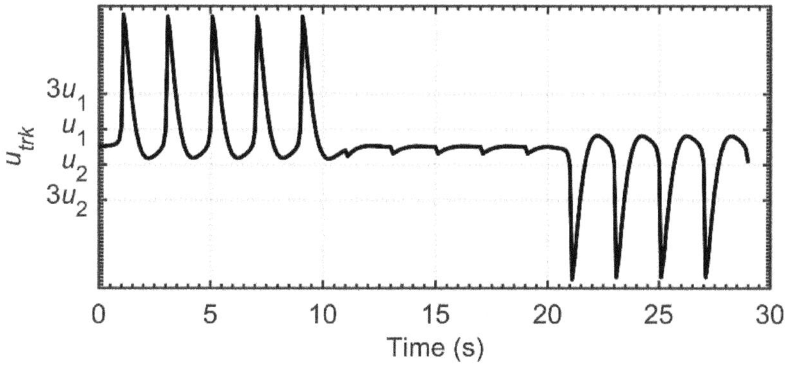


**Figure 7.9** Result of Example 3.

[1] As a note: $\Delta f$ will always lie between ± ½ the track rate (±1/2$T$) since the tracker is a discrete time system. Even if the actual $\Delta f$ is some other value, the tracker will treat it as if it is between ± ½ the track rate. Thus, for example, if the track rate is 10 Hz and the actual $\Delta f$ is 10,000 Hz, the effective, or folded, or ambiguous, $\Delta f$ will be zero Hz. If the actual $\Delta f$ is 10,001 Hz, the effective $\Delta f$ will be 1 Hz.

During the first 10-second interval, the angle track tended toward Target 1 but oscillated over a significant interval. During third 10-second interval, the angle track tended toward Target 2, but again oscillated over a significant interval. During the second interval, the angle track somewhat followed the centroid of the two targets but still oscillated. Although not obvious, the average angles over the three 10-second intervals are $u_1$, $(u_1 + u_2)/2$, and $u_2$. Thus, the average behavior follows the theoretical prediction of (7.8).

The time between the peaks of Figure 7.9 is 2 seconds, which is the reciprocal of the Doppler difference between the two targets. Thus, the tracker is following the beat frequency between the two targets.

The behavior illustrated in Figure 7.9 is well explained by theory [1, 2, 4] . If we substitute (7.16) and (7.17) into (7.11), and perform appropriate manipulations, we obtain the equation

$$e_u = \frac{(\sigma_1 - \sigma_2)\Delta u - (\sigma_1 + \sigma_2)u_{trk} - 2\sqrt{\sigma_1\sigma_2}\cos(2\pi\Delta f t)u_{trk}}{\sigma_1 + \sigma_2 + 2\sqrt{\sigma_1\sigma_2}\cos(2\pi\Delta f t)}. \tag{7.20}$$

If we assume the track loop is fast enough to maintain $e_u = 0$, we obtain

$$u_{trk} = \frac{(\sigma_1 - \sigma_2)\Delta u}{(\sigma_1 + \sigma_2) + 2\sqrt{\sigma_1\sigma_2}\cos(2\pi\Delta f t)}. \tag{7.21}$$

If $\sigma_1 > \sigma_2$, $u_{trk}$ will be proportional to $+\Delta u$. That is, the tracker will tend to follow Target 1. If $\sigma_2 > \sigma_1$, $u_{trk}$ will be proportional to $-\Delta u$. That is, the tracker will tend to follow Target 2. This is what we observed in Figure 7.9. However, (7.21) also indicates $u_{trk}$ can deviate significantly from $\pm\Delta u$. Specifically, $u_{trk}$ can vary, in magnitude, from a minimum of

$$|u_{trkmin}| = \left|\frac{(\sigma_1 - \sigma_2)}{(\sigma_1 + \sigma_2) + 2\sqrt{\sigma_1\sigma_2}}\right|\Delta u, \tag{7.22}$$

to a maximum of

$$|u_{trkmax}| = \left|\frac{(\sigma_1 - \sigma_2)}{(\sigma_1 + \sigma_2) - 2\sqrt{\sigma_1\sigma_2}}\right|\Delta u. \tag{7.23}$$

In this example, $\sigma_1 = 1$ dBsm and $\sigma_2 = 0$ dBsm during the first 10 second interval. This means $u_{trk}$ could, theoretically, vary between $u_{trkmin} = 0.06\Delta u$ and $u_{trkmax} = 17.4\ \Delta u$. This did not occur because the track loop bandwidth was not large enough to maintain $e_u = 0$.

It is also noted ,from (7.21), that $u_{trk}$ varies in proportion to $\Delta f$. This is what led to the 2-second spacing between peaks in Figure 7.9.

As an interesting point, we note that, as $\sigma_1$ and $\sigma_2$ get closer together, but not equal, $|u_{trkmax}|$ tends toward $\infty$ and $|u_{trkmin}|$ tends toward 0, which means $u_{trk}$ could vary significantly. This idea that $u_{trk}$ can vary significantly is sometimes referred to as the *glint* problem [8–13]. However, in that application, the "targets" would be scattering surfaces on a single target.

As a note, the variations in $u_{trk}$ could be reduced by averaging over several track updates. However, this could have the negative effect of making the tracker less sensitive to target maneuvers, since the averaging would essentially reduce the bandwidth of the tracker.

### 7.3.4 Example 4: Example 3 with a Realistic Antenna

The antenna pattern used in Example 3 was ideal in that $G_\Sigma(u) = 1$ and $G_\Delta(u) = u$. In this example, we use a more realistic antenna pattern. In particular, we use the 50-element, uniformly illuminated, linear, space-fed phased array discussed in Section 5.4, and used in the examples of Sections 5.9 and 5.10. That antenna had a beam width of about 2.2 degrees. We also use the modified exact discriminator discussed in Section 5.8.2, with $k_\Delta = 0.2$.

The targets were 0.5 degrees apart to be sure both were located in the main beam of the sum pattern. As with Example 3, we initialized the angle tracker to $[u_{trk}, \dot{u}_{trk}] = [0, 0]$ and began the simulation with Target 1 one dB larger than Target 2. After 10 seconds, we set the target sizes equal and, after another 10 seconds, we made Target 2 one dB larger than Target 1.

The result of this example is shown in Figure 7.10. The plot of $u_{trk}$ resembles that of Figure 7.9 for the cases where one target was larger than the other. However, when the target sizes were equal, the response is quite different. This is due to the "glint" phenomenon discussed at the end of Section 7.3.3. Specifically, as the antenna moved toward the centroid of the two targets, the effective values of $\sigma_1$ and $\sigma_2$ were close, but not exactly equal. As a result, the denominator of (7.20) became a very small positive or negative number over the time period of 10 to 20 seconds. This caused large spikes in the error signal, $e_u$, out of the discriminator, which the angle tracker tried to null by increasing the variations in $u_{trk}$.

Figure 7.11 contains a plot of the simulation output when noise was included. For this example, the SNR on both targets, when their RCSs were equal, was held constant at 10 dB.

In this case, it is difficult to tell which target the tracker is following. The deviations in $u_{trk}$ appear to be larger in the 10 second to 20 second interval. However, they are also large in the 0 second to 10 second interval. In other simulation runs, the response was quite different than that shown in Figure 7.11. Also, in about 30% of the runs, the tracker either lost track or produced excessively large errors.

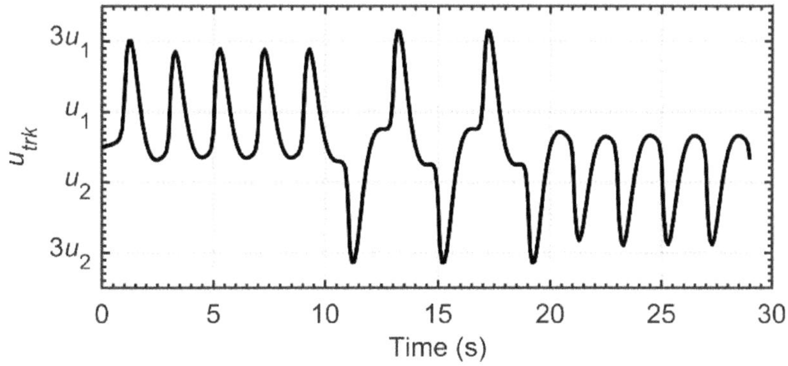


**Figure 7.10** Example 4 Results

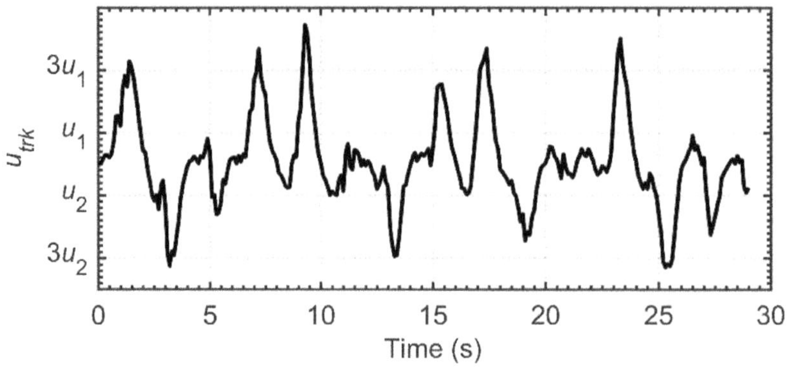


**Figure 7.11** Result of Example 4, with noise.

## 7.4 CROSSING TARGET EXAMPLES

For the next examples, we examine how the range and angle tracker of Section 5.10 behaves in the presence of two targets that approach each other, and then diverge. The general geometry we consider is illustrated in Figure 7.12. At the beginning of the simulation, the targets are located 20 km down range, relative to the radar, and at cross range locations of ±3 km. We assume the targets are at an altitude of zero. This is unrealistic. However, it allows us to avoid having to model two angle trackers. We only need to model the azimuth, or $u$, angle tracker.

Both targets are approaching the radar at 140 m/s in the y, or down range direction and 60 m/s in the x, or cross range direction. Initially, we use nominally equal radar cross sections (RCSs) of 1-$m^2$ but later will vary the RCS of one of the targets to investigate the impact of different RCSs on tracking behavior.

We initialize the range and angle trackers to the location of the left target, which we designate as Target 1. We assume perfect initialization of the positon and velocity states of the range tracker, and perfect initialization of the position state of the angle tracker. We set the initial velocity state of the angle tracker to zero.

We adjust the parameters of the radar range equation, specifically $K_{ref}$, so the initial SNR on Target 1 is 20 dB (see Chapters 4, 5, and 6).

We use a track rate of 10 Hz, and both the angle and range trackers use Benedict-Bordner α-β filters with cutoff frequencies of $f_c$ = 1 Hz. The resulting filter coefficients are

$$\alpha = 0.3230 \text{ and } \beta = 0.0622 . \tag{7.24}$$

The range tracker uses the sampling gate discriminator with $q$ = ½. The waveform is a 15 μs, LFM pulse with a bandwidth of 1 MHz (see Chapter 4). We match the Doppler frequency of the matched filter to the initial Doppler frequency of Target 1 and keep it at that value throughout the simulation run.

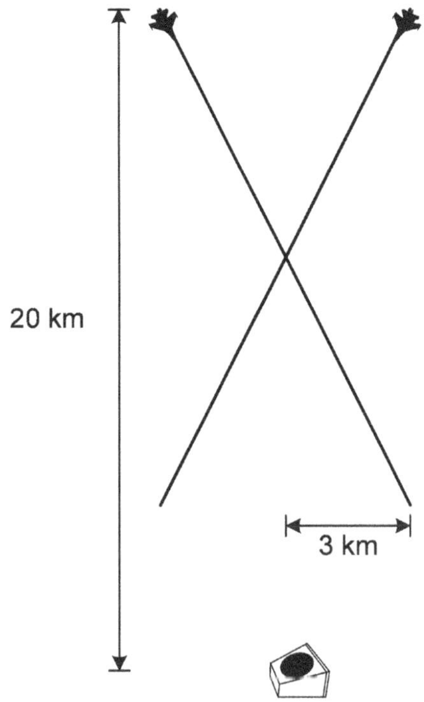


**Figure 7.12** Geometry used in the crossing target examples.

To properly capture the effect of Doppler frequency differences between the two targets, we modify the matched filter model to include the effects of target motion. The matched filter model used in Chapter 4 is (see (4.20))

$$MF(\tau,f)=\sqrt{P_s}e^{j\psi_R}e^{j\pi f(\tau_R+\tau)}\left(\tau_p-|\tau-\tau_R|\right) \times \text{sinc}\left(\left[f-\alpha(\tau-\tau_R)\right]\left[\tau_p-|\tau-\tau_R|\right]\right)\text{rect}\left[\frac{\tau-\tau_R}{2\tau_p}\right]. \quad (7.25)$$

To include the effects of target motion, we modify the model to

$$MF(\tau,f)=\sqrt{P_s}e^{-j2\pi f_{RF}\tau_R}e^{j\pi f(\tau_R+\tau)}\left(\tau_p-|\tau-\tau_R|\right) \times \text{sinc}\left(\left[f-\alpha(\tau-\tau_R)\right]\left[\tau_p-|\tau-\tau_R|\right]\right)\text{rect}\left[\frac{\tau-\tau_R}{2\tau_p}\right] \quad (7.26)$$

by replacing $\psi_R$ with $-2\pi f_{RF}\tau_R$. For a target with a constant range rate of $\dot{R}$, we have

$$\begin{aligned}\psi_R &= -2\pi f_{RF}\frac{2}{c}\left(R_0 + \dot{R}t\right) = -4\pi R_0/\lambda - 2\pi\left(2\dot{R}/\lambda\right)t \\ &= -\phi_{R0} - 2\pi f_d t\end{aligned} \tag{7.27}$$

In (7.27), $f_d$ is the target Doppler frequency.

For the antenna, we use the squinted beam model of Section 5.4, which had a beam width of about 2.2°. We use the modified exact discriminator of Section 5.8.2 with $k_\Delta = 0.2$. With this model, the error signal is given by

$$e_u = \text{Re}\left[\frac{V_\Delta\left(\tau_{trk}\right)V_\Sigma^*\left(\tau_{trk}\right)}{\left|V_\Sigma\left(\tau_{trk}\right)\right|^2 + 0.2\left|V_\Delta\left(\tau_{trk}\right)\right|^2}\right]S_D \tag{7.28}$$

where $V_\Sigma(\tau_{trk})$ is the output of the sum channel matched filter evaluated at the tracked range, and $V_\Delta(\tau_{trk})$ is the output of the difference channel matched filter at the tracked range. The discriminator scaling factor, $S_D$, is 0.0318 sines/sine. We use the Re operator in (7.28) because the antenna model we use implies the radar is using amplitude comparison monopulse.

### 7.4.1 Example 5: Equal Doppler Frequencies and Target Sizes

With the geometry of Figure 7.12, the radar is at the same cross-range as the point where the two target trajectories cross, i.e., at $x = 0$. Because of this, and the symmetry of the target trajectories, the range from the radar to the two targets is the same. Also, the range rate associated with each target is the same. This means the Doppler frequency difference between the target returns is zero, and we expect the radar will track the centroid of the targets, once they are close enough to both be in the beam. Figure 7.13 illustrates that this is what happened.

The left plot of Figure 7.13 is a down range vs. cross range plot of the two target trajectories, and a plot of a trajectory created from the outputs of the range and angle trackers. The $y$ (down range) and $x$ (cross range) coordinates of the track are given by

$$\begin{aligned}y_{trk} &= R_{trk}\sqrt{1 - u_{trk}^2} \\ x_{trk} &= R_{trk}u_{trk}\end{aligned} \tag{7.29}$$

where $R_{trk}$ is the tracked range and $u_{trk}$ is the tracked angle.

The upper right figure contains plots of the error between the actual target ranges and the tracked target range. The plots appear as one because the ranges to the two targets are the same. The lower right figure contains plots of the actual target angles and the tracked angle.

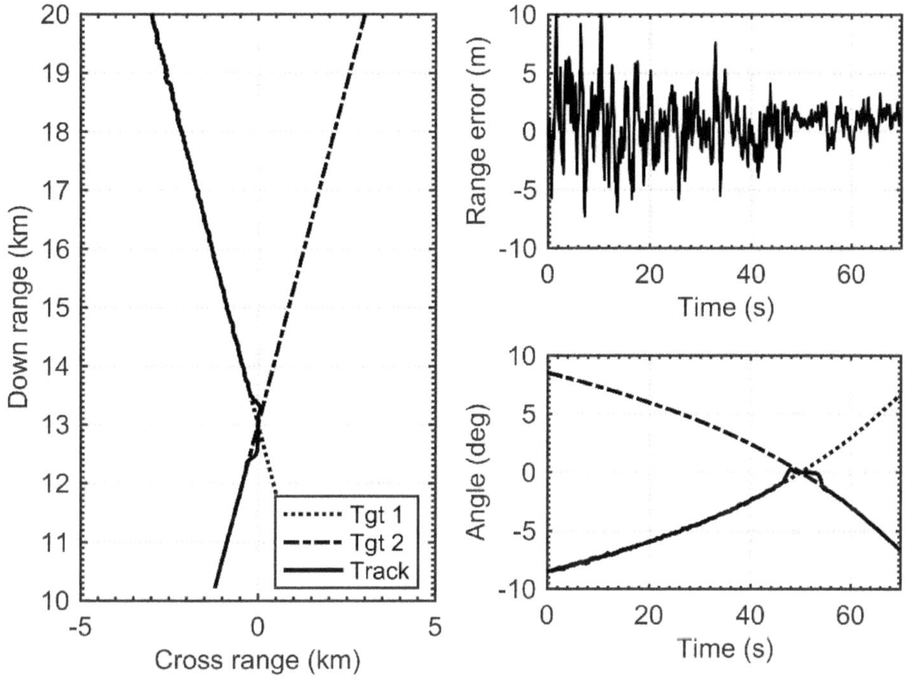


**Figure 7.13** Crossing targets—equal Doppler frequencies and equal target RCSs.

As expected, the angle track stays on Target 1 until the targets are separated by approximately 2°, the approximate antenna beam width. At that time, Target 2 starts moving into the sum beam (the antenna is still nominally centered on Target 1). When this happens, the tracker starts tracking a weighted centroid of the target angles. This continues until the beam is centered between the two targets. At that time, the angle tracker settles on the centroid of the two target angles. This happens because the signal amplitudes of the two targets are the same (recall that the target RCSs are equal). The slight overshoot in the track angle, before it settles on zero, is due to track filter transients. As a note, range does not enter into the picture because the range to the two targets is the same.

The tracked angle remains at zero until the targets separate to about 2°. At that point, both targets are moving off the main beam of the antenna, and the angle track loop becomes influenced by noise. The noise influence tends to move the track in one direction or the other, until the beam moves on to one of the targets. At that point, the angle tracker starts tracking that target. In the particular example of Figure 7.13, the angle tracker switched to Target 2. In other simulation runs, with different noise sequences, the angle tracker would return to Target 1. If one of the targets was made slightly larger than the other, the track tended to follow the larger target, after the targets crossed.

The scenario of this example is ideal because, in practice, it is very unlikely the Doppler frequencies of the two target will be exactly the same. However, it is still of value because it illustrates the centroid tracking properties discussed in Section 7.3.

### 7.4.2 Example 6: Different Doppler Frequencies and Target Sizes

For this example, we move the radar 500 m to the right. The result of this is, the ranges to the two targets are not the same, and the range rates of the two targets are also not the same. We also make the RCS of Target 2 one dB larger than the RCS of Target 1. We keep all of the other simulation parameters the same as in Example 5, including the noise samples.

Because of the Doppler frequency difference between the two targets, and the fact that Target 2 is 1 dB larger than Target 1, we expect the track to switch from Target 1 to Target 2 somewhat abruptly. As can be seen by Figure 7.14, this is what happened. As the two targets approached each other, they both moved into the antenna beam, and the main lobe of the matched filter response. When this happened, the track changed from Target 1 to Target 2.

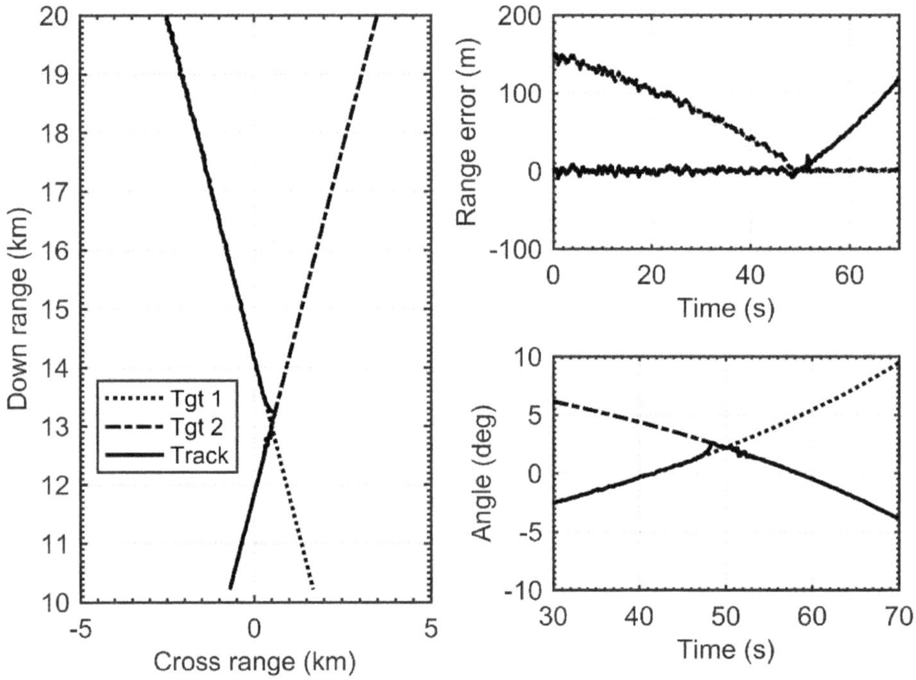


**Figure 7.14** Crossing targets—different Doppler frequencies and target RCSs.

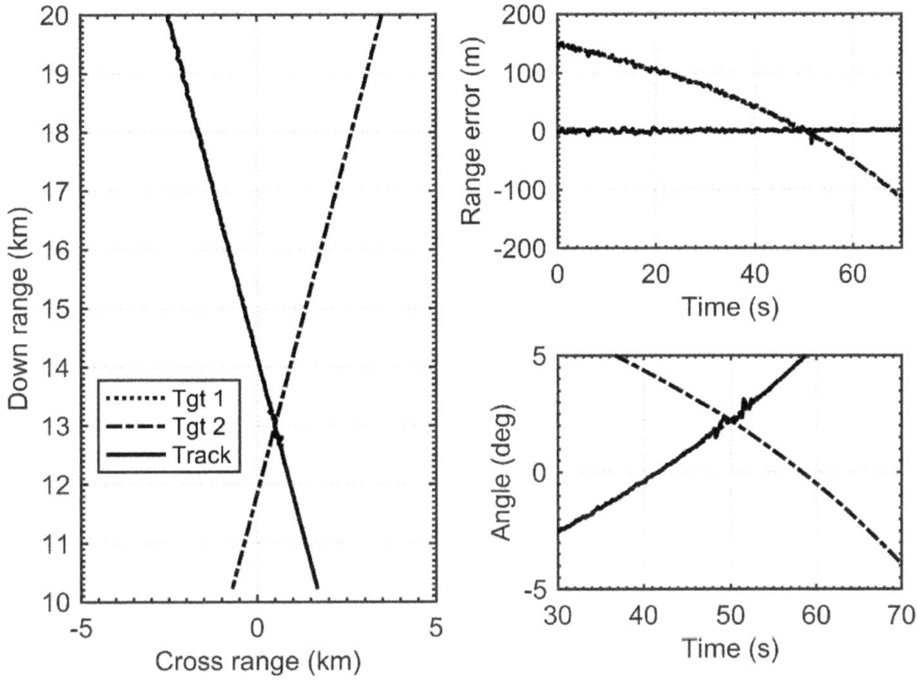


**Figure 7.15** Crossing targets—different Doppler frequencies and target RCSs—Target 1 larger than Target 2.

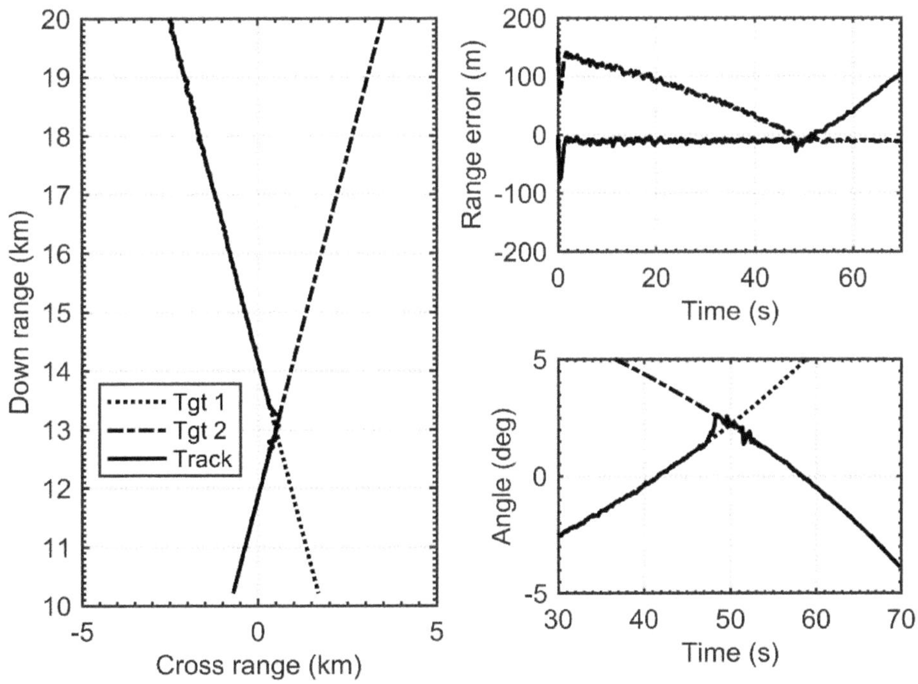


**Figure 7.16** Crossing targets—Different Doppler frequencies and target RCSs—target 1 larger than Target 2.

Figure 7.15 contains plots of the simulation results for the case where the RCS of Target 1 was 1 dB larger than the RCS of Target 2. In this case, the track stayed with Target 1.

Figure 7.16 contains an interesting set of simulation plots that were the result of an error. Specifically, it was generated by the same script used for Figure 7.14, except that the matched filter was tuned to zero Doppler frequency, instead of the initial Doppler frequency. The result was the range bias error indicated in the upper right plot. In this case, the lower range error plot is not centered on zero, as in Figures 7.14 and 7.15. Instead, the range tracker had a small bias error. This bias error is due to the LFM range-Doppler coupling discussed in Section 6.4.

### 7.4.3 Example 7: A Different Geometry—Collinear Target Trajectories

In this example, we want to see if the range tracker exhibits the same track behavior as the angle tracker. To do so, we change the geometry of the previous examples to focus on the performance of the range tracker.

We have the targets fly toward the radar along a line that passes directly through the radar. We start tracking Target 1 when it is 20 km down range (and 0 km in cross range). At that time, Target 2 is 20.5 km down range. Target 1 is approaching the radar at 140 m/s, and Target 2 is approaching the radar at 155 m/s. After about 33 seconds of track, Target 2 overtakes and passes Target 1. The targets have the same RCS.

Because of the choice of target velocities, the difference in Doppler frequency between the two targets is 500 Hz. Since this is an integer multiple of the 10 Hz track rate, the apparent, or ambiguous, Doppler frequency is zero. As such, we expect the tracker will exhibit a centroid track behavior. Figure 7.17 illustrates this is what happened.

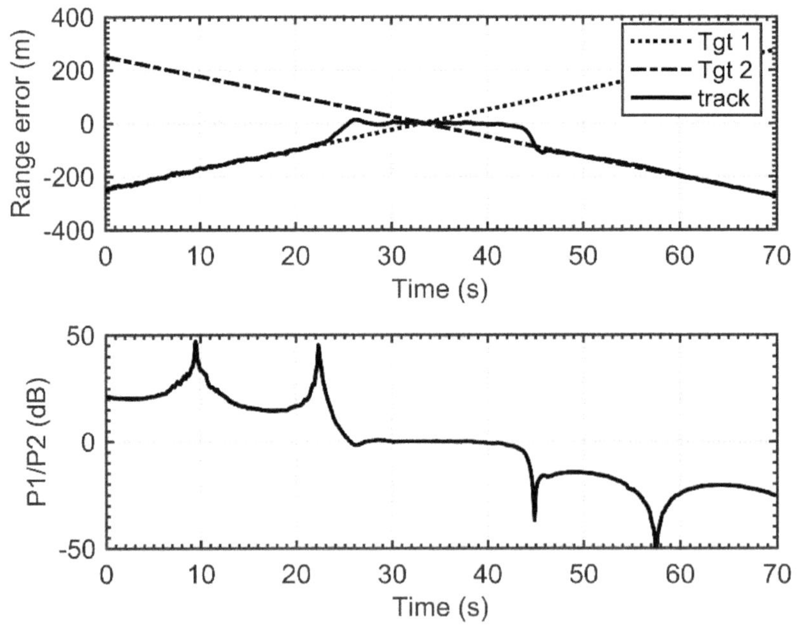


**Figure 7.17** Collinear targets—different Doppler frequencies, but zero apparent Doppler frequency difference, same target RCSs.

The top portion of Figure 7.17 contains plots of the ranges of the two targets and the tracked range, all referenced to the centroid between the target ranges. The bottom portion contains a plot of the relative powers (P1 and P2) of the two target signals at the output of the matched filter.

The top plot indicates the radar tracks the centroid of the two targets once their separation approaches the range resolution of the matched filter, which is 150 m. At that time, the ratio of the powers (P1/P2) very close to 0 dB. At about 45 seconds, the ranges separate by more than the range resolution, which means the range tracker has to follow one of the targets. It chooses Target 2 because, at that time, it will be closer to the radar than Target 1 and will thus have a higher signal amplitude, due to the $R^4$ variation of SNR.

The scallops in the lower plot below about 20 seconds and above about 45 seconds are due to the fact that one of the targets is in the side lobes of the matched filter response.

As a variation on this example, we changed the velocity of Target 2 to $(-150 - 1.5\pi)$ m/s. This resulted in an ambiguous Doppler frequency of 0.41 Hz, which is less than the 1 Hz cutoff frequency of the range tracker. As a result, the range tracker exhibited the oscillatory behavior discussed in Sections 7.3.3 and 7.3.4. This is illustrated in Figure 7.18. When the targets approached each other, their sizes were close enough for the discriminator to exhibit an oscillatory behavior similar to that discussed in Sections 7.3.3 and 7.3.4. Since the frequency of the oscillations was less than the track loop band width, the range tracker was able to follow them for a while. Eventually, the targets became symmetrically located about the matched filter response peak, and the tracker settled to tracking the centroid. It maintained the centroid track until the targets separated far enough in range for the tracker to track only one of them.

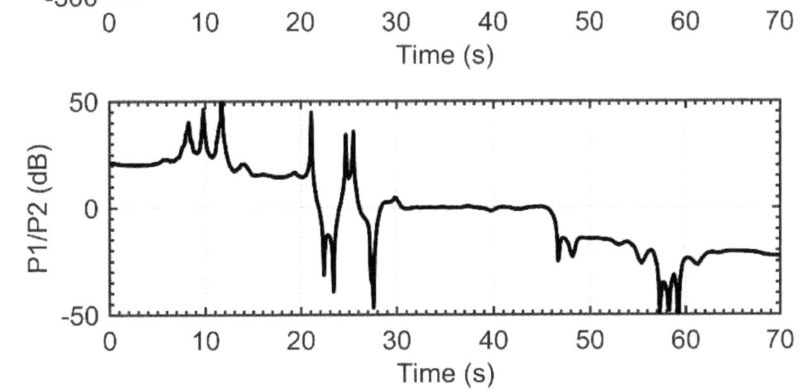


**Figure 7.18** Collinear targets—nonzero apparent Doppler frequency difference, same target RCSs.

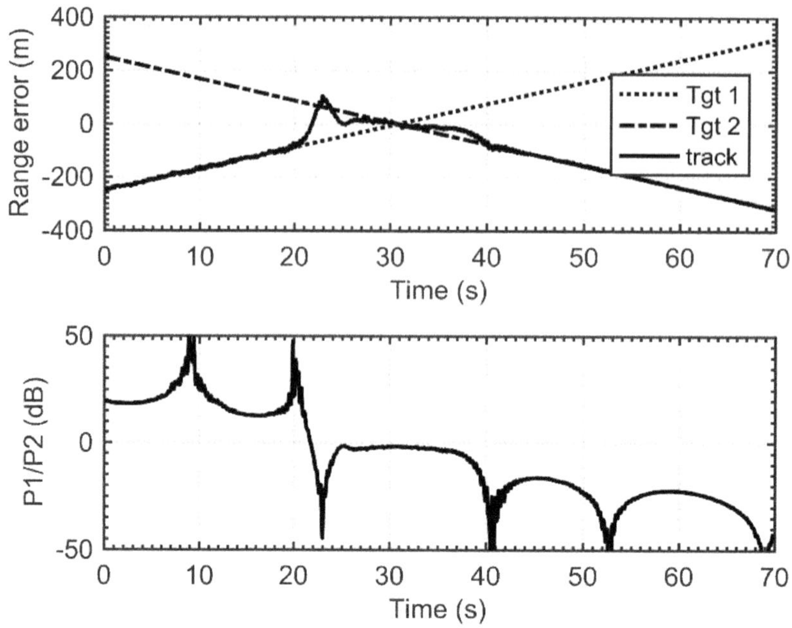


**Figure 7.19** Collinear targets—nonzero apparent Doppler frequency difference, different target RCSs.

To close out this example, we change the RCS of Target 2 to 1.5 m$^2$, and adjust its velocity to $(-150 - 2\pi)$ m/s. This results in an apparent Doppler frequency difference of about 2.77 Hz, which is larger than the 1 Hz cutoff frequency of the range tracker.

The result of this simulation run is contained in Figure 7.19. In this case, the tracker tracked Target 1 until about the 21 second time point. The reason for the switch is not clear. However, around the 21 second time point, the range error at the discriminator output had a small bias error. It is suspected that this caused the track to drift toward, and eventually overshoot the Target 2 range. At about the 22 second time point, the combination of RCS differences and locations on the matched filter response caused the track to follow a weighted centroid of the two target ranges. This continued until the targets separated far enough for the return from Target 1 to fall into a null of the matched filter response. At that point, the tracker tracked Target 2.

## 7.5 CROSSING TARGET EXAMPLES—FLUCTUATING TARGET RCS

In the previous set of simulations, the RCS of the targets was constant throughout the simulation runs. In this section, we consider the case where the RCSs of the targets are fluctuating.

### 7.5.1 Example 8: Same Set Up as Example 5 with Fluctuating RCSs

For this example, we use the simulation set up from Example 5 (Section 7.4.1) but use fluctuating RCSs instead of constant RCSs. In Example 5, the RCSs were equal, and the Doppler frequencies of the targets were the same. As a result, the angle tracker tended to track the centroid of the two targets, when both targets were in the main beam of the antenna pattern (see Figure 7.13).

The result from one simulation run of this example is contained in Figure 7.20. In this case, the track did not follow the centroid of the targets. It stayed on Target 1 until just before the targets crossed. At that time, it moved to Target 2, and stayed there until just after the targets crossed, at which time the track moved back to Target 1 for a while. Finally, the track switched back to Target 2. This result is not unexpected. Because of the RCS fluctuation of the two targets, it is possible that one or the other is the larger at any one time. Since they are both in the main beam of the antenna, the radar could just as likely track one or the other, or it could track their angle centroid at any one time. As the targets separate in angle, the radar will tend to track one of them, since the other will be attenuated by the antenna beam pattern, once the angle separation becomes large enough.

As indicated above, after the targets crossed, the tracker was as likely to follow one target as the other. This is illustrated in Figure 7.21. This figure is a plot like the left plot of Figure 7.20 for the case where we ran the simulation five times with a different RCS fluctuation used on each of the runs. As expected, near the point where the targets crossed, the radar was just as likely to track one or the other, or their centroid. After the targets crossed, the radar tended to track one or the other.

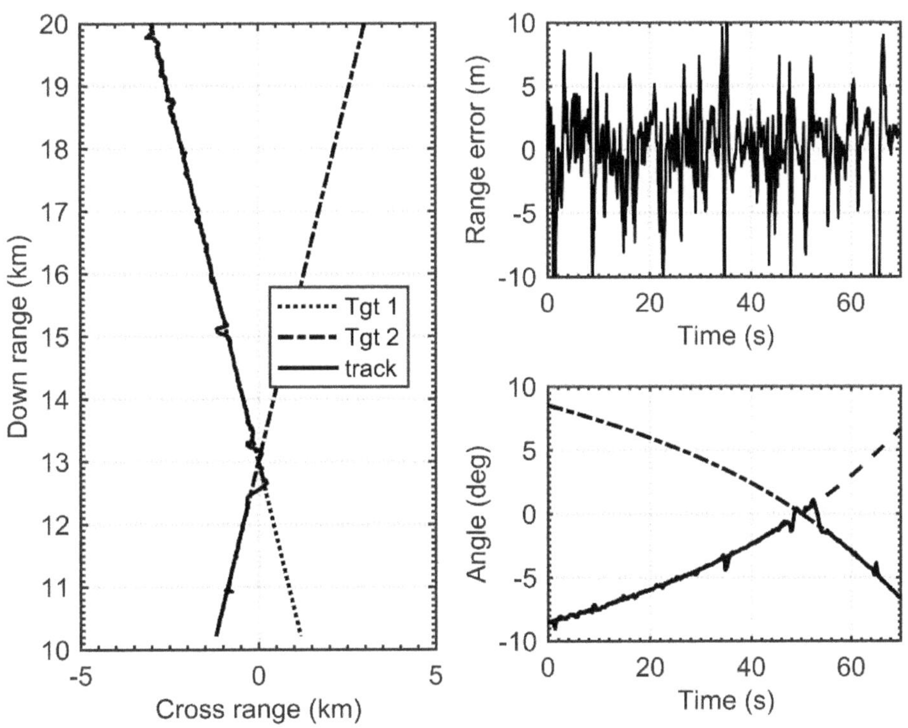


**Figure 7.20** Crossing targets—equal Doppler frequencies and equal average target RCSs—fluctuating target RCSs.

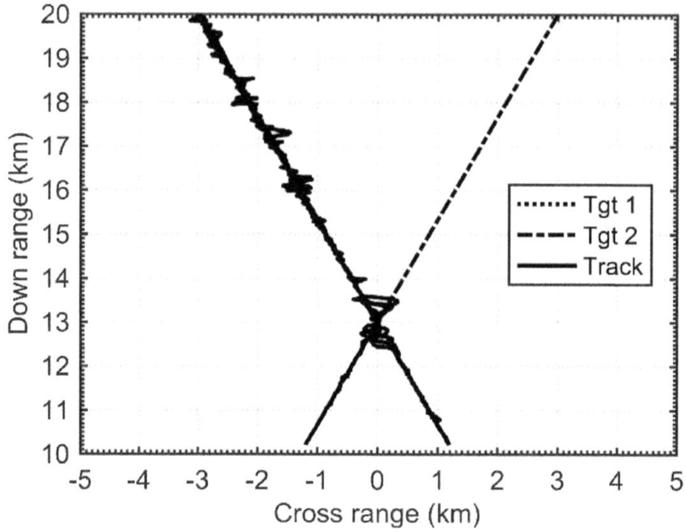


**Figure 7.21** Crossing targets—equal Doppler frequencies and equal average target RCSs—fluctuating target RCSs—five simulation executions.

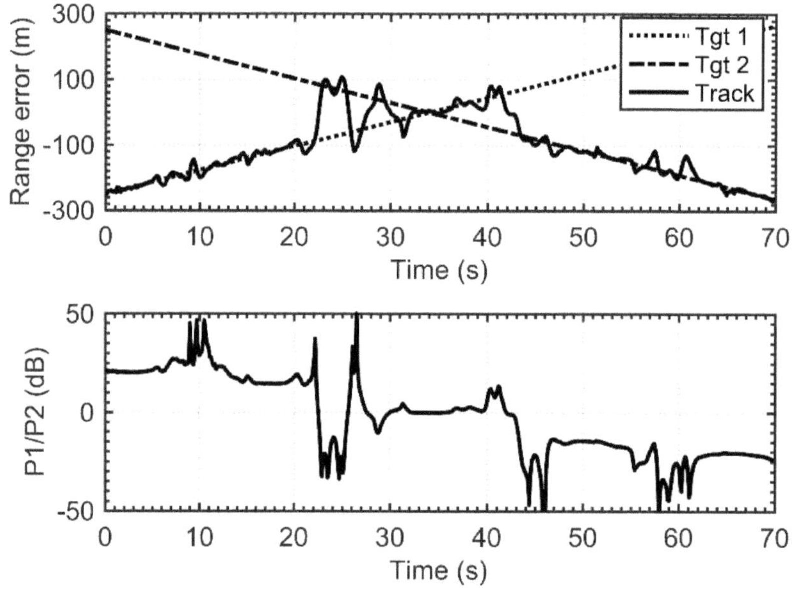


**Figure 7.22** Collinear targets—non-zero apparent Doppler frequency difference and same average RCSs—fluctuating target RCSs.

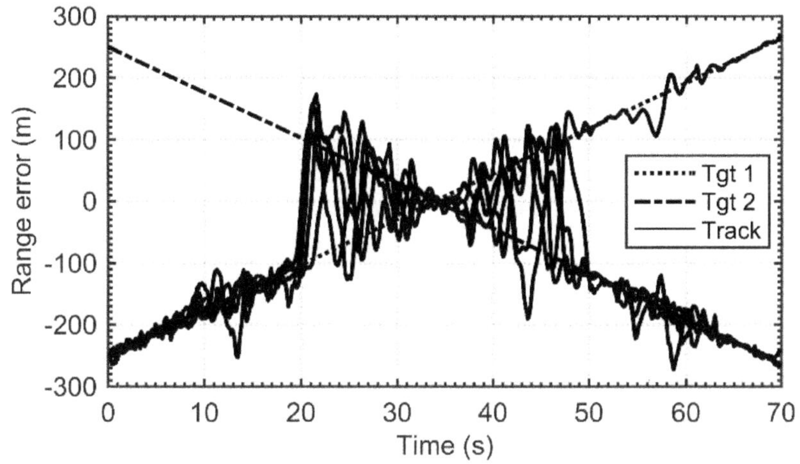


**Figure 7.23** Collinear targets—nonzero apparent Doppler frequency difference and same average RCSs—fluctuating target RCSs—five simulation runs.

### 7.5.2 Example 9: Same Set Up as Example 7 with Fluctuating RCSs

For this example, we use the simulation set up from Example 7 that produced Figure 7.18 (Section 7.4.3) but use fluctuating RCSs instead of constant RCSs. The result of this example is shown in Figure 7.22. The response is similar to Figure 7.18 in that the track moves between targets 1 and 2 during the 20 second to 30 second time interval. However, there is more variation in the track because of the fluctuating RCS. In Figure 7.18, the track followed the centroid for a short time after the targets were at the same range. When the fluctuation was added, the track stayed with Target 1 for a short time. Eventually, the signal strength of Target 2 became larger than that of Target 1, because Target 2 was closer to the radar than Target 1. At that time, the radar followed Target 2.

Figure 7.23 contains a plot like the top plot of Figure 7.22 for the case where we executed the simulation 5 times. As we saw with Figure 7.21, the track moved between the two targets when they were close in range. After the targets separated in range, the radar tended to track one or the other. In this case, the radar tended to favor Target 2, after separation, because Target 2 was closer to the radar than Target 1, and thus had a higher, average, SNR.

## 7.6 MULTIPATH EXAMPLES

To close this chapter, we consider tracking in the presence of multipath [13, p. Ch 11; 14–16]. Tracking in the presence of multipath can be considered a two-target problem where one "target" is the actual target, and the second "target" is its image. This is illustrated in Figure 7.24. That figure illustrates that energy reaches the target by two paths, and reradiated energy reaches the radar by the same two paths. The path directly to and from the target is termed the *direct path*, and the other path is termed the *reflected path*. If the earth is smooth, there will be only one *reflected path* between the target and the radar. One part of the reflected path is a line from the target to the earth, and the other part is a line from the earth to the radar. The two lines intersect at a reflection point, which is located such that the angles of the lines relative to the earth are equal. This point is termed the *specular reflection* point. The idea that the energy between the radar and the target travels by more than one path is most likely the origin of the word *multipath*. When one of the paths is due to specular reflection, the commonly used term is *specular multipath*.

When the reflecting surface is not perfectly smooth, there will be several (infinitely many, in theory) reflection points. The locations of these reflection points depend on the height and shape of the surface irregularities. The idea that target energy can travel between the radar and target by multiple, non-specular, paths is termed *diffuse multipath* [17–19]. The term diffuse probably originates from the fact that the reflected energy is distributed, over several, diffuse, points on the (rough) surface. Figure 7.25 contains a simple example of diffuse multipath, where the dash-dot lines are paths of diffuse paths . For the diffuse points, the angle of incidence and reflection are not equal, when referenced to a horizontal surface.

The accepted practice of analyzing the effects of multipath is to assume the multipath is divided between specular and diffuse multipath in a proportion that depends on surface roughness and operating frequency, or wavelength, of the radar. For calm water and smooth surfaces, such as concrete or asphalt, the predominant multipath is specular multipath. For complex surfaces, such as choppy water and surfaces with shrubbery, the predominant multipath is diffuse multipath [17–19]. The apportionment of multipath contributions between specular and diffuse is characterized by parameters termed specular and diffuse *reflection coefficients*, $\rho$ [14, 20–22] . The reflection coefficient is a complex number with a magnitude between 0 and 1. For specular reflection, a reflection coefficient magnitude of $|\rho_S| = 1$ means all of the energy incident on the reflection point leaves the reflection point. Values between 0 and 1 are used when a surface is not smooth but still exhibits some degree of specular multipath.

The phase of $\rho_D$, the reflection coefficient for diffuse multipath, is usually taken to be random. Several authors discuss the phase of $\rho_S$ and indicate it can take on values from 0 to $\pi$ depending on polarization and elevation angle [23]. However, $\pi$ is the most commonly used value [21].

Multipath, and reflection coefficients in particular, are treated extensively in the literature [14] [17-19] [23-26]. It appears as if the original basis for most of the discussion of reflection coefficient is Kerr's work in the MIT Radiation Laboratory Series [22].

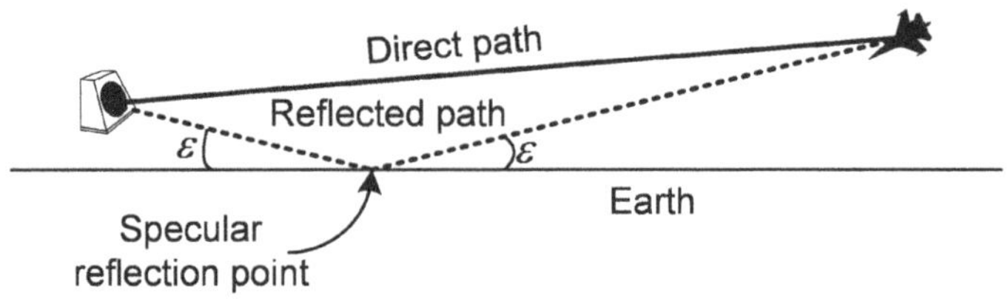


**Figure 7.24** Illustration of multipath.

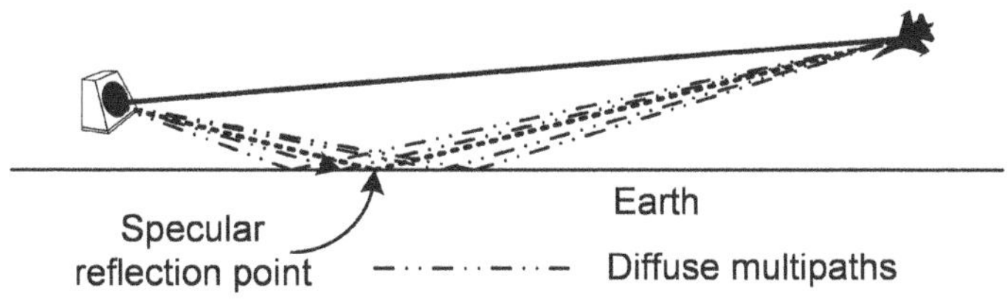


**Figure 7.25** Illustration of diffuse multipath.

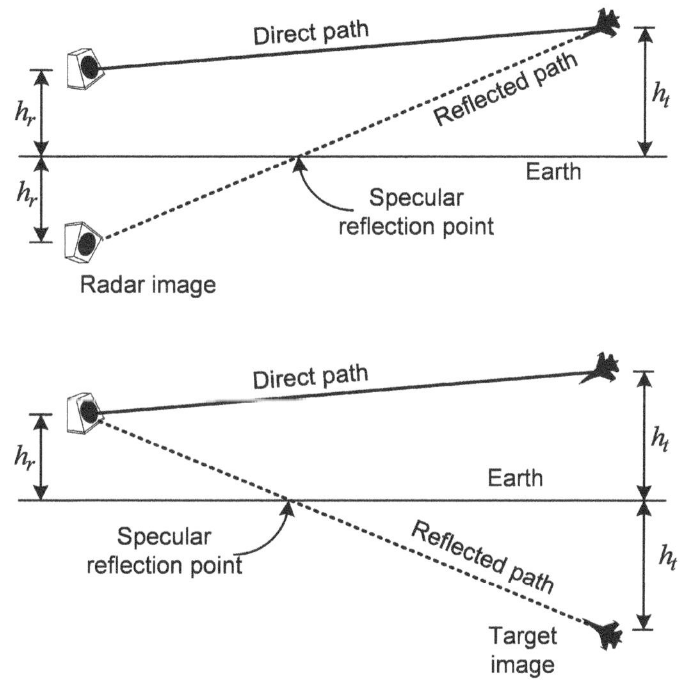


**Figure 7.26** Two methods of modeling specular multipath.

### 7.6.1 Specular Multipath Modeling

There are two common methods of modeling specular multipath effects [13, p. 273; 14; 18; 20, p. 290; 27]. These are illustrated in Figure 7.26. With one method, an image of the radar is placed below the ground at a distance of $h_r$, where $h_r$ is the height of the phase center of the radar antenna. By doing this, we can consider two straight lines between the radar(s) and target, instead of one straight line for the direct path, and one bent line for the reflected path.

The other, more common, method is to place an image of the target below the ground at a distance of $h_t$, where $h_t$ is the target altitude. Again, the reason for doing this is that it is easier to base analyses, and simulation, on two straight lines, than on one straight line and one bent line. We use the target image approach.

### 7.6.2 Target Model for Multipath

We are now in a position to develop a target model we can use in multipath simulations. We consider the case where the actual target is flying toward the radar at a constant velocity and altitude over a flat earth. With this, the down-range positon, $x(t)$, and altitude, $y(t)$, of the actual target are

$$x_a(t) = x_0 - Vt \text{ and } y_a(t) = h_t \tag{7.30}$$

where $x_0$ is the initial down-range position, $V$ is the target speed, and $h_t$ is the target altitude. The range to the target is

$$R_a(t) = \sqrt{x_a^2(t) + (y_a(t) - h_r)^2} \, . \tag{7.31}$$

The elevation angle[2] to the target is

$$\varepsilon_a(t) = \sin^{-1}\left(\frac{y_a(t) - h_r}{R_a(t)}\right) \tag{7.32}$$

or, in sine space

$$v_a(t) = \frac{y_a(t) - h_r}{R_a(t)} \, . \tag{7.33}$$

The associated parameters for the image target due to specular multipath are

$$x_I(t) = x_a(t) = x_0 - Vt \text{ and } y_I(t) = -h_t, \tag{7.34}$$

$$R_I(t) = \sqrt{x_I^2(t) + (y_I(t) - h_r)^2} \tag{7.35}$$

and

$$v_I(t) = \frac{y_I(t) - h_r}{R_I(t)} \, . \tag{7.36}$$

These parameters are illustrated in Figure 7.27.

---

[2] In the examples up to now, we were tracking in range and azimuth, or $u$. For the multipath examples, we switch to range and elevation, or $v$.

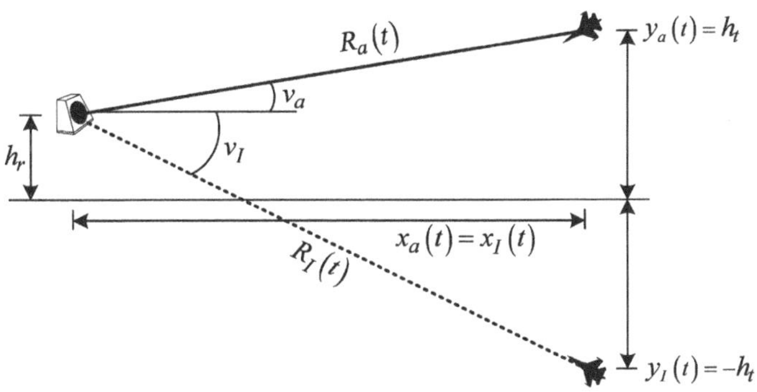


**Figure 7.27** Specular multipath target model.

To generate the signals out of the matched filters in the sum and difference channels, we need to account for the fact that there are two signal paths from the radar to the target, and two signal paths from the target back to the radar, for a total of four signal paths. The signal paths are:

1. The direct path from the radar to the target and the direct path back to the radar.
2. The direct path from the radar to the target and the reflection path from the target back to the radar.
3. The reflection path from the radar to the target and the direct path from the target back to the radar.
4. The reflection path from the radar to the target and the reflection path from the target to the radar.

There will be different range delays associated with the paths

1. The range delay for path 1 is $2R_a(t)/c$.
2. The range delay for path 2 is $(R_a(t)+R_I(t))/c$.
3. The range delay for path 3 is $(R_I(t)+R_a(t))/c$.
4. The range delay for path 4 is $2R_I(t)/c$.

Since the ranges associated with the actual and image targets are slightly different functions of time, the actual and image target returns will exhibit slightly different Doppler frequencies. Also, since the differences in range can be large relative to the wavelength of the transmit signal, the multipath could cause signal fading, and enhancement, as discussed in text books that discuss the impact of multipath on SNR and detection performance [13, p. 276; 14; 28].

The signals on paths 1 and 2 will experience a transmit antenna voltage directivity of $G_\Sigma(v_a)$ V/V, and the signals on paths 3 and 4 will experience a transmit antenna voltage directivity of $G_\Sigma(v_I)$ V/V.

On receive paths 1 and 3, the signal will be modified by the receive antenna directivity of $G_\Sigma(v_a)$ V/V in the sum channel and $G_\Delta(v_a)$ V/V in the difference channel. On receive paths 2 and 4, the signal will be modified by the receive antenna directivity of $G_\Sigma(v_I)$ V/V in the sum channel and $G_\Delta(v_I)$ V/V in the difference channel.

The signals in paths 2 and 3 are modified by the reflection coefficient, $\rho_S$, and the signal on path 4 is modified by $\rho_S^2$. The reflection coefficient is not applied to the path 1 since it is a direct path from the radar to the target and back to the radar.

### 7.6.3 Example 10: Tracking in the Presence of Specular Multipath

For this example, we use a radar height of $h_r = 10$ m and a target altitude of $h_t = 500$ m. We start the target 20 km down range and fly it toward the radar at a speed of 150 m/s. We assume a specular reflection coefficient of $\rho_S = -1$. We use the same radar model as in the previous examples (see Section 7.4). We adjust the radar parameters and target RCS to provide an average SNR of 20 dB at the start of tracking. We use the same Swerling 1 - like RCS fluctuation model used in previous examples.

The result of one execution of the simulation is contained in Figure 7.28. In this particular run, the track transitioned from the actual target to the image at about 35 seconds. The exact reason the tracker transitioned between the target and its image is not clear, since the behavior of the tracker is influenced by several factors. However, an examination of Figure 7.29 might provide some insight. This figure is a plot of the total signal amplitude of the sum signal in the angle channel before noise is added. A small amount of the variation is caused by fading due to the Doppler frequency difference of the target and its image. However, the majority of the variation is due to RCS fluctuation. At about the 35 second mark, the time the track transitioned from the target to the image, the sum signal of the angle channel faded to a very small value. This meant the angle error signal was dependent mainly on noise.

It is likely noise pushed the track toward the image before the signal fade. During the fade, the signal level into the angle tracker was small, so the tracker most likely continued to drift toward the image. By the time the sum signal amplitude increased, the beam had moved closer to the image than the target. This made the amplitude of the image return larger than the amplitude of the actual target return, which caused the angle track to transition to the image. As time progressed, the angle between the actual target and the image increased to the point where the actual target return was in the side lobe region of the sum antenna pattern, and thus attenuated relative to the image.

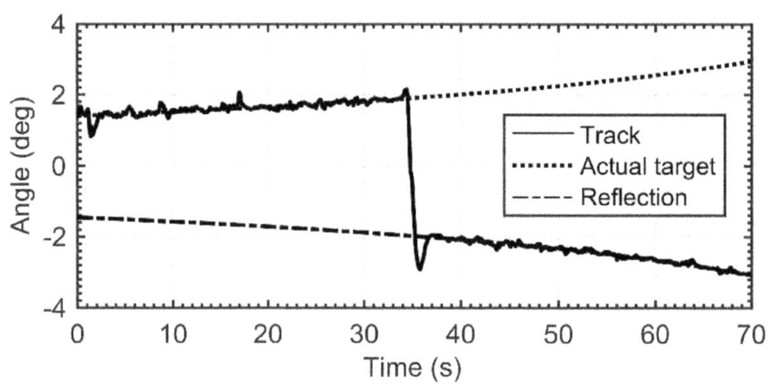


**Figure 7.28** Multipath tracking results.

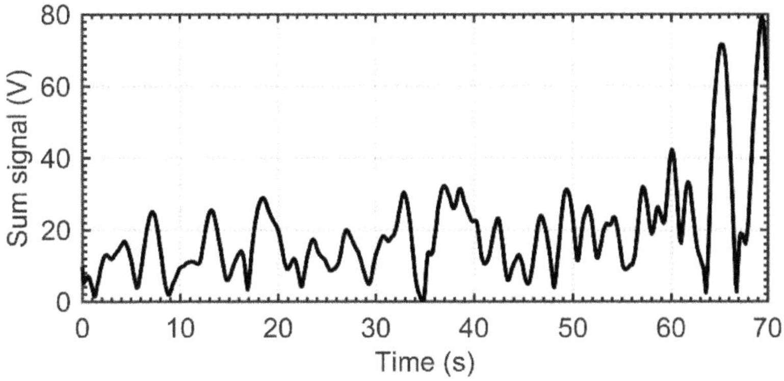


**Figure 7.29** Signal fade that may have initiated the transition of the track from the target to the image.

To check the assertion the fade was due to RCS fluctuation, the simulation was rerun with a constant RCS target. The other simulation parameters, and the noise, were the same as for the case that produced Figures 7.28 and 7.29. Since the RCS of the target was constant, the only variation in the angle channel sum signal amplitude (before noise is added) was a small ripple caused by the Doppler mismatch, or multipath fading, depending on interpretation. This is illustrated in Figure 7.30, which is a plot like Figure 7.29. In this case, there is no signal fade due to RCS fluctuation. As a result, the radar tracked the direct path signal, as illustrated in Figure 7.31.

In most executions of the simulation (with the fluctuating RCSs), the track stayed on the actual target. This was the expected behavior since the track was initiated on the actual target, and the initial angle between the target and image was larger than the beam width of the antenna, which caused the image return to be in the side lobe region of the sum pattern. However, on some simulation executions, the track did transition to the image; probably because of a fading phenomenon similar to the one that resulted in Figure 7.28. In still other cases, the tracker completely lost track. It is expected this was caused by a long period of signal fade due to RCS fluctuation.

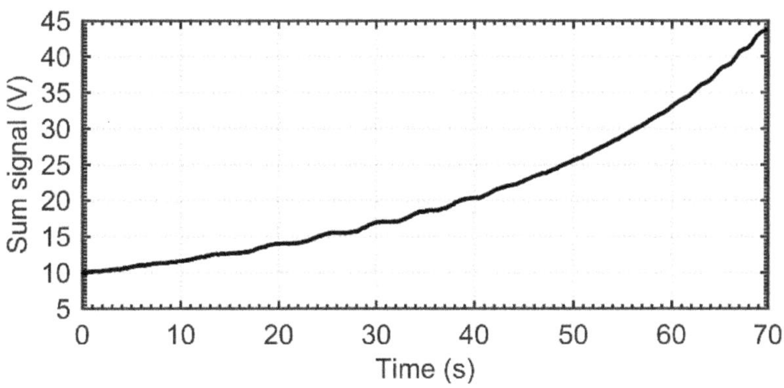


**Figure 7.30** Signal amplitude variation when the target RCS is constant.

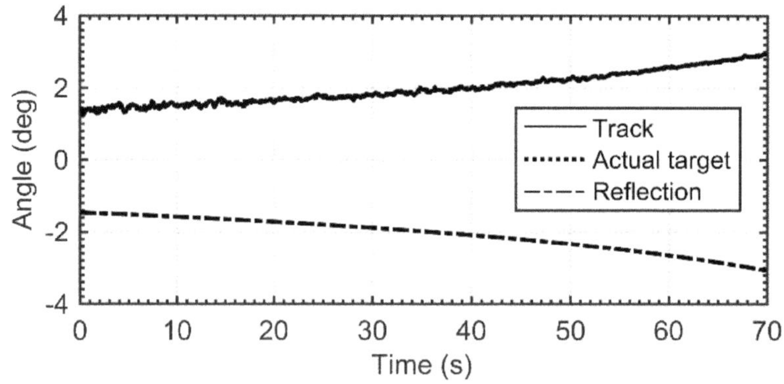


**Figure 7.31** Multipath tracking results for the same conditions as Figure 7.30, but with constant a RCS target.

When we reduced the target altitude from 500 m to 200 m, the track results became more interesting, as illustrated in Figure 7.32. The simulation conditions (i.e., noise, RCS fluctuation, etc.) that produced Figure 7.32 were the same as those that produced Figure 7.28, except for the lower target altitude. At the lower target altitude, the target and image were separated, in angle, by about the beam width. This meant both the target and its image were in the main beam of the sum pattern. As a result, the track was able to transition between the target and its image, as illustrated in the earlier, two target tracking examples. Toward the end of the simulation run, the angle separation increased to over a beam width. Because of this, the tracker tended to follow either the target, or its image. In Figure 7.32, the tracker followed the image. Several executions of the simulation indicate it is equally probable the track will settle on the target or image. In some cases, the radar lost track of both the target and its image.

Figure 7.33 is a plot like Figure 7.30. It was generated using the same simulation that produced Figure 7.32, except that the RCS of the target was constant. The fluctuation is more pronounced than in Figure 7.30 because both targets are in the main beam of the sum pattern. As indicated in the discussion related to Figure 7.30, we can say the fluctuation is due to the difference in Doppler frequencies between the target and its image. Figure 7.34 contains a plot of the Doppler frequency between the target and its image. This figure illustrates there is a direct relation between the Doppler frequency difference and the lobing structure. As the Doppler frequency difference increases, the distance between the lobes of Figure 7.33 decreases. Thus, we can conclude that the oscillatory, or lobing, structure of Figure 7.33 is due to the frequency difference (the beat frequency) between the direct target return and the return from its image.

We could also interpret the lobing structure as being due to multipath fading [13; 14; 20; 29; 30, p. 103]. As the target approaches the radar, the ranges to the target and its image will change. From (7.30), (7.31), (7.34), and (7.35), the range difference is (see Problem 22)

$$\Delta R(t) = R_I(t) - R_a(t) = \frac{2h_r h_t}{x_a(t)}. \tag{7.37}$$

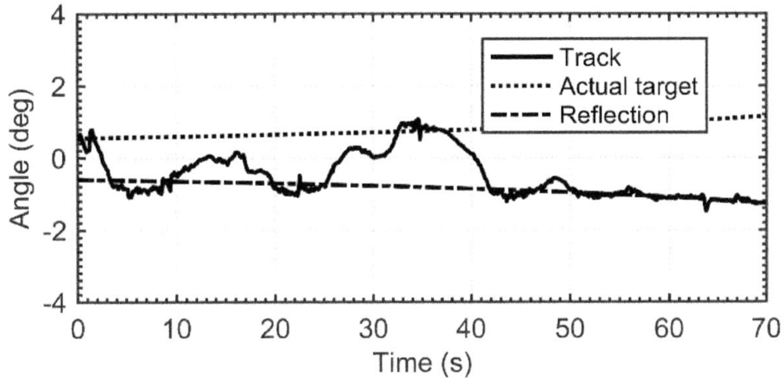


**Figure 7.32** Simulation results when the target altitude was decreased to 200 m.

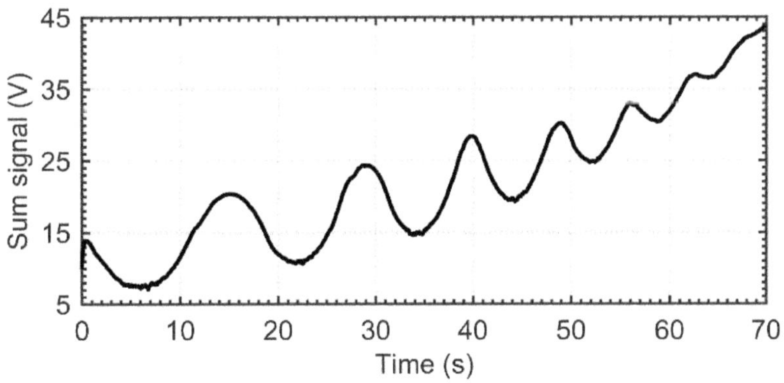


**Figure 7.33** Signal amplitude variation when the target RCS is constant—target altitude of 200 m.

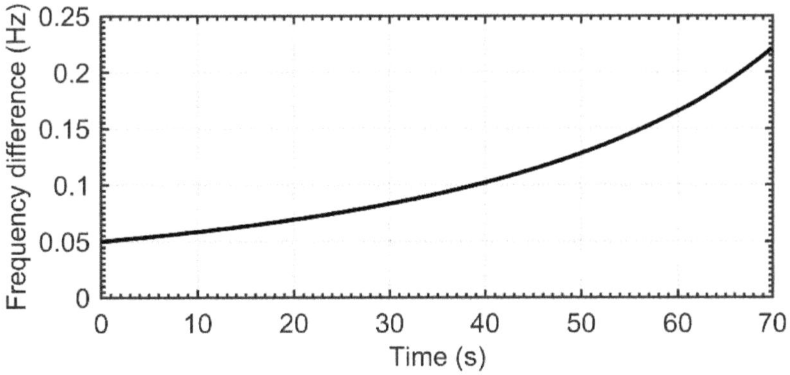


**Figure 7.34** Frequency difference between the signals from the target and its image—target altitude of 200 m.

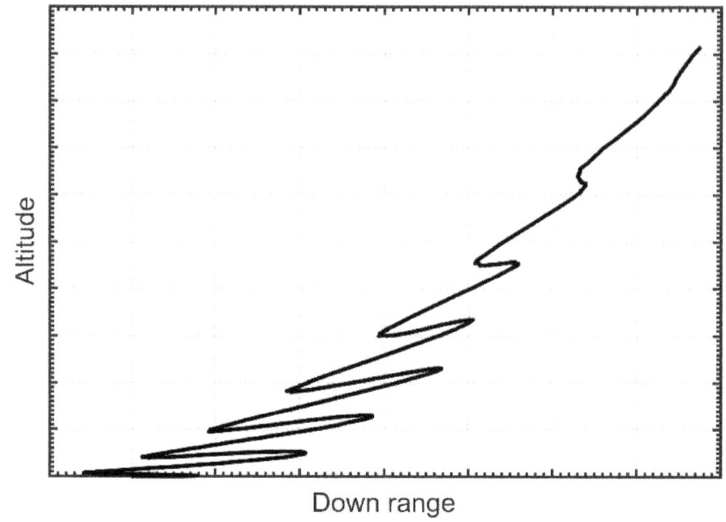


**Figure 7.35** Multipath lobing structure interpretation of Figure 7.33.

This range difference will cause a change in the phase between the two signals. If the signal amplitudes are almost the same, as is the case in this example, they will add when the phase difference is zero, and cancel when the phase difference is 180°. That is, the total signal will periodically fade, as indicated in Figure 7.33. As the target approaches the radar, $\Delta R(t)$ will increase (because $x_a(t)$ decreases), and the time between fades will become shorter. This is also demonstrated in Figure 7.33, where the spacing between signal fades becomes shorter. That the signal doesn't fade to a value of zero is due to the fact that we considered four paths from the radar to the target and back to the radar. Had we considered only the receive problem, which has two paths, the amplitude would have periodically faded to zero.

Figure 7.35 contains the data in Figure 7.33 plotted as a polar plot. In this plot, the radial direction is sum signal amplitude (the vertical axis of Figure 7.33), and the angular direction is the target elevation angle, $\varepsilon_a(t)$, (see (7.32)), referenced to the minimum value of $\varepsilon_a(t)$. The plot of Figure 7.35 demonstrates the lobing structure found in several references that discuss multipath fading in connection with search radars [13; 14; 20; 29; 30, p. 103]. The fact that it doesn't exhibit the well behaved lobing structure found in many references is due to the four-path discussion of the previous paragraph.

## 7.7 EXERCISES

1. Implement the fluctuating RCS script of Figure 7.1 and plot the amplitude and phase of the RCS for filter bandwidths of 0.5, 1.5, 3.0, and 6.0 Hz.
2. Repeat Example 1 and produce plots like those in Figures 7.2. Try different filter bandwidths in the fluctuating RCS script.

3. Create plots like Figure 7.5 for the case where there is no noise in the radar. How do the plots compare to Figure 7.5? Do they support the statement in the paragraph before Figure 7.4?
4. Reproduce Figure 7.6.
5. Derive equations (7.14) through (7.18).
6. In the sentence below equation (7.18), we claimed $S_u = 1.0$. Verify or refute, this claim. See the related discussions in Chapter 5.
7. Implement the tracker discussed in Section 7.3.2 and reproduce Figure 7.8.
8. In foot note 1, we claimed $\Delta f$ will always lie between $\pm 1/2$ the track rate. Explain why the claim is correct.
9. Modify your script from Problem 7, as discussed in Section 7.3.2, and generate a plot like Figure 7.9.
10. In the second paragraph of Section 7.3.3, we claimed the average values of $u_{trk}$ over the three 10-second intervals of Figure 7.10 were $u_1$, $(u_1 + u_2)/2$, and $u_2$. Verify that is, indeed, approximately the case.
11. Derive equations (7.20) and (7.21), and plot $u_{trk}$ versus $\theta(t) = 2\pi\Delta f t$ over the range of $-\pi$ to $\pi$. Generate plots $\sigma_1 = 1$ and $\sigma_2 = 10^{1/10}$ (a 1-dB difference in RCS), $\sigma_2 = 1$ and $\sigma_1 = 10^{1/10}$ and $\sigma_1 = \sigma_2 = 1$.
12. Modify your script from Problem 9, as discussed in Section 7.3.4, and reproduce Figure 7.10.
13. Add noise to your script for Problem 12 and generate plots like Figure 7.12. Run you script several times to see if you observe the phenomenon discussed in the paragraph above Figure 7.11.
14. Set up the example discussed in Section 7.4 and generate plots like Figure 7.13. Verify the statements in the second paragraph above Section 7.4.2.
15. Modify your Problem 14 script as discussed in Section 7.4.2 and generate a plot like Figures 7.14 and 7.15.
16. Implement the tracker discussed in Section 7.4.3 (Example 7) and generate plots like Figures 7.17, 7.18, and 7.19.
17. Implement the tracker discussed in Section 7.5.1 and generate plots like Figure 7.20. Run the simulation several times to generate a plot like Figure 7.21.
18. Add the fluctuating RCS to the targets in your script for Problem 16 and generate plots like Figures 7.22 and 7.23.
19. Derive (7.30) through (7.36).
20. Generate a script to model multipath, as described in Sections 7.6.2 and generate plots like Figures 7.28 through 7.31.
21. Reduce the target altitude to 200 m and generate plots like Figure 7.32 through 7.35.
22. Derive (7.37).

## References

[1] M. C. Budge Jr. and S. R. German, *Basic Radar Analysis*, Norwood, MA: Artech House, 2015.

[2] E. F. Knott, J. F. Shaeffer and M. T. Tuley, *Radar cross Section: its Prediction, Measurement, and Reduction*, Artech House, 1985.

[3] G. T. Ruck, D. E. Barrick, W. D. Stuart and C. K. Krichbaum, Eds., *Radar Cross Section Handbook*, New York: Plenum Press, 1970.

[4] D. K. Barton and P. Hamilton, Eds, ANRO Engineering, Inc., *Radar Evaluation Handbook*, Norwood, MA: Artech House, 1991.

[5] L. V. Blake, *Radar Range-Performance Analysis*, Norwood, MA: Artech House, 1986.

[6] M. I. Skolnik, *Introduction to Radar Systems*, 3rd ed., New York: McGraw-Hill, 2001.

[7] Swerling, P., "Probability of Detection for Fluctuating Targets," *RAND Corp.*, Santa Monica, CA, *Res. Memo.* RM-1217, Mar. 17, 1954. Reprinted: IRE Trans. Inf. Theory, vol. 6, no. 2, Apr. 1960, pp. 269–308

[8] D. K. Barton, *Radar Equations for Modern Radar*, Boston: Artech House, 2013.

[9] Norton, K. A., and A. C. Omberg, "The Maximum Range of a Radar Set," *Proc. IRE*, vol. 35, no. 1, Jan. 1947, pp. 4–24. First published Feb. 1943 by U.S. Army, Office of Chief Signal Officer in the War Department, in Operational Research Group Report, ORG-P-9-1.

[10] D. K. Barton, Radars, *Volume 2, The Radar Range Equation*, Dedham, MA: Artech House, 1975.

[11] A. Papoulis, *Probability, Random Vairables, and Stochastic Processes*, 3rd ed., New York: McGraw-Hill, Inc., 1991.

[12] P. Z. Peebles Jr., *Probability, Random Variables, and Random Signal Principles*, 3rd ed., New York: McGraw-Hill, Inc., 1993.

[13] H. Urkowitz, *Signal Theory and Random Processes*, Dedham, MA: Artech House, 1983.

[14] P. Swerling and W. L. Peterman, "Impact of target RCS fluctuations on radar measurement accuracy," *IEEE Transactions on Aerospace and Electronic Systems,* vol. 26, no. 4, pp. 685-686, July 1990.

[15] S. S. Blackman, *Multiple-Target Tracking with Radar Application*, Norwood, MA: Artech House, 1986.

[16] Y. Bar-Shalom, Ed., *Multitarget-Multisensor Tracking: Applications and Advances*, vol. II, Norwood, MA: Artech House, 1992.

[17] S. S. Blackman, "Multiple hypothesis tracking for multiple target tracking," *IEEE Aerospace and Electronic Systems Magazine,* vol. 19, no. 1, pp. 5-18, 2004.

[18] L. D. Stone, R. L. Streit, T. L. Crown and K. L. Bell, *Bayesian Multiple Target Tracking*, 2 ed., Norwood, MA: Artech House, 2014.

[19] S. S. Blackman and R. Popoli, *Design and Analysis of Modern Tracking Systems*, Norwood, MA: Artech House, 1999.

[20] J. T. Andrews, *Examination of Dichotomous and Centroid Tracking of a Monopulse Angle Tracking unit,* Ann Arbor, MI: ProQuest, 2017.

[21] I. E. Kliger and C. F. Olenberger, "Multiple Target Effects on Monopulse Signal Processing," *IEEE Transactions on Aerospace and Electronic Systems,* vols. AES-11, pp. 795-804, September 1975.

[22] I. E. Kliger and C. F. Olenberger, "Phase-Lock Loop Jump Phenomenon in the Presence of Two Signals," *IEEE Transactions on Aerospace and Electronic Systems,* vols. AES-12, pp. 55-56, January 1976.

[23] I. Kanter, "Varieties of Average Monopulse Responses to Multiple Targets," *IEEE Transactions on Aerospace and Electronic Systems,* vols. AES-17, January 1981.

[24] C. Melino, "Average Monopulse Angle Tracking Response to Two Unresolved Sources," *IEEE Transactions on Aerospace and Electronic Systems,* vols. AES-23, September 1987.

[25] S. Charland, "An Investigation of the Transfer of Monopulse Tracking Between Two," Defence Research Establishment, Ottowa, Canada, 1989.

[26] R. M. Tillman, *Centroid and Dichotomous Tracking for Monopulse Radar Systems,* The University of Alabama in Huntsville Master's Thesis, 1998.

[27] T. Benedict and G. Bordner, "Synthesis of an optimal set of radar track-while-scan smoothing equations," *IRE Transactions on Automatic Control,* vol. 7, no. 4, pp. 27-32, 1962.

[28] R. H. Delano, "A Theory of target glint or angular scintillation in radar tracking," *Proc. of the IRE,* pp. 1778-1784, December 1953.

[29] D. D. Howard, "Radar target glint in tracking and guidance system based on echo signal phase distortion," *Proceedings of the National Electronim Conference*, 1959.

[30] J. E. Lindsay, "Angular glint and the moving, rotating, complex radar target," *IEEE Transactions on Aerospace and Electronic Systems,* vols. AES-4, no. 2, pp. 164-173, March 1968.

[31] H. Yin and P. Huang, "Unification and comparison between two concepts of radar target angular glint," *IEEE Transactions on Aerospace and Electronic Systems,* pp. 778-783, 1995.

[32] S. M. Sherman and D. K. Barton, *Monopulse Principles and Techniques*, 2nd ed., Norwood, MA: Artech House, 2011.

[33] D. K. Barton, *Radars, Volume 4, Radar Resolution & Multipath Effects*, Dedham, MA: Artech House, 1975.

[34] P. Beckmann and A. Spizzichino, *The Scattering of Electromagnetic Waves from Rough Surfaces*, Artech House, 1987.

[35] D. K. Barton, "Low Angle Radar Tracking," *Proc. IEEE,* vol. 62, no. 6, pp. 687-704, June 1974.

[36] D. K. Barton, "Low-Altitude Tracking over Rough Surfaces, I: Theoretical Predictions," in *IEEE Eascon-79*, Washington, 1979.

[37] D. K. Barton, *Modern Radar System Analysis*, Norwood, MA: Artech House, 1988.

[38] D. K. Barton and H. R. Ward, *Handbook of Radar Measurement*, Englewood Cliffs, New Jersey: Prentice Hall, 1969.

[39] D. E. Kerr, *Propagation of Short Radio Waves*, vol. 13, New York: McGraw-Hill, 1951.

[40] M. I. Skolnik, Ed., *Radar Handbook*, New York: McGraw-Hill, 1970.

[41] C. R. Burrows and S. S. Attwood, *Radio Wave Propagation*, New York: Academic Press, Inc., 1949.

[42] W. S. Ament, "Toward a Theory of Reflection by Rough Surface," in *Proc. IRE*, 1953.

[43] L. V. Blake, "Reflection of Radio Waves from a Rough Sea," in *Proc. IRE*, 1950.

[44] G. C. Evans, "Influence of Ground Reflections on Radar Target-Tracking Accuracy," *Proc. IEE,* vol. 113, no. 8, pp. 1281-1286, August 1966.

[45] S. Kingsley and S. Quegan, *Understanding Radar Systems*, Raleigh, NC: Scitech Publishing, 1999.

[46] J. L. Eaves and E. K. Reedy, *Principles of Modern Radar*, New York: Van Nostrand Reinhold, 1987.

[47] G. M. Ewing, *Calculus of Variations with Applications*, New York: Dover Publications, Inc., 2016.

[48] D. Tenne and T. Singh, "Characterizing performance of α-β-γ filters," *IEEE Transactions on Aerospace and Electronic Systems,* vol. 38, no. 3, pp. 1072-1087, July 2002.

[49] J. Sklansky, "Optimizing the dynamic parameters of a track-while-scan system," *RCA Review,* vol. 18, no. 2, pp. 163-185, June 1957.

[50] G. M. Siouris, *An Engineering Approach to Optimal Control and Estimation Theory*, New York: Wiley-Interscience, 1996.

[51] M. C. Budge Jr., "EE 706 - Kalman Filtering Techniques," 2010. [Online]. Available: http://www.ece.uah.edu/courses/material/EE706-Merv/index.htm. [Accessed 28 February 2017].

[52] R. J. Polge and B. K. Bhagavan, "A Study of the G-H-K Tracking Filter," UAH Research Report No. 176, MICOM Report No. RE-CR-76-1, DTIC ADA021317, 1975.

[53] C. H. Houpis and S. N. Sheldon, *Linear Control System Analysis and Design with MATLAB*, 6th ed., Boca Raton, FL: CRC Press, 2013.

[54] R. C. Dorf and R. H. Bishop, *Modern Control Systems*, 13th ed., Pearson Education, Inc., 2016.

[55] R. E. Kalman, "A new approach to linear filtering and prediction problems," *Transactions of the ASME, Journal of Basic Engineering, Series D,* vol. 82, no. 1, pp. 35-45, March 1960.

[56] R. E. Kalman and R. S. Bucy, "New results in linear filtering and prediction theory," *Transactions of the ASME, Journal of Basic Engineering,* vol. 83, no. 3, pp. 95-107, December 1961.

[57] B. K. Bhagavan and R. J. Polge, "Performance of the g-h Filter for Tracking Maneuvering Targets," *IEEE Transactions on Aerospace and Electronic Systems,* Vols. AES-10, no. 6, pp. 864-866, Nov. 1974.

[58] H. R. Simpson, "Performance measures and optimization condition for a third-order sampled-data tracker," *IEEE Transactions on Automatic Control,* vol. 8, no. 2, pp. 182-183, April 1963.

[59] D. Tenne and T. Singh, "Optimal design of α-β-(γ) filters," in *American Control Conference, 2000. Proceedings of the 2000*, Chicago, IL.

[60] S. Neal, "Parametric relations for the α-β-γ filter Predictor," *IEEE Transactions on Automatic Control,* vol. 12, no. 3, pp. 315-317, June 1967.

[61] D. G. Schultz and J. L. Melsa, *State Functions and Linear Control Systems*, New York, NY: McGraw-Hill, 1967.

[62] J. L. Melsa and D. G. Schultz, *Linear Control Systems*, New York, NY: McGraw-Hill, 1969.

[63] S. C. Gupta, Transform and State Variable Methods in Linear Systems, Huntington, New York: Robert E. Kreiger Publishing Co., Inc., 1971.

[64] E. Kreyszig, *Advanced Engineering Mathematics*, 7th ed., New York: John Wiley & Sons, 1993.

[65] J. L. Lagrange, *Mecanique Analytique*, Paris: Courcier, 1811.

[66] I. Grattan-Guinness, Ed., *Companion Encyclopedia of the History and Philosophy of the Mathematical Sciences*, New York: Routledge, 1993.

[67] L. P. Eisenhart, *Riemannian Geometry*, Princeton, New Jersey: Princeton University Press, 1997.

[68] J. A. Cadzow and H. R. Martens, *Discrete-Time and Computer Control Systems*, F. F. Kuo, Ed., Englewood Cliffs, New Jersey: Prentice Hall, 1970.

[69] R. E. Zeimer, W. H. Tranter and D. R. Fannin, *Signals & Systems Continuous and Discrete*, 4th ed., Upper Saddle River, NJ: Prentice Hall, 1984.

[70] W. R. Ahrendt, *Servomechanism Practice*, New York: McGraw-Hill Book Company, Inc., 1954.

[71] S. A. Davis, *Outline of Servomechanisms*, New York: Unitech, 1966.

[72] H. M. James, N. B. Nichols and R. S. Phillips, Eds., *Theory of Servomechanisms*, New York: McGraw-Hill Book Company, Inc., 1947.

[73] M. Williamson, *Dictionary of Space Technology*, Cambridge University Press, 2010.

[74] H. D. Young and R. A. Freedman, *Sears and Zemansky's University Physics*, 10th ed., Addison-Wesley, 1999.

[75] S. J. Orfanidis, *Introduction to Signal Processing*, Englewood Cliffs, New Jersey: Prentice Hall, 1996.

[76] R. G. Lyons, *Understanding Digital Signal Processing*, 3rd ed., New York: Prentice Hall, 2011.

[77] A. V. Oppenheim and R. W. Schafer, *Discrete-Time Signal Processing*, Englewood Cliffs: Prentice-Hall, 1989.

[78] E. O. Brigham, *The Fast Fourier Transmform and Its Applications*, Englewood Cliffs, NJ: Preintice-Hall, Inc., 1988.

[79] G. R. Cooper and C. D. McGillem, *Continuous & Discrete Systems*, 3rd ed., Philadelphia: Saunders College Publishing, 1991.

[80] 徐振来 (Xu Zhenlai), 相控阵雷达数据处理 (Phased Array Radar Data Processing), 北京: 国防工业出版社 (Beijing: National Defense Industry Press), 2009.

[81] 蔡慶宇 (Cai Qingyu) and 张伯彦 (Zhang Boyan), 相控阵雷达数据处理及其仿真技术 (Phased Array Radar Data Processing and Its Simulation Techniques), 北京: 国防工业出版社 (Beijing: National Defense Industry Press), 1997.

[82] 何友 (He You), 修建娟 (Jian Juan)and 关欣 (Guan Xin), 雷达数据处理及应用 (Radar Data Processing and Applications), 北京: 电子工业出版社 (Beijing: Publishing House of Electronics Industry, 2013.

[83] J. G. Proakis and D. G. Manolakis, *Digital Signal Processing*, 4th ed., Upper Saddle River, New Jersey: Pearson Prentice Hall, 2007.

[84] А. Б. Сергиенко (A. B. Sergienko), Цифровая обработка сигналов (*Digital Signal Processing*), Москва: ЗАО Издательский дом «Питер» (Moscow: ZAO Peter Publishing House), 2002.

[85] D. K. Barton, Radars, Volume 7, *CW and Doppler Radar*, Dedham, MA: Artech House, 1978.

[86] S. Butterworth, "On the Theory of Filter Amplifiers," *Experimental Wireless & the Wireless Engineer,* vol. 7, pp. 536-541, October 1930.

[87] C. L. Dolph, "A Current Distribution for Broadside Arrays Which Optimizes the Relationship Between Beam Width and Side-Lobe Level," *Proceedings IRE,* vol. 34, no. 6, pp. 335-348, June 1946.

[88] H. J. Riblet, "Discussion on Dolph's Paper," *Proc. IRE,* vol. 35, no. 5, pp. 489-492, May 1947.

[89] R. W. Hamming, *Digital Filters*, 2nd ed., Englewood Cliffs, NJ: Prentice-Hall, Inc., 1983.

[90] T. T. Taylor, "Design of Line-Source Antennas for Narrow Beamwidths and Low," *IRE Transactions on Antennas and Propagation,* vols. AP-3, pp. 16-28, January 1955.

[91] N. Levanon and E. Mozeson, *Radar Signals*, Hoboken, New Jersey: Wiley-Interscience, 2004.

[92] P. M. Woodward, *Probability and Information Theory with Applications to Radar*, 2nd ed., New York: Pergamon Press, 1953.

[93] H. J. Blinchikoff and A. I. Zverev, *Filtering in the Time and Frequency Domains*, New York: John Wiley & Sons, Inc., 1976.

[94] A. I. Zverev, *Handbook of Filter Synthesis*, John Wiley & Sons, 1967.

[95] R. Schaumann and . M. E. V. Valkenburg, *Design of Analog Filters*, New York: Oxford University Press, 2001.

# Chapter 8

# Acquisition and Track Initiation

## 8.1 INTRODUCTION

In the examples of previous chapters, we initialized the various trackers with either perfect target information, or target information with small errors. In actual radars, the various trackers use data gathered by what we will term a *search data source*. We will refer to data provided by the search data source as *search data*. The search data can come from a remotely located search radar by way of a communications, command and control ($C^3$) network, a closely located search radar through a direct feed or $C^3$ network, a search radar collocated with the track radar, or the search function of the track radar. In general, the accuracy of the search data degrades as the source moves further away from the track radar. That is, the data provided by the search function of the track radar will be more accurate than data provided by a collocated search radar, which will be more accurate than data provided by a closely located search radar, and so forth.

The degradation in accuracy can also be due to the fact that the range and angle resolution of the search data source is usually poorer than the range and angle resolution of the track radar. Also, the data generated by the search data source is usually in a different coordinate system than the one used in the track radar. Thus, the data must be converted from the source coordinate system to the track radar coordinate system. Those conversions introduce errors that degrade the accuracy of the data the track radar must use.

Because of degradation in accuracy of the search data, the track radar may need to perform some type of *acquisition search* to find the target. Hereinafter, we will term this *acquisition* and refer to an *acquisition volume*. The size of the acquisition volume, in angle and range, and possibly Doppler frequency, will depend on the accuracy of the search data received by the track radar. The size of the acquisition volume is also affected by the time it takes the search data to reach the track radar. Specifically, there will be target motion during the time period from when the search data source obtained the search data and when it reaches the track radar. This is termed *data latency*. Basically, the target motion causes the target location to be different than when the search data was gathered. Thus, knowledge of the range and angle degrades.

If the search data is accurate, and arrives at the track radar quickly, the acquisition volume could be as small as one beam in angle and very few range resolution cells. This might be the case when the search data comes from the search function of the track radar. In contrast, if data is coming from a remotely located search radar, through a $C^3$ network, the track radar acquisition volume could consist of a somewhat large number of beams and several range cells.

## 8.2 BACKGROUND

The exact design of an acquisition and track initiation process is dependent on the type of search data available to the track radar, and the type of data needed by the track filters. However, in general, the process would include

- Designing the acquisition volume;
- Developing a method for searching the volume;
- Developing a method of using the results of the search to estimate target parameters;
- Developing a method of eliminating, or reducing, false detection due to noise;
- Defining a track filter initialization approach.

### 8.2.1 Acquisition Volume Design

The design of an acquisition volume depends on the nature of the data from the search data source. As an example, suppose the search data source is the acquisition function of the track radar, and that the radar implements the search function by scanning a pencil beam through a preset angular region, in sine space. Suppose further the $u$ and $v$ beam widths of the search beam are the same as the $u$ and $v$ beam widths of the beam used in track. Finally, suppose the acquisition and track initiation process can be scheduled very soon after the radar records a detection in search, so that the target moves very little between search and acquisition. In this case, the acquisition volume could consist of a single beam and very few range cells. The acquisition process would consist of transmitting a waveform in the beam position and searching a few range cells to look for a target return. The results of this single waveform acquisition would be used to provide the initial range, $u$ and $v$ to the track filters. The transmission of the single waveform also serves as a verify process since, if the search detection was caused by noise, there is a high probability the radar would not record a detection during the acquisition process.

In cases where the search data is less accurate, the acquisition volume will need to be larger than one beam and a few range cells, and the acquisition process will require more steps. As an example, suppose the search data comes from a remotely located search radar through a $C^3$ network. Suppose further the search radar is a 2-D radar [1, p. 37] that uses a broad beam in elevation and a narrow beam in azimuth. In this case the elevation, or $v$, dimension of the acquisition volume would need to contain enough beams to cover the search radar elevation beam width, plus any uncertainty caused by geometry, and any other errors caused by target motion, $C^3$ time delays, and other such factors. The azimuth, or $u$, dimension of the acquisition volume could be smaller than the $v$ dimension, but would need to account for similar uncertainties.

The range dimension of the acquisition volume will be influenced by the accuracy of the range data provided by the search data source, plus the uncertainties indicated in the previous paragraph. It can also be influenced by the location of the search data source relative to the track radar. An example of this is illustrated in Figure 8.1. In the left part of the figure, the search and track radars are almost collinear. The rectangular block denotes the possible down-range and cross-range target locations that will cause them to be in a particular range cell and beam of the search radar (the pairs of lines emanating from the radars denote their beam widths). Since the radars are almost collinear, their range coordinates are close. Thus, the range extent

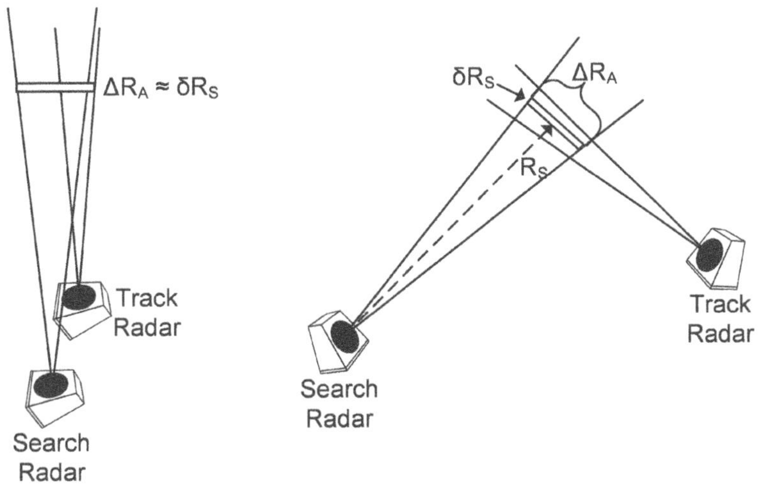


**Figure 8.1** Radar geometry requiring an expanded acquisition volume.

of the acquisition volume, $\Delta R_A$, would be determined almost completely by the range accuracy of the search radar, $\delta R_S$.

In the right image, the search and track radars are horizontally separated. In this case, the range dimension of the acquisition volume must accommodate the cross-range dimension of the search radar. Suppose the range to the target is $R_S$, and the azimuth beam width of the search radar is $\theta_S$, radians. If the target is anywhere in the range extent $\Delta R_S = R_S\theta_S$, it will be detected by the search radar. From the track radar perspective, the target could be anywhere in the interval $\Delta R_S$. The range dimension of the acquisition volume would need to be selected to account for this uncertainty.

### 8.2.2 Acquisition and Track Initiation Process

The fact that the acquisition volume could span several beam positions and range cells means the previously discussed acquisition and track initiation process will need to include the extra steps needed to find the target in range, $u$ and $v$. That is, it must perform a search of the acquisition volume. It must then verify any detections found during the search, to eliminate false detections (false alarms), measure $R$, $u$ and $v$, and then initialize the track filters. A logic diagram of this process is contained in Figure 8.2.

The process starts with the $R$, $u$ and $v$ handed over from the search data source to the track radar. We assume these parameters are relative to the phase center of the track radar. Had the search data source also provided range-rate, it could be used to tune the matched filter used during acquisition. As a note, the track radar would probably want to use an LFM waveform during acquisition, since it is a Doppler insensitive waveform. If the radar used a pulsed Doppler waveform, the process discussed herein becomes somewhat more complicated because of the need to also measure Doppler frequency during the acquisition process, and to account for things such as range and Doppler ambiguities [2].

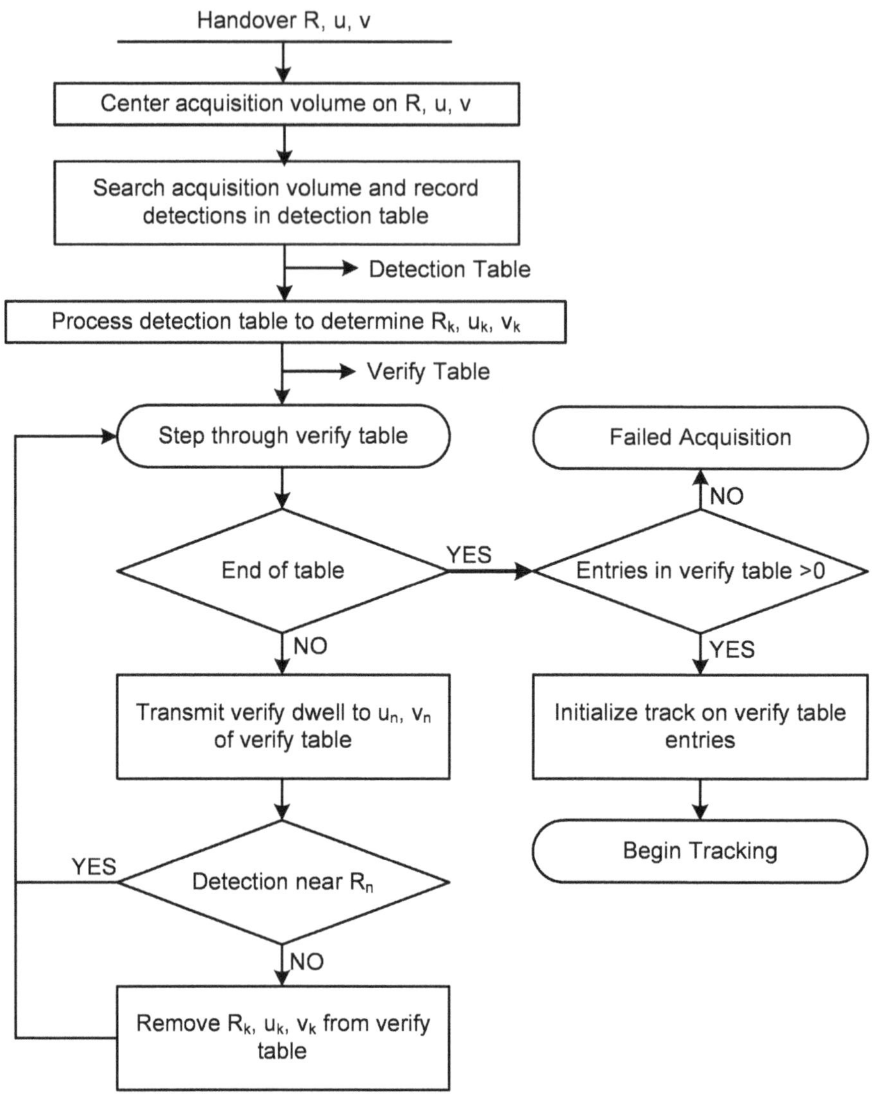


**Figure 8.2** Acquisition and track initiation process.

### 8.2.2.1 Search Acquisition Volume

To start the acquisition phase, the radar would center the acquisition volume on the value of $R$, $u$ and $v$ handed over from the search data source. It would then search the volume. That is, it would transmit waveforms to each of the beam locations of the acquisition volume, receive the return signal and search over the acquisition range window ($\Delta R_A$) for detections. The results of this process would be stored in what we term a *detection table* for later processing. The detection table would contain the beam number, range cell number, and amplitude of each detection.

As an illustrative example, we consider the situation where the acquisition volume contains 9 beams as illustrated in Figure 8.3. The range dimension of the volume contains 11 range cells centered on *R*, the range handed over from the search data source. The nine circles represent the beams of the search volume, and the numbers in the circles are beam numbers we will use shortly. Beam 5 is centered on the handover *u* and *v* and is the center of the acquisition volume. The dot between beams 4 and 5 denotes the location of the target.

In this example, we assume the target causes detections in beams 4 and 5, and in range cells 5 and 6, in each beam. We also assume noise causes a detection in beam 5, range cell 9, and in beam 9, range cell 6. The resulting detection table is illustrated in Table 8.1. The table entries in the first two columns are the beam and range cell numbers just indicated. The entries in the third column are the signal amplitudes associated with the detections. They are used to determine range and angles during the next step of the acquisition process.

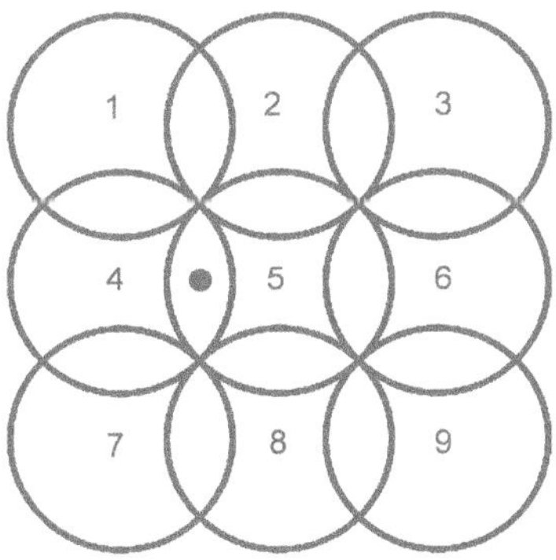


**Figure 8.3** Example acquisition volume—9-beam rectangular raster.

**Table 8.1**
Detection Table Example

| *Beam Number* | *Range Cell Number* | *Signal Amplitude* |
|---|---|---|
| 4 | 5 | 10 |
| 4 | 6 | 12 |
| 5 | 5 | 14 |
| 5 | 6 | 16 |
| 5 | 9 | 4 |
| 9 | 6 | 3 |

### 8.2.2.2 Process Detection Table

The next step in the process (see Figure 8.2) is to process the detection table to produce measurements of range and angles ($R_k$, $u_k$, $v_k$) to associate with the various table entries. This process produces what we term a *verify table*.

The first step in this process is to step through the detection table and combine measurements in adjacent range cells of a beam. The premise of this algorithm is that the adjacent measurements are from a single target, since a target will typically produce measurements in one, or more, range cells. When it produces returns in more than one range cell, they will be adjacent. Noise almost always causes a detection in only one range cell.

The range cell numbers and amplitudes are used to form a weighted average range cell, using the weighted range measurement technique of Chapter 4. Specifically, if the adjacent range cell numbers are $N_m$ and their associated signal amplitudes are $V_m$, the weighted range cell number is

$$N_R = \frac{\sum_{m=1}^{M} N_m V_m}{\sum_{m=1}^{M} V_m} \tag{8.1}$$

where the sum is taken over the total number of adjacent range cells.

If this procedure is applied to Table 8.1, the result is as indicated in Table 8.2. In this case, the pairs of detection in beams 4 and 5 have been combined into single detections, and the size of the detection table has been reduced from 6 entries to 4 entries. The two noise detections are carried over to the reduced detection table because they are not adjacent, in range, to any other detections.

The signal amplitude associated with the two, merged, detection is the total of the signal amplitudes of the two detections. This choice was somewhat arbitrary. However, if spacing between range cells is equal to the range resolution of the waveform used during acquisition, the sum would represent the peak of the matched filter output (see Chapter 4), which is located very close to the weighted average range of the two cells.

**Table 8.2**

Reduced Detection Table

| *Beam Number* | *Range Cell Number, $N_R$* | *Signal Amplitude, $V_{Rn}$* |
|---|---|---|
| 4 | 5.55 | 22 |
| 5 | 5.53 | 30 |
| 5 | 9 | 4 |
| 9 | 6 | 3 |

The next step in the detection table reduction is to combine entries that are close in range, but in different beams. This step is a *beam splitting* operation aimed at providing a better estimate of the angular position of the target. It also serves to reduce the size of the table, and thus limit the number of beams necessary in the verify stage of the acquisition and track initiation process.

For this step, the algorithm searches the reduced detection table to find range cell numbers that are less than a specified distance apart, $\Delta N_R$. It then uses these to compute range and angles to be passed to the verify stage.

The value of $\Delta N_R$ will affect how many entries are used to determine the final range and angles passed to the verify stage. It also affects how many detections are declared targets. If the SNR is reasonably high, the combined range cell numbers, in range cells that contain target returns, will be very close. Thus, $\Delta N_R$ could be set to a small value. For the example of Table 8.2, a $\Delta N_R$ of 0.1 ($1/10^{th}$ of a range resolution cell) would work well.

In low SNR cases, the $V_m$ could contain a significant noise component, which could cause the $N_R$ entries of Table 8.2 to be separated by almost a range resolution cell. Based on this, a compromise $\Delta N_R$ close to 0.5 might be appropriate. As a note, these observations are based on the assumption the range cell spacing is equal to the range resolution. If the cells are spaced closer, this logic would need to be modified.

For the example of Table 8.2, the use of $\Delta N_R = 0.5$ would mean the first, second, and fourth entry would be included in the set used to compute the angles and range included in the verify table. The $N_R$ spacing between the third entry and the others of the table is greater than 0.5. Thus, it would be retained as a separate entry in the verify table.

The fourth entry of Table 8.2 is the result of a false detection and should not be combined with the other two entries. However, it passes the $\Delta N_R$ test and must be included. This is a case where noise can cause not only a false detection, but also a measurement error. In this particular case, inclusion of the fourth entry will not significantly affect the final range and angle measurements since the amplitude associated with it is much smaller than the amplitudes associated with the first and second entries.

The algorithm would compute the final range cell number as a weighted average of the $N_R$ that are combined. It would compute the final angles as a weighted average of the locations of the beams. Equation (8.1) would be used to compute the final ranges, and the final angles would be computed from

$$u_c = \frac{\sum_n u_n V_{Rn}}{\sum_n V_{Rn}} \quad \text{and} \quad v_c = \frac{\sum_n v_n V_{Rn}}{\sum_n V_{Rn}} \tag{8.2}$$

where $(u_n, v_n)$ is the u-v location of beam $n$, and the $V_{Rn}$ are the amplitude entries of Table 8.2.

The output of this stage is the verify table that is passed on to the verify stage. An example of such a table is contained in Table 8.3. In this table, the $u$ and $v$ entries are the measured angles, and the third column contains the measured range. The measured range is computed from the final, weighted average, range cell number.

### 8.2.2.3 Verify and Track Initiation

The main purpose of the verify stage is to eliminate false detections during acquisition. The basis of this statement is, it is highly unlikely noise will cause detections in the same beam and in range cells that are close. Thus, if noise caused a detection in a specific beam position and range cell during acquisition, it is very unlikely it will cause a detection in the same beam position and range cell during verify.

As indicated in Figure 8.2, the radar steps through the verify table and transmits waveforms to each of the beam locations of the table. It then looks for detections range cells close to the ranges in the verify table (usually the two cells closest to the range from the verify table). If no detections are found, that row of the verify table is deleted. This process continues until the end of the table.

If there are no entries left in the table after the verify stage, the acquisition is deemed a failure. If that happens, the computer would need to determine what to do: schedule another acquisition attempt with the same size acquisition volume, schedule another attempt with a larger acquisition volume, wait for operator interventions, or some other action.

The radar would initiate track on any entries left in the verify table by using these entries to initialize the position states of the $R$, $u$ and $v$ track filters. The velocity state of these filters would be set to zero. The exception to this might be the range filter. If range rate was supplied as part of the handover data, it could be used to initialize the velocity state of the range filter.

An alternative to using the entries in the verify table to initialize the track filters would be to enable the monopulse channels during verify and use them to determine angle. The range could be measured by using the range measurement technique discussed in Chapter 4.

During the initial track phase, the filter bandwidths should be made large so filter transients will settle quickly. In α-β and α-β-γ filters, this can be done by the appropriate selection of $\alpha$. $\beta$ and $\gamma$ could be selected using the Benedict-Bordner and Polge-Bhagavan techniques discussed in Chapter 3 [3, 4]. A good choice for bandwidth might be about half the track rate (cutoff frequency, $f_c$ = ¼ the track rate).

If the trackers are Kalman filters, the wide bandwidth could be obtained by using a large $Q$ matrix to force the filter to emphasize the measurements (see Chapter 3). Unfortunately, selection of a suitable $Q$ is not as clear as selection of $\alpha$. This is due to the fact that the effective bandwidth of a Kalman filter is also a function of the initial state covariance matrix ($P$, see Chapter 3) and the measurement variance ($R$, see Chapter 3).

**Table 8.3**
Verify Table

| *u Beam Location* | *v Beam Location* | *Measured Range* |
|---|---|---|
| $u_c$ | $v_c$ | $R_c$ |
| $u_5$ | $v_5$ | $R_9$ |

## References


[1] D. K. Barton, *Radar Equations for Modern Radar*, Boston: Artech House, 2013.

[2] M. C. Budge Jr. and S. R. German, *Basic Radar Analysis*, Norwood, MA: Artech House, 2015.

[3] T. Benedict and G. Bordner, "Synthesis of an optimal set of radar track-while-scan smoothing equations," *IRE Transactions on Automatic Control,* vol. 7, no. 4, pp. 27-32, 1962.

[4] R. J. Polge and B. K. Bhagavan, "A Study of the G-H-K Tracking Filter," UAH Research Report No. 176, MICOM Report No. RE-CR-76-1, DTIC ADA021317, 1975.

# Acronyms and Abbreviations

| | |
|---|---|
| ADC | Analog-to-digital converter |
| AESA | Active electronically scanned antenna |
| AGC | Automatic gain control |
| BPF | Bandpass filter |
| BT | Time-bandwidth |
| CPI | Coherent processing interval |
| CW | Continuous wave |
| dB | Decibel |
| DC | Direct current |
| ENU | East north up |
| ESRS | Electronic scanning radar systems |
| FFT | Fast Fourier transformer |
| FIR | Finite impulse response |
| HOT | Higher order terms |
| GHz | Gigahertz |
| Hz | Hertz |
| I&Q | Inphase and quatrature |
| IEEE | Institute of Electrical and Electronics Engineers |
| IF | Intermediate frequency |
| K | Degrees Kelvin |
| kHz | Kilohertz |
| km | Kilometer |
| LFM | Linear frequency modulation |
| LO | Local oscillator |
| LPF | Lowpass filter |
| m | Meter |
| m/s | Meter per second |
| MF | Matched filter |
| MHz | Megahertz |
| mm | Millimeter |

| | |
|---|---|
| NCO | Numerically controlled oscillator |
| PRF | Pulse repetition frequency |
| PRI | Pulse repetition interval |
| PRN | Pseudo random noise |
| R-C | Resistor-capacitor |
| RCS | Radar cross section |
| RF | Radio frequency |
| rms | Root mean square |
| ROC | Region of convergence |
| s | Second |
| SNR | Signal-to-noise ratio |
| SP | Signal processor |
| SPDT | Single pole double throw |
| T/R | Transmit/receive |
| THAAD | Theater high altitude area defense |
| TWS | Track while scan |
| UAH | University of Alabama in Huntsville |
| W | Watt |
| WSS | Wide-sense stationary |
| μs | Microsecond |

# Variables

| | |
|---|---|
| $A$ | Input magnitude |
| $A$ | State variables |
| $a_k$ | kth element weight |
| $a_n$ | Denominator coefficients |
| $Az$ | Azimuth angle |
| $Az_T$ | Azimuth angle to target |
| $Az_{trk}$ | Azimuth location of antenna beam |
| $B$ | State variables |
| $B_{AGC}$ | AGC bandwidth |
| $b_k$ | kth element weight |
| $B_{LFM}$ | LFM bandwidth |
| $b_m$ | Numerator coefficients |
| $B_n$ | Effective noise bandwidth |
| $BW$ | Antenna beam width |
| $C$ | State variables |
| $c$ | Speed of light |
| $c_n$ | Polynomial coefficients |
| $D$ | Transient performance measure (distance) |
| $d$ | Element spacing |
| $d_1$ | Distance from phase center of feed to reflector |
| $d_2$ | Distance from reflector to aperture plane |
| $\Delta Az$ | Azimuth angle error |
| $\Delta El$ | Elevation ange error |
| $\delta_f$ | BPF offsets |
| $\Delta\phi_{rk}$ | Additional phase shift |
| $\delta\phi_{rk}$ | Phase shift differential |
| $\Delta N_R$ | Specified range cell separation |
| $\Delta R$ | Pulse range resolution |
| $\Delta r_k$ | Additional path length |
| $\delta R_S$ | Range accuracy of the search radar |

| | |
|---|---|
| $dt$ | Integration step size |
| $\Delta t$ | Update period |
| $\Delta\tau$ | Range delay error |
| $\Delta_\tau$ | Range error |
| $\Delta u$ | Direction cosine error |
| $\Delta V$ | Error signal |
| $\Delta v$ | Direction cosine error |
| $\Delta V_{avg}$ | Early/late sample average |
| $\Delta V_{norm}$ | Normalized error signal |
| $\delta y$ | Feed offset |
| $e$ | Error signal |
| $E(k)$ | Elliptic integral of the second kind |
| $e(n)$ | $k$-element input vector |
| $E(s)$ | Error - continuous time transfer function |
| $e(t)$ | Error - continuous time domain |
| $E(z)$ | Error - discrete time transfer function |
| $e(z)$ | Error - discrete time domain |
| $E_0$ | E-field magnitude |
| $\varepsilon_a(t)$ | Target elevation angle |
| $E_{ak}(t)$ | E-field at array element |
| $e_{AZ}$ | Azimuth error voltage |
| $E_{bk}(t)$ | E-field at array element |
| $E_{Bk}(t)$ | E-field radiated by the kth element on the back of the array |
| $e_{El}$ | Elevation error voltage |
| $e_f$ | Doppler error signal |
| $E_k(t)$ | E-field at the kth element |
| $El$ | Elevation ange |
| $El_T$ | Elevation ange to target |
| $El_{trk}$ | Elevation location of antenna beam |
| $e_R$ | Range error signal |
| $ess$ | Steady state error |
| $e_u$ | $u$ error signal |
| $e_v$ | $v$ error signal |
| $e_\theta$ | Angle error signal |
| $F$ | $k \times k$ element state transition matrix |
| $F(n-1)$ | State transition matrix |
| $\phi(t)$ | Phase of the received signal |
| $\phi(t)$ | Phase modulation |
| $\phi(y)$ | Phase gradient |

| | |
|---|---|
| $F(z)$ | Standard input z-transform |
| $f_B$ | Doppler filter bandwidth |
| $f_c$ | Closed loop track bandwidth (Hz) |
| $f_c$ | One-sided, half-power bandwidth of the α-β and α-β-γ filters |
| $f_c$ | Carrier frequency |
| $f_c$ | Filter cutoff frequency |
| $f_d$ | Target Doppler frequency |
| $f_d(n)$ | Bandpass filter center frequencies |
| $f_{dinit}$ | Initial Doppler for track filter |
| $f_{dmeas}$ | Measured Doppler frequency |
| $f_{ds}(n)$ | Unambiguous smoothed Doppler frequency |
| $f_{dtrk}$ | Predicted Doppler frequency |
| $f_{dtrk}(n)$ | Unambiguous tracked Doppler frequency |
| $f_{dtrkambig}$ | Ambiguous Doppler frequency |
| $f_m$ | Mixer tuning frequency |
| $f_M$ | Matched filter Doppler frequency |
| $\phi_r$ | Phase of electric field on circular arc |
| $f_r$ | Scan frequency in Hz |
| $f_{RF}$ | Carrier frequency |
| $f_s$ | Track rate (Hz) |
| $G$ | $k \times r$ element input distribution matrix |
| $g(n)$ | Impulse response of the (closed loop) α-β filter |
| $G(s)$ | Continous time system frequency response |
| $G(z)$ | Discrete time system frequency response |
| $G(\theta_T)$ | Normalized directive gain pattern |
| $G_C(s)$ | Controlled element - continuous time transfer function |
| $G_C(z)$ | Controlled element - discrete time transfer function |
| $G_{CL}(s)$ | Closed loop continuous time transfer function |
| $G_{CL}(z)$ | Closed loop discrete time transfer function |
| $G_F(s)$ | Track filter - continuous time transfer function |
| $G_F(z)$ | Track filter - discrete time transfer function |
| $G_F(z)$ | Open loop discrete time filter transfer function |
| $G_{OL}(s)$ | Open loop continuous time transfer function |
| $G_{OL}(z)$ | Open loop discrete time transfer function |
| $h(n)$ | Arbitrary but realizable function |
| $h(t)$ | Matched filter impulse response |
| $h_r$ | Height of the phase center of the radar antenna |
| $H^T$ | $s \times k$ element output distribution matrix |
| $h_t$ | Target altitude |

| | |
|---|---|
| $I$ | Identity matrix |
| $I_E$ | Early gate integral output |
| $I_L$ | Late gate integral output |
| $J(g(n),\lambda)$ | Overall performance measure (Benedict and Bordner) |
| $J_1(x)$ | First order Bessel function of the first kind |
| $K$ | Parameter used to establish type 3 servo bandwidth (real valued) |
| $K$ | Elements in subarray |
| $K(k)$ | Elliptic integral of the first kind |
| $K(n)$ | Weighting or gain matrix |
| $K(n)$ | Kalman filter gain matrix |
| $K_\phi$ | Phase constant |
| $K_{OL}$ | Tracker closed loop gain |
| $K_R$ | Tracker opern loop gain |
| $K_R$ | Scaling constant |
| $L$ | A real, lower triangular matrix |
| $\lambda$ | Wavelength associated with the radar RF |
| $M$ | Input parameter - integer |
| $MF(\tau)$ | Matched filter output |
| $MF(\tau, f)$ | Matched filter output |
| $N$ | System type |
| $N$ | Number of poles |
| $n$ | Track stage |
| $n$ | Effective number of pulses noncoherently integrated by tracker |
| $n(t)$ | Noise into matched filter |
| $N_{BPF}$ | Number of bandpass filters |
| $N_{chip}$ | Number of chips |
| $N_{FFT}$ | Number of points in FFT |
| $n_k$ | Samples from a WSS random process |
| $N_o$ | Gaussian noise power spectral density |
| $n_o(t)$ | Noise out of matched filter |
| $N_R$ | Weighted range cell number |
| $P_N$ | Noise power |
| $P_p(n)$ | Kalman filter predicted covariance matrix |
| $P_S$ | Return signal level scale at MF output to provide proper SNR |
| $P_S$ | Signal power |
| $P_s(n)$ | Kalman filter smoothed covariance matrix |
| $q$ | Spacing variable (in terms of waveform resolution), between early/late samples ($0 < q < 1$) |
| $q$ | Range gate offset |

| | |
|---|---|
| $Q(n)$ | Kalman filter covariance matrix of input disturbance $W(n)$ |
| $\theta_k$ | Asymptote angle |
| $\theta_s$ | Squint angle |
| $\theta_S$ | Azimuth beam width of the search radar |
| $\theta_T$ | Target angle |
| $R$ | Autocorrelation (and autocovariance) matrix for the vector $\eta$ |
| $\rho$ | Parameter based on detection threshold relative to SNR corresponding to the average target RCS |
| $\rho$ | Reflection coefficient |
| $r(n)$ | Filter output |
| $R(n)$ | Kalman filter covariance matrix of noise on measurements $V(n)$ |
| $\dot{R}(t)$ | Range-rate of target |
| $R(z)$ | Filter output z-transform |
| $R(\tau)$ | Autocorrelation function |
| $r_{12}$ | Noise correlation coefficient |
| $R_a(t)$ | Range to target |
| $R_{act}(n)$ | Actual range |
| $r_{alt}$ | Alternate solution (Benedict and Bordner) |
| $\rho_D$ | Diffuse reflection coefficient |
| $R_e(n)$ | Range track error |
| $R_I(t)$ | Range to target image |
| $R_n$ | True target range at the beginning of the n[th] track update period |
| $R_o(\tau)$ | Autocorrelation of the noise at the matched filter output |
| $R_p(n)$ | Tracked range |
| $R_{ref}$ | Reference range |
| $R_\sigma$ | Variance ratio |
| $\rho_S$ | Specular reflection coefficient |
| $R_S$ | Range to target |
| $\Sigma$ | Sum signal |
| $\sigma_{asy}$ | Asymptote real axis intercept |
| $S_D$ | Discriminator scaling factor |
| $S_f$ | Doppler discriminator scale factor |
| $SNR_{ref}$ | Reference SNR |
| $\sigma_o^2$ | Output noise variance |
| $\sigma_{ref}$ | Reference RCS |
| $\sigma_{Rmeas}$ | root-mean-square range measurement error |
| $S_S$ | Squint angle |
| $S_\tau$ | Discriminator scale factor |
| $S_\tau$ | Saturation value of the discriminator |

| | |
|---|---|
| $S_\tau$ | Range discriminator scale factor |
| $\sigma_{tgt}$ | Target RCS |
| $s_{trk}$ | Predicted angle from tracker |
| $\sigma_y^2$ | Noise variance |
| $\sigma_{ys\eta}^2$ | Variance of $y_{s\eta}(n)$ |
| $T$ | Sample period |
| $T$ | Track rate in Hz |
| $t$ | Time |
| $\tau_c$ | Chip width |
| $\tau_{chip}$ | Chip, or subpulse, width |
| $T_{CPI}$ | Coherent processing interval |
| $T_{dwell}$ | Pulsed Doppler waveform dwell duration |
| $\tau_E$ | Early gate sample time |
| $\tau_{init}$ | Initial range delay for the track filter |
| $\tau_L$ | Late gate sample time |
| $T_{n-1}$ | Update period of closed loop tracker |
| $\tau_p$ | Pulsewidth |
| $\tau_r$ | Range delay to target |
| $\tau_R$ | Range delay to target |
| $\tau_{res}$ | Waveform range resolution |
| $\tau_{Rest}$ | Range delay estimate |
| $T_{sim}$ | Simulation duration |
| $\tau_{trk}$ | Range delay provided by tracker (track point, predicted range delay) |
| $\tau_{trkactual}$ | Actual range delay |
| $\tau_{trkambig}$ | Ambiguous range delay |
| $u$ | sine space angle |
| $U(n)$ | Discrete time unit step function |
| $U(t)$ | Unit step function |
| $u(t)$ | LFM pulse |
| $u_{init}$ | Initial $u$ for the track filter |
| $u_T$ | $u$ direction cosine to target |
| $u_T, v_T$ | Target angles in sine space |
| $u_{trk}$ | $u$ location of antenna beam in sine space |
| $u_{trk}, v_{trk}$ | Track angles of antenna beam in sine space |
| $u_{trk}, v_{trk}$ | Predicted target angles in sine space |
| $v$ | sine space angle |
| $V(n)$ | Kalman filter noise on measurements |
| $V(t)$ | Voltage at the feed output |

| | |
|---|---|
| $V(t)$ | Normalized output of the matched filter |
| $v_1(t)$ | Port 1 input signal - Rat Race |
| $v_1(t)$ | Port 1 input signal - 3-dB Coupler |
| $v_2(t)$ | Port 2 input signal - 3-dB Coupler |
| $v_{21}(t)$ | Signal at ports 2, due to $v_1(t)$, rat race |
| $v_{23}(t)$ | Signal at ports 2, due to $v_3(t)$, rat race |
| $v_3(t)$ | Port 3 input signal, rat race |
| $v_3(t)$ | Port 3 output signal, 3-dB coupler |
| $v_4(t)$ | Port 4 output signal, 3-dB coupler |
| $v_{41}(t)$ | Signal at ports 4, due to $v_1(t)$, rat race |
| $v_{43}(t)$ | Signal at ports 4, due to $v_3(t)$, rat race |
| $V_a(t)$ | Voltage at summer output |
| $v_a(t)$ | Target elevation angle in sine space |
| $V_{ak}(t)$ | Voltage at array element |
| $V_b(t)$ | Voltage at summer output |
| $V_{bk}(t)$ | Voltage at array element |
| $V_\Delta(t)$ | Difference voltage |
| $v_\Delta(t)$ | Signal an port 4, the Δ port, rat race |
| $V_{diff}$ | Difference signal |
| $V_E$ | Early gate sample value (detector output) |
| $v_I(t)$ | Target image elevation angle in sine space |
| $v_{inN}(t)$ | Noise voltage at receiver input |
| $v_{inT}(t)$ | Target voltage at receiver input |
| $V_k$ | Above threshold envelope detector outputs |
| $V_L$ | Late gate sample value (detector output) |
| $v_{MF}(t)$ | Matched filter output |
| $V_R$ | Matched filter peak value |
| $v_r(t)$ | Received signal |
| $V_{RF\Delta}(t)$ | Difference signal from the monopulse combiner |
| $V_{RF\Sigma}(t)$ | Sum signal from the monopulse combiner |
| $V_{RF\Delta}$ | Resolver output |
| $V_{RF\Delta u}$ | Azimuth signal |
| $V_{RF\Delta v}$ | Elevation signal |
| $V_\Sigma(t)$ | Sum voltage |
| $v_\Sigma(t)$ | Signal an port 2, the Σ port, rat race |
| $v_T$ | $v$ direction cosine to target |
| $v_t(t)$ | Normalized transmit pulse |
| $v_{trk}$ | $v$ location of antenna beam |
| $W(n)$ | Kalman filter input disturbance |

| | |
|---|---|
| $\omega_c$ | Closed loop bandwidth (rad/s) |
| $\omega_d$ | Damped frequency |
| $\omega_n$ | Undamped natural frequency |
| $X$ | $N$ element vector of zero-mean, unit variance, independent, complex, Gaussian random variables |
| $X(n)$ | $k$-element state vector |
| $X(n)$ | System state variable |
| $x(t)$ | Down-range positon of target |
| $X_a(s)$ | Target parameter, continuous time transfer function |
| $x_a(t)$ | Target parameter, time domain |
| $x_a(t)$ | Tracker input, continuous time domain |
| $x_a(t)$ | Actual down-range positon of target |
| $X_a(z)$ | Target parameter, discrete time transfer function |
| $x_I(t)$ | Down-range positon of target image |
| $Xnew$ | New state vector |
| $Xold$ | Initial state vector |
| $X_p$ | Predicted track filter state |
| $X_p(n)$ | Predicted state (vector) |
| $X_p(s)$ | Predicted target parameter, continuous time transfer function |
| $x_p(t)$ | Predicted target parameter, time domain |
| $X_p(z)$ | Predicted target parameter, discrete time transfer function |
| $X_s$ | Smoothed state vector |
| $X_s(n)$ | Smoothed state (vector) |
| $x_{trk}$ | Cross range track coordinate |
| $y(n)$ | $s$-element input vector |
| $y(n)$ | Measurement |
| $y(t)$ | Altitude of target |
| $y_a(t)$ | Actual altitude of target |
| $y_I(t)$ | Altitude of target image |
| $y_p(n)$ | Predicted measurement |
| $y_{s\eta}(n)$ | Filter output with zero-mean, unit-variance, white noise input |
| $y_{trk}$ | Down range track coodinate |
| $\zeta$ | Damping coefficient or damping ratio |
| $z_0$ | Parameter used to establish type 3 servo transient response |
| $\Delta\tau$ | Spacing between noise samples |
| $\Delta$ | Difference signal |
| $\Delta f$ | Doppler error |
| $\Delta R$ | Range error |
| $\Delta R_A$ | Range extent of the acquisition volume |

| | |
|---|---|
| $\Delta\alpha$, $\Delta\varepsilon$ | Target location relative to the track axis |
| $\Delta\theta$ | Solid angle of the target relative to the track axis |
| $\Delta\theta$ | Angle error |
| $\alpha$ | Track filter coefficient |
| $\alpha$ | LFM slope |
| $\alpha_b$, $\varepsilon_b$ | Center of the beam relative to the track axis |
| $\beta$ | Track filter coefficient |
| $\chi(\tau, f)$ | Ambiguity function |
| $\delta(n)$ | Kronecker delta |
| $\delta(t)$ | Drica delta function |
| $\delta\tau$ | desired spacing between samples |
| $\phi(t)$ | Phase modulation |
| $\gamma$ | Track filter coefficient |
| $\eta$ | $N$ element vector of the noise samples at the matched filter output |
| $\eta(n)$ | Zero-mean, unit-variance, white noise |
| $\lambda$ | Lagrange multiplier |
| $\theta_b$ | Squint angle |
| $\theta_T$ | Actual target angle |
| $\theta_T(t)$ | Location of the target relative to the center of the beam |
| $\theta_{trk}$ | Tracked target angle |
| $\sigma$ | Target RCS |
| $\sigma_{angmeas}$ | Expected angle measurement error |
| $\sigma_{fdmeas}$ | Expected Doppler measurement error |
| $\sigma_{ref}$ | Reference RCS |
| $\sigma_{Rfluct}$ | Estimate RCS fluctuation on rms tracking error |
| $\sigma_{tgt}$ | Target RCS |
| $\tau_{ak}$ | Propagation time to array element |
| $\tau_{bk}$ | Propagation time to array element |
| $\tau_{Rmeas}$ | Measured range delay |
| $\tau_{Rn}$ | Range delay |
| $\tau_{Rn}(t)$ | Target range delay at the beginning of the $n^{th}$ track update period |
| $\tau_{trk}$ | Predicted range delay |
| $\omega_r$ | Scan rate in radians per second |
| $\psi_{Rn}$ | Received signal phase term |

# About the Authors

**Dr. Mervin C. Budge**, Ph.D., is chief scientist of Dynetics where he is responsible for overall technical quality. He also conducts research in the area of radars and air defense systems. He conducts analysis and design of all aspects of ground-based and airborne radars including antennas, transmitters, receivers, signal processors, tracking, data processing and displays and controls. This includes performance analyses and specific radar designs.

Dr. Budge has served as an adjunct professor at the University of Alabama Huntsville (UAH) since 1973 and previously served as a part-time professor at the Southeastern Institute of Technology, an instructor at Texas A&M University, and both a graduate and undergraduate teaching assistant at the University of Louisiana in Lafayette. He currently teaches courses in radar, tracking and Kalman filters at UAH.

He holds a Ph.D. in electrical engineering from Texas A&M University and master's and bachelor's degrees in electrical engineering from the University of Louisiana in Lafayette. Dr. Budge's graduate research included "Prediction of Clear Air Turbulence," sponsored by NASA Marshall Space Flight Center; "Space Shuttle Aeroelastic Stability and Control Properties," sponsored by NASA Johnson Space Center; "Parameter Identification for Linear Systems," a Themis Grant; "Electric Power System Control," sponsored by Philco-Ford; and "Analytical Techniques for Predicting Flow Fields in Oil Wells." His graduate thesis and dissertational work was in the areas of the effect of interaction in multivariable control systems and the design of recursive and nonrecursive digital filters. He has authored more than three dozen publications.

Dr. Budge's honors include Outstanding Educator Award from the Huntsville Section of IEEE; the Dynetics R. Duane Hays Award for Technical Excellence; the Outstanding Engineer Award from the Huntsville Section IEEE; Outstanding PhD Student, Electrical Engineering Department, Texas A&M University; and Outstanding PhD Student, Texas A&M University. He holds memberships in Eta Kappa Nu, Tau Beta Pi, Sigma Xi, Phi Kappa Phi, the Institute of Electrical and Electronic Engineers, Phi Mu Alpha Sinfonia, and Kappa Delta Pi.

**Shawn R. German** is a senior principal engineer at Dynetics, where he has been employed since 1995. He is responsible for performing radar system design and analysis. His current areas of research include radar receivers, advanced signal processing methodologies, CFAR and detection theory. He is responsible for generating assessments of ground based and airborne radars, seekers, and integrated air defense systems, including performance analyses and specific radar designs. Mr. German has served as a subject matter expert in a number of radar disciplines including: system test and measurement, RF and IF system testing, and performance analysis. He has served as a consultant on various radar related technologies and theories.

Mr. German's expertise includes the following areas: radar system repair and alignment, radar system and subsystem performance analysis, environment and target modeling, waveform analysis and design, signal processor analysis, radar system design, radar subsystem design, signal processor hardware prototyping, development of mathematical models from measured data, radar system and subsystem simulation and emulation, radar ECM and ECCM, and phased array theory.

Mr. German is a senior member of the Institute of Electrical and Electronic Engineers and holds memberships in Eta Kappa Nu, Tau Beta Pi, Phi Kappa Phi, and the Association of Old Crows. He holds bachelor's and master's degrees in electrical engineering from Mississippi State University and an associate's degree in electronics technology from the Mississippi Gulf Coast Community College. He is completing a PhD in electrical engineering from the University of Alabama Huntsville (UAH). His undergraduate research in "Error Control Coding in Satellite Communication" was sponsored by NASA.

Mr. German has served as a graduate teaching assistant in the Electrical Engineering Department at Mississippi State University and teaches internal classes at Dynetics. He has coauthored several publications.

# Index


- $\alpha$
  - calculation of, 57
  - equation for, 58
  - tracker example, 72
- $\alpha$-filter, 61
  - matrices, 61
- $\alpha$-$\beta$, 2, 3, 7, 84
  - bandwidth, 57
  - block diagram, 47
  - closed loop transfer function, general, 62
  - gain matrix, 43
  - noise reduction measure, 54
  - open loop transfer function, general, 61
  - performance measure, 55
  - relation, 56
  - track filter, 46
  - tracker example, 72
  - transient performance measure, 54
  - values, general, 64
- $\alpha$-$\beta$ filter, 41, 141
  - Doppler Tracker, 207
- $\alpha$-$\beta$-$\gamma$, 2, 3, 141
  - bandwidth, 57
    - equal zeros, 69
  - block diagram, 47
  - closed loop transfer function
    - critically damped, 66
    - equal zeros, 68
    - general, 65
  - coefficients
    - critically damped, 67
    - equal zero form, 69
  - critically damped example, 130
  - critically damped form, 65, 66
  - equal zeros form, 68
  - gain matrix, 43
  - open loop transfer function
    - equal zeros, 68
    - general, 65
  - Polge-Bhagavan, 46
  - relation, 56
  - rcsponscs
    - equal zeros, 69
    - critically damped, 67
  - state transition matrix, 42
  - track filter, 46
  - tracker example, 72
- $\alpha$-$\beta$-$\gamma$ filter, 41
- 12-horn feed, 164
- acceleration
  - constant, model, 45
- acquisition
  - process, 286
  - search, 285
  - volume, 285, 286
- Acquisition, 285
- acquisition volume
  - influence of geometry on, 286
- AGC, 95
  - for normalization, 182
- ambiguity function
  - LFM, 219
  - phase coded, 219
- ambiguous
  - Doppler, 215, 218
  - range, 214
- amplitude comparison, 142

difference voltage, 159
monopulse, 253
sum and difference voltage, 159
sum voltage, 159
angle
azimuth, 139
discriminator, 181
elevation, 139
error
functional level model, 202
information, 150
position, 139
predicted, 141
squint, 158
tracked, 139
tracker, 7, 139
closed loop, 139
angle tracker, 42
example 1, 193
example 2, 198
antenna
ideal, 253
aperture plane, 145
array
linear
phase comparison, 146
space fed, 154
weights, 146
association, 3
multiple hypothesis, 3
nearest neighbor, 3
probabilistic data, 3
asymptotes, 14
angles, 14
intersect, 14
automatic gain control (AGC), 95
azimuth, 139
bandpass limiter
for normalization, 182
bandwidth
closed loop, 32
effective noise, 53
noise, 52
Type 1 servo, 60
α-β, 57
α-β filter, general, 63
α-β-γ, 57
α-β-γ, equal zeros, 69
α-filter, 60
beam
squinted, 140
beam splitting, 291
beams
simultaneous, 175
squinted, 159
beat frequency, 256, 276
Benedict-Bordner, 46, 84
example, 123
α
equation for, 58
α-β design, 53
bilinear transform, 27
block diagram
Kalman, 47
z-transform example, 31
α-β, 47
α-β-γ, 47
BPF, 209
response, 210
Butterworth
filter, 211
frequency response, 211
$C^3$, 285
calculation of α
Benedict-Bordner, 57
Polge-Bhagavan, 57
calculus of variations, 55, 84
centroid
elliptic, 251
track, 260, 261, 267, 269
track, range, 264
tracking, 251
Chebychev
filter, 212
parameters, 212, 213
closed loop
poles, 13
closed loop bandwidth, 32, 33
closed loop range tracker, 91
closed loop transfer function
Type 1 servo, 59

α -filter, 59
α-β-γ, critically damped, 66
α-β-γ, equal zeros, 68
α-β-γ, general, 62
closed loop transfre function
α-β-γ, general, 65
coefficients
α-β-γ, critically damped, 67
α-β-γ, equal zero form, 69
coherent processing interval, 169
comparison
amplitude, 142
phase, 142
configuration
pulsed radar, 208
conical
scan, 141
on receive only (COSRO), 141
conical scan, 176
description, 176
constrained
feed array, 142
continuous wave
tracker, 207
control theory approach, 59
Type 1 servo, 59
α -filter, 59
α-β filter, 61
α-β-γ filter, 65
controlled element, 2, 142
antenna, 2
range gate, 2
coordinae
ENU, 139
coordinate
transformation, 1
coordinate system
Cartesian, 139, 209
spherical, 139
corporate
feed array, 142
correction
spherical, 143, 155
phase shift for, 156
COSRO
description, 170
cost function
Benedict-Bordner, 84
coupler, 3-dB, 163
coupling
range-Doppler, 225
covariance
matrix, 83
predicted, 71
propagation, 82
smoothed, 71
critically damped, 17
α-β-γ, 65, 66
crossing target
tracking, 258
in angle, 258
in range
Doppler effect, 265
RCS fluctuation, 266
damped
critically, 17, 32
over, 16, 17, 18, 19, 32
under, 18, 19, 32
damped frequency, 32
damping
coefficient, 31
ratio, 31
damping ratio, 33
data latency, 285
DC gain, 11, 21
decay time, 32
delay
range, 92
detection
false, 291
detection table, 289
determining the open loop parameters, 30
dichotomous
track, range, 264
tracking, 252
difference
voltage
amplitude comparison, 159
phase comparison, 150
diffuse
multipath, 270

direct
path, 269
direct measurement
Doppler, 217
discrete time servos, 19
discriminator
angle, 181
exact, 3
log, 3
curve, 98
low PRF, 225
Doppler, 207, 209
CW, 208
low PRF radar, 218
pulsed Doppler, 208
equation
low PRF, 224
exact, 142, 182, 252
gain, 95
LFM, 100
indeterminate condition, 97
log, 142, 190
modifie exact, 142
modified exact, 187
monopulse
exact, 252
pulsed Doppler, 214
range, 91
block diagrams, 102
integrating gate, 3
sampling gate, 3, 92
summing gate, 101
sampling gate
equation for, 95, 100
gain, 95
slope
CW, 211
low PRF, 224
summing gate, equation for, 102
discriminator curve, 8, 72
integrating gate
LFM pulse, 107
integratng gate
unmodulsted pulse, 107
sampling gate
phase coded pulse, 102
unmodulated pulse
equation for
sampling gate, 98
integrating gates, equation for, 106
Doppler
ambiguous, 218
direct measurement, 217
discriminator, 207, 209
low PRF radar, 218
error
signal, 207
frequency, 207, 208
matched, 219
tuning
matched filter, 218
Doppler frequency, 260
difference, 264
mismatch, 264
Doppler tracking
functional level model
comparison to actual discriminator, 239
Dual target tracking, 251
dwell, 169
elevation, 139
elliptic
centroid, 251
integral, 251
equal zeros
$\alpha$-$\beta$-$\gamma$, 68
equation
predicted measurement, 47
prediction, 44, 47
smoothing, 47
equations
Kalman filter, 71
error, 43
bias, 12, 22, 93
bias, range, 264
equation, 210
formation block, 210
range
equation for, 94
signal
Doppler, 207
steady state, 9, 10

steady state analog, 12
steady state digital, 22
error sensor, 1, 139
error signal
range, 93
exact
discriminator
amplitude comparison, 182
phase comparison, 182
exact processor
constrained feed array, 185
space fed array, 186
example
results, case 1, 73
results, case 2, 75
target scenario, 72
tracker parameters, 73
Example
CW tracker, 226
tracker
low PRF, 232
α, α-β, α-β-γ trackers, 72
example plots, 34
examples
range tracker, 121
simulation iterative loop, 124
simulation setup, 124
Examples
Sampling gate discriminator, α–β filter, unmodulated pulse, 122
fading
multipath, 273, 278
feed, 143
12-horn, 165
amplitude comparison, 143
constrained, 142
corporate, 142
four horn, 143, 164
multimode, 164
sidelobe control, 165
space, 142
space fed array, 154
feed, constrained
sidelobe control, 165
FFT, 217
filter
band pass, 209
Benedict-Bordner, 3
Butterworth, 211
Chebychev
Type I, 212
Type II, 212
FIR, 215
g-h, 7
g-h-k, 7
initialization, 292
Kalman, 2, 3, 7
low pass, 211
open loop, track, 217
Polge-Bhagavan, 3
track, 2, 7, 92
α-β, 2
α-β-γ, 2
Filter
Kalman, 41
α-β, 41
α-β-γ, 41
final value theorem
s-domain, 11
z-domain, 21
finite impulse response
filter, 215
FIR, 215
Fourier Transform, 87
frequency
beat, 256
damped, 32
Doppler, 113, 123, 207, 208
matched filter, 113
radio, 208
residula, 113
undamped, 31
frequency response
continuous time, 24
discrete time, 24
functional error model, 202
gain
DC, 11, 21
open loop, 13
gate mode
narrow, 98

normal, 98
wide, 98
geometric progression, 151
glint, 257
behavior, 265
H matrix, 43
higher order terms, 45
HOT, 45
as model uncertainties, 46
impulse invariant, 27
impulse response
α-β, 53
indeterminate condition, 97
information
angle, 150, 152
initialization
Kalman filter, 72
α filter, 72
α-β filter, 72
α-β-γ filter, 72
input
constant, 11, 21, 22
constant acceleration, 11
constant velocity, 11
quadratic, 11
ramp, 11, 21
step, 11, 22
z-domain, 21
input distribution matrix, 35
instantaneous
normalization, 252
integration step size, 29
Kalman, 2, 3, 41, 141
block diagram, 47
state transition matrix, 42
Kalman filter, 70
extended, 71
initialization, 72
Kalman Filter
equations, 71
Lagrange multiplier, 55
Laplace transform
of derivative, 25
LFM
pulse, 98
range discriminator, 98
slope, 221
lobing, 276, 278
multipath, 276
log processor, 182, 190
loot locus
and transient response analog, 13
and transient response digital, 22
LPF, 211
magic tee, 160, 161
matched
range, 219
matched filter
impulse response, 113
output, 223
reaponse
matched range, 224
response, 114
general, 114
LFM pulse, 114
matched range, 223
phase coded pulse, 115
unmodulated pulse, 114
sidelobes, 265
tuning, 220
matched filter output
LFM pulse, 99
unmodulated pulse, 99
matrix
gain, 41
state transition, 41
measurement, 41
predicted, 41
range-rate, 43
measurement sensor, 1
metric, 1
acceleration, 1
angle, 1
position, 1
range-rate, 1
velocity, 1
model
constant acceleration, 45
modeling
range tracker, 111
state variable method, 24
z-transform method, 24, 27, 28

modified exact processor, 187
monopulse
  amplitude comparison, 140, 154, 253, 260
  combiner, 140, 160
    block diagram, 160
  comparator, 140, 160
  phase comparison, 146, 253
  receiver
    two channel, 170
      COSRO, 170
  receivers
    three channel, 167
  sensing, 142
  three channel, 141
monopulse combiner, 141
monopulse comparator, 141
monopulse ratio, 183
multipath
  diffuse, 270, 271
  effect of target altitude, 276
  experiments, 269, 272
    specular, 274
  fading, 276
  lobing, 276
  modeling, 271
  reflection coefficient, 270
  specular, 269
multiple beams, 141
noise
  autocorrelation
    matched filter input, 115
  bandwidth, 52
  considerations, 115
  correlation, 226
    matched filter output, 116, 117
      LFM pulse, 117
      phase coded pulse, 117
      unmodulated pulse, 117
  effective bandwidth, 53
  generation, 118
    algorithm, 119
    autocorrelation matrix, 118
    Cholesky decomposition, 118
    Toeplitz matrix, 118
  matched filter input, 115
  modeling, 226
  scaling, 120
noise model, 115
noise power, equation for, 120
noise reduction measure, 53
  α-β, 54
normalization
  error, in angle tracking, 141
  instantaneous, 252
open loop parameters
  determining, 30
open loop transfer function
  Type 1 servo, 59
  α filter, 59
  α-β-γ, equal zeros, 68
  α-β-γ, general, 65
  α-β, general, 61
Open loop transfer function
  α-β, 47
  α-β-γ, 48
oscillate
  from sample to sample, 23
output distribution matrix, 35
output vector, 35
overdamped, 16, 18, 19
parameter. See metric
path
  direct, 269
  reflected, 269
performance measure
  α-β, 55
phase
  range delay, 113
phase comparison, 142
  difference voltage, 150
  sum voltage, 152
phase variable, 25
pilot pulse, 169
plane
  wave, 146
pole
  at s = 0, 20
  at z = 1, 20
poles
  closed loop, 13

complex, 13
open loop, 13
Polge-Bhagavan, 46, 89
$\alpha$
equation for, 58
$\alpha$-$\beta$-$\gamma$ design, 56
predicted
measurement, 41, 42
measurement equation, 47
positon, 42
range, 93
predicted covariance, 71
predicted state, 41
prediction
equation, 44, 47
stage, 41, 44
prediction equation, 47
propagate, 43
PSK signal model, 233
pulse
LFM, 219
phase coded, 219
unmodulated, 92
width, 92
pulsed Doppler
discriminator, 214
radar
multifunction, 1
search, 1, 285
track, 1, 285
track-while-scan, 1
radar range equation, 120
range
ambiguous, 214
delay, 91, 92
predicted, 93
direct measurement, 92, 109
several samples, 111
direct measurement equation
two samples, 110
discriminator, 91
LFM pulse, 98
error, 92
error signal, 93
estimator, optimum, 91
gates, 91
matched, 219
rate, 209
resolution, 93
straddling loss, 111
tracker, 42
closed loop, 91
modeling, 111
range delay, 92
phase, 113
range discriminator
block diagrams, 102
integrating gate, 3
sampling gate, 3, 92
summing gate, 101
unmodulated pulse, 103
range error
equation for, 94
range gates
narrow, 98, 103
normal, 98
wide, 98, 103
range measurement, 109
range measurement equation, 110
range resolution, 93
range tracker, 42
track filter implementation, 122
range tracker examples, 121
range tracking
example 1
sampling gate, $\alpha$–$\beta$ filter, LFM pulse, 129
example 1 plots, 125
example 2
integrating gate discriminator, $\alpha$–$\beta$ filter, unmodulated pulse, 128
sample plots, 131
sequence of events, 128
example 3
direct range measurement, $\alpha$–$\beta$–$\gamma$ filter, unmodulated pulse, 129
sample plots, 11 dB initial SNR, 133
functional level model, 133, 239
range-Doppler
coupling, 225
range-rate, 209
rat race, 161

RCS
  fluctuation, effect on tracking, 247
realizability constraint
  on g(n), 53
  on h(n), 56
reference
  range, 122
  RCS, 122
  SNR, 122
reflected
  path, 269
reflection
  point, 269
    specular, 269
region of convergence, 87
regions
  s- and z-plane, 23
residual, 43
response
  BPF, 210
responses
  α-β filter, general, 64
  α-β-γ, critically damped, 67
  α-β-γ, equal zeros, 69
  α-filter, 60
RF, 208
ROC, 87
root locus, 3
  algorithm, 13, 14, 22
  analog, 13
  asymptotes, 14
  branches, 14
  definition, 14
  digital, 22
  Example 1, 15
  Example 2, 16
  Example 3, 18
  Example 4, 19
  Example 5 - digital, 22
  sketching, 14, 22
Routh-Hurwitz, 80, 81
sample period, 24, 28
search data, 285
search data source, 285
sensor
  angle, 3
    conical scan, 3
    COSRO, 3
    LORO, 3
    monopulse, 3
    simultaneous beams, 3
  error, 1, 139
  frequency, 3
  measurement, 1
  range, 3
servo
  block diagra, analog, 9
  block diagram, 13
  block diagram, digital, 20
  digital, 20
  discrete time, 19, 20
  modeling, 24
  modeling – analog, 24
  modeling – digital, 35
  Type 0, 11
  Type 1, 9, 23
  Type 2, 9
  Type 3, 9
  unity feedback, 13
servomechanism, 3, 7
  servos, 7
  unity feedback, 7
servomechanisms, 7
signal
  scaling, 120
signal and noise generation
  algorithm, 121
signal and noise scaling, 120
signal model, 112
signal power, equation for, 120
signal processor considerations, 121
  coherent integration, 121
  noncoherent integration, 121
  proper tuning, 121
signals
  difference, 141
  error, 141
simulatin setup
  low PRF tracker, 234
simulation example, 28
  state variable, 28
  z-transform, 30

simulation examples, 28
simulation parameters, 123
simulation results
  low PRF tracker
    LFM waveform, no Doppler tuning, 238
    PSK waveform, 5 dB SNR, 237
simulation setup
  example
    CW tracker, 227
simultaneous beams, 175
sine space, 139
smoothed covariance, 71
smoothed state, 41
smoothing
  equation, 47
  stage, 41
smoothing equation, 47
spherical
  correction, 155
  E-field, 146
spherical correction, 143
squint, 140, 143
  angle, 158
squinted
  beams, 159
stability, 12
  triangle, 48, 51, 79
standard quadratic form, 31
state
  predicted, 41
  smoothed, 41
  transition, 41
  vector, 42
state transition, 41
state transition matrix, 35
  α-β, 42
state variable, 26
  equation, 27, 35
  equation - example, 28, 33
state vector, 35
  α-β, 42
steady state error, 10
  analog, 12
  and system type, analog, 9
  and system type, digital, 20
  digital, 22
sum
  voltage
    amplitude comparison, 159
    phase comparison, 152
superpositon
  equation for, 112
Swerling, 245
  approximation, 246
  RCS model, 245
system
  continuous time, 9
  type, 9
  type, discrete, 20
system type, 3
target scenario
  example, 72
Taylor series, 44
time constant, 32
track, 1
  filter, 91
track filter
  alpha-beta, 254
  α-β, 46
  α-β-γ, 46
Track Initiation, 285
track point, 93
track rate, 42
track update period. See sample period
tracker, 7, 139
  analog, 7
  angle, 7, 42, 139
    closed loop, 139
    example 1, 193
    example 2, 198
  bias in, 12
  closed loop, 1, 46
  combined
    low PRF, 222
  combined range and angle, 198
  continuous, 1
  digital, 7
  example
    CW, 226
    low PRF, 232
  hybrid, 8

open loop, 1
combined, 215
range, 12, 42
structure, 46
type, 1
tracker parameters
example, 73
tracking, 1
centroid, 251
crossing target, 258
dichotomous, 252
effect of RCS on, 247
elliptic centroid, 252
range and angle, 251
two target, 4, 251
track-while-scan, 1
transfer function
closed loop, 13
definition, 25
derivation, 35
open loop, 9, 13, 20, 22
track filter, 47
z-domain, 35, 36
$\alpha$-$\beta$, 47
$\alpha$-$\beta$-$\gamma$, 48
transient performance measure, 53
$\alpha$-$\beta$, 54
transient response
root locus analog, 13
root locus digital, 22
underdamped, 13
Type 1, 9
Type 1 servo
$\alpha$ filter, 59
Type 2, 9
Type 2 servo
Kalman, 48
$\alpha$-$\beta$, 48
Type 3 servo
Kalman, 48
$\alpha$-$\beta$-$\gamma$, 48
undamped frequency, 31
underdamped, 17, 18, 19
unity feedback, 7
unstable, 19
update period, 42
variance
ratio, 48, 51, 79, 82
reduction, 48
verify table, 290, 292
wavefront
planar, 143
spherical, 143
wavelength, 148
weight
amplitude, 146, 149
phase, 146, 149
weight function, 215
weights
phase, 154
zero
closed loop, 18
zeros
open loop, 13

www.ingramcontent.com/pod-product-compliance
Ingram Content Group UK Ltd.
Pitfield, Milton Keynes, MK11 3LW, UK
UKHW050726260926
475579UK00005B/32

9 781630 813352